CAMBRIDGE COMPARATIVE PHYSIOLOGY

GENERAL EDITORS:

J. BARCROFT, C.B.E., M.A., F.R.S.
Fellow of King's College and Professor of
Physiology in the University of Cambridge
and
J. T. SAUNDERS, M.A.
Fellow of Christ's College and Lecturer in
Zoology in the University of Cambridge

THE ELEMENTS OF EXPERIMENTAL EMBRYOLOGY

THE ELEMENTS OF
EXPERIMENTAL EMBRYOLOGY

BY

JULIAN S. HUXLEY, M.A.

Honorary Lecturer in Experimental Zoology,
King's College, London

AND

G. R. DE BEER, M.A., D.Sc.

Fellow of Merton College, and Jenkinson Lecturer
in Embryology, Oxford

CAMBRIDGE
AT THE UNIVERSITY PRESS
1934

CAMBRIDGE
UNIVERSITY PRESS

University Printing House, Cambridge CB2 8BS, United Kingdom

Cambridge University Press is part of the University of Cambridge.

It furthers the University's mission by disseminating knowledge in the pursuit of education, learning and research at the highest international levels of excellence.

www.cambridge.org
Information on this title: www.cambridge.org/9781107502420

© Cambridge University Press 1934

First published 1934
First paperback edition 2015

A catalogue record for this publication is available from the British Library

ISBN 978-1-107-50242-0 Paperback

To

ROSS HARRISON and HANS SPEMANN

CONTENTS

PREFACE

A few words are needed to explain the scope of this book. The study of the developmental processes of animals is an enormous field, of which only a small fraction can be dealt with in a volume of this size. The observational and comparative study of embryology falls outside the boundaries of this series; in any case, it has already been treated in numerous authoritative works. Even on the experimental and physiological side, however, there remains the difficulty of selection from the vast mass of somewhat heterogeneous material which many lines of research have provided for consideration and synthesis.

In the first place, development is not merely an affair of early stages; it continues, though usually at a diminishing rate, throughout life. The processes of amphibian metamorphosis or of human puberty; the form-changes accompanying growth; senescence and natural death itself—these are all aspects of development; and so, of course, is regeneration.

We feel that it is impossible to treat the whole life-cycle in a single volume, and have accordingly set an arbitrary limit to our material. We have deliberately restricted ourselves to the early period of development, from the undifferentiated condition up to the stage at which the main organs are laid down and their tissues histologically differentiated—in other words, to Wilhelm Roux's "prefunctional period". Growth, absolute and relative; the effects of function on structure and on size; the morphogenetic effects of hormones—the details of these and of other related topics we have deliberately omitted, and we have contented ourselves with the addition of a final chapter in which the main peculiarities of the functional period are contrasted with those of the pre-functional period of primary differentiation. Any satisfactory treatment of the latter portion of the developmental cycle would require a separate volume.

In the second place, within the period of early development, we have exercised a further selection. In a new field of biology such as this, there are always two levels of approach. One of these is broadly biological, while the other is physiological in the stricter sense. The prime aim of the worker approaching the problem on the physiological level will always be to analyse the processes involved in terms of physics and chemistry. The worker on the biological level will aim at discovering general rules and laws which he is content to leave to his physiological colleague for future analysis in more fundamental terms, but which, meanwhile, will give coherence and a first degree of scientific explanation to his facts. Both methods are necessary for progress; and while most biologists hope and expect that one day their laws will, thanks to the labours of their physiological colleagues, be made comprehensible in the most fundamental physico-chemical terms, they can reflect that it is they who must first reveal the existence of these laws before the pure physiologist can hope to begin his analysis. The biologist can also remember that these laws have their own validity on their own level, whether they be physico-chemically analysed or not.

We may take a salient example from the contents of this book. Spemann's discovery of "organisers" in the process of gastrulation of Amphibia, and the extension of the concept to other stages of development and to other groups of organisms, have made it possible to understand on the biological level many processes of development which were previously obscure. At the moment we can only throw out crude guesses as to the underlying physiology of organisers and their effects, but the discovery opens a new field of research to physiologists, which they themselves would not have been likely to hit upon for many years. And even if and when the physiological analysis has been made, the empirical biological laws concerning organisers will not lose their validity or their interest; they will merely have been extended and deepened.

At the present moment, research into developmental problems is being actively prosecuted on both the biological and the physiological levels. Following up the early work of Roux, Hertwig, Driesch, Herbst, Jenkinson, Delage, Brachet, Morgan, and Wilson, a flourishing school of *Entwicklungsmechanik* has grown up in Germany, and another, no less successful, in the United States. Meanwhile, on the physiological side, the advance has also been striking, and we may perhaps cite as particular examples such works as Fauré-Fremiet's *Cinétique du Développement*; Gray's *Experimental Cytology*; Dalcq's *Bases Physiologiques de la Féconda-tion*; and Needham's classic book on *Chemical Embryology*.

So far, however, little progress has been made in equating the results of the two lines of approach, and it seems clear that a con-siderable time must elapse before it will be possible to do so satis-factorily. At the moment the two fields are almost as unrelated as were, through most of the nineteenth century, the cytological and the experimental-genetic approaches to the problem of heredity, which are now inseparable.

That being so, we have not attempted to include the results of the purely physiological study of development in this survey. This means that we have deliberately excluded such topics as the physiology of fertilisation, the mechanics of cleavage, and the bio-chemistry of the egg and embryo, save where they have a specific bearing on the biological problems involved.

In other words, what we have attempted to do is to give some account of the results of the experimental attack on the problem of the biology of differentiation—the production of an organised whole with differentiated parts out of an entirely or relatively un-differentiated portion of living material. Almost the only short books on this subject since Jenkinson's *Experimental Embryology* and his (posthumous) *Lectures* are Brachet's *L'Œuf et les Facteurs de l'Ontogénèse*, Dürken's *Grundriss der Entwicklungsmechanik*, Weiss' *Entwicklungsphysiologie der Tiere*, and de Beer's *Introduction to Experimental Embryology*; and each of these treats the subject along rather different lines. Among larger works, Wilson's *The Cell*, Morgan's *Experimental Embryology*, Dürken's *Lehrbuch der Experimentalzoologie*, and Schleip's *Determination der Primitivent-wicklung* are the most important which have appeared since the pioneer works on the subject. A perusal of them will suffice to show the extreme diversity of their lines of approach. What we have felt is that at present there exists in the subject a vast body of facts and a relative paucity of general principles. We have accordingly aimed at marshalling the facts under the banner of general prin-ciples wherever possible, even when the principle seemed to be only provisional.

Many of the illustrations have been drawn specially for this book by Miss B. Phillipson, to whose care and skill we wish here to make acknowledgments. Particular thanks are due to Miss P. Coombs for her help with typing and many other details of preparing the book for press. Acknowledgment is hereby made to those authors and publishers of the journals whose names appear in the legends to the figures, by whose courtesy they are here reproduced.

We wish to express our thanks to Prof. E. S. Goodrich, F.R.S., to Mr and Mrs R. Snow, and to Mr C. H. Waddington, who not only read part or all of the manuscript but also made several helpful suggestions. We are under a particular debt of gratitude to Dr Sven Hörstadius, and to Professor J. Runnström, who very kindly permitted us to make use of some as yet unpublished results, and also to Professor M. Hartmann who has kindly enabled us to reproduce a figure from the (as yet unpublished) 2nd edition of his *Allgemeine Biologie*.

In conclusion, we should like to acknowledge our debt to the late Dr J. W. Jenkinson, an Oxford man, and the pioneer of Experimental Embryology in this country, and to express our deep appreciation of the care and skill which the Cambridge University Press has expended on the production of this volume.

J. S. H.

G. R. DE B.

January, 1934

ACKNOWLEDGMENTS

Acknowledgment is due and hereby gratefully made to the following, for permission to reproduce figures:
Akademische Verlagsgesellschaft m.b.H., Leipzig (*Verh. d. Deutscher Zool. Gesells.*, *Zeitschrift f. wissenschaftl. Zoologie*); Cambridge University Press (*Anatomy and the Problem of Behaviour*, Coghill; *Biological Reviews*; *Experimental Cytology*, Gray); Chicago University Press (*Botanical Gazette*; *General Cytology*, Cowdry; *Individuality in Organisms*, Child; *Physiological Zoology*); Columbia University Press (*Experimental Embryology*, Morgan); MM. Gaston Doin et Cie, Paris (*Archives de Morphologie*); Herren Gustav Fischer, Jena (*Allgemeine Biologie*, *Verh. d. Anatomischen Gesells.*, *Zoologische Jahrbücher*); Messrs Henry Holt & Co., U.S.A. (*Physiological Foundations of Behavior*, Child); The Lancaster Press, Inc., U.S.A. (*Biological Bulletin*); The Marine Biological Laboratory, Woods Hole, Mass., U.S.A. (*The Biological Bulletin*); Messrs Methuen & Co., Ltd. (*Problems of Relative Growth*, Huxley); Le Muséum d'Histoire Naturelle, Genève (*Revue Suisse de Zoologie*); National Academy of Sciences, Washington, U.S.A. (*Proceedings*); Oxford University Press (*Introduction to Experimental Embryology*, de Beer; *Experimental Embryology*, Jenkinson; *Quarterly Journal of Microscopical Science*); The Council of the Royal Society, London (*Proceedings, Philosophical Transactions*); The Science Press, U.S.A. (*The American Naturalist*); Herren Julius Springer, Berlin (*Archiv f. Entwicklungsmech. d. Organismen*, *Archiv f. Mikros. Anat. u. Entwicklungsmech.*, *Naturwissenschaften*, *Ergebnisse d. Biologie*, *Handb. d. norm. u. pathol. Physiol.*, *Zeitschrift f. vergleich. Physiol.*); Herr Georg Thieme, Leipzig (*Biologisches Zentralblatt*); The Waverley Press (*The Science of Life*, Wells, Huxley and Wells); The Waverley Press, Inc., U.S.A. (*Journal of Experimental Medicine*); The Williams and Wilkins Company, U.S.A. (*Quarterly Review of Biology*); The Wistar Institute of Anatomy and Biology, U.S.A. (*Journal of Experimental Zoology*).
Acknowledgment to the authors of the works from which these illustrations are taken is made in the legends.

Chapter I

HISTORICAL INTRODUCTION TO THE PROBLEM OF DIFFERENTIATION

§ 1

The production of the adult living organism with all its complexity out of a simple egg (or its equivalent in the terminology of the ancients) is a phenomenon and a problem which has attracted the attention of philosophers as well as of scientists for over two thousand years. To give a brief account of the history of ideas relating to this problem is no easy matter, but the task is fortunately facilitated by the fact that Dr E. S. Russell and Prof. F. J. Cole, F.R.S. have recently devoted volumes to certain aspects of this subject, and to the reader who desires to become better acquainted with it, no better advice can be given than to refer him to *The Interpretation of Development and Heredity*, and to *Early Theories of Sexual Generation*. The historical section of Dr Needham's *Chemical Embryology*, and various works of Dr Charles Singer also provide much valuable information.

Meanwhile, a brief attempt will be made in the following few pages to outline the essential features of the chief schools of thought concerning problems of development, in order to show how the modern science of experimental embryology came into being, and to present it in its proper historical setting.

The kernel of the problem is the appearance during individual development of complexity of form and of function where previously no such complexity existed. In the past, there have existed two sharply contrasted sets of theories to account for it. One view accepts the phenomenon as essentially a genesis of diversity, a new creation, and attempts to understand it as such. This coming into existence of new complexity of form and function during development is styled *epigenesis*.

The difficulties which other thinkers experienced in trying to understand how epigenesis may be brought about led them to deny that it exists: i.e. to say that there is no fresh creation of diversity in development from the egg, but only a realisation, expansion, and rendering visible of a pre-existing diversity. *Preformation* is the fundamental assumption of views of this type, and they are classed together as preformationist theories. But the doctrine of preformation, however, met with even graver obstacles, both logical and empirical, than the opposite view, and biological opinion is now united in maintaining the existence of a true epigenesis in development. In recent years, however, the discoveries of genetics have reintroduced certain elements of the preformationist theory, but in more subtle form. As will be seen later, the modern view is rigorously preformationist as regards the hereditary constitution of an organism, but rigorously epigenetic as regards its embryological development.

To a large extent, the preformationist view assumes as already given that which the epigenetic attempts to study and to explain; and the problem is complicated by the fact that notions of embryonic development have been confused with concepts of heredity. This is evident in the attempt, on the part of the author of *Peri Gones* in the Hippocratic corpus, to explain development by assuming a part-to-part correspondence between the parts of the body of the parent and those of the offspring: the corresponding parts being related to one another via the "semen", or, as would now be said, via the germ-cells. By assuming that the embryo at its earliest stage is a minute replica of the adult, its parts having been "preformed" by representative particles coming from the corresponding parts of the parent, the preformationist hypothesis attempts to solve at one stroke both the problem of hereditary resemblance between generations and the problem of development within each generation.

This view was in reality shattered by Aristotle's criticism, but it was revived and widely held during the seventeenth and eighteenth centuries, largely owing to the fact that mechanistic explanations had come into vogue, and it seemed impossible to understand epigenesis on mechanistic lines. One of the foremost exponents of the preformationist hypothesis was Charles Bonnet. His views were freed from the crude idea that the preformation in the egg was spatially identical with the arrangement of parts in the adult and fully developed animal, or that the "homunculus" in the sperm, with the head, trunk, arms and legs which it was supposed to have (and which certain over-enthusiastic observers claimed to have seen through their microscopes; see Cole's *Early Theories of Sexual Generation*) only required to increase in size, as if inflated by a pump, in order to produce development. Instead of regarding the rudiments of the organs as being preformed in their definitive adult positions, Bonnet imagined them as "organic points" which subsequently had to undergo considerable translocation and re-arrangement. He was thus able to reconcile his belief in preformation with the empirical fact that the germ or blastoderm of the early chick showed no resemblance to a hen.

Bonnet's theories were ahead of his facts, and, indeed, he seems to have been proud of it, for he refers to the preformationist view as "the most striking victory of reason over the senses". The hypothesis of such an invisible and elastic preformation was perhaps permissible in Bonnet's day, but later observational and experimental evidence has rendered it utterly untenable. Further, a rigid preformationist view which asserts that the egg is a miniature and preformed adult, necessarily implies that the egg must also contain the eggs for the next generation; the latter eggs must therefore also contain miniature embryos and the eggs for their subsequent generations. Bonnet realised that an *emboîtement* or encasement of this kind *ad infinitum* would be an absurdity. (Incidentally, it may be noticed that if it were true, phylogenetic evolution—unless it too were preformed and predetermined—would be an impossibility.) But then, if *all* subsequent generations are not preformed in miniature now, there must come a time when they *are* determined and preformed. Before this time they were neither determined nor preformed, and this making of a new determination, albeit pushed into the future, is the antithesis of preformation.

If pushed to its extreme conception of infinite encasement, then preformation is absurd; if not pushed to this extreme, preformation will not account for the determination of ultimate future generations; and if it did apply, preformation would be an unsatisfactory view in that it assumes that the diversity which is progressively manifested in development is ready-made at the start, and in no way attempts to explain it causally or to interpret it in simpler terms.

Logically, the preformationist view is associated with the notion of separate particles being transmitted from parent to offspring, though the converse does not hold. In preformationist theory, the hypothetical particles establish the one-to-one link between the corresponding organs and parts of parent and offspring, whereas the modern view, which combines an epigenetic outlook on development with the particulate theories of neo-Mendelism, denies any such simple correspondence between hereditary germinal unit and developed adult character. Darwin's theory of pangenesis resembles that of the Hippocratic writer in this respect, the pangens being supposed to come from all parts of the body of the parent and to be transmitted, via the germ-cells or "semen", to the offspring whose development they mould. Embryologically, however, Darwin's theory is vague, and leaves the question of preformation open. Weismann's theory of the germ-plasm, in which the *determinants* are regarded as representing the predetermined but not spatially preformed diversity of the future embryo, differs from that of previous preformationists in that the particles are regarded as coming, *not* from the corresponding parts of the body of the parent, but from the germ-plasm, of which each generation of individual organisms is held to be nothing but the life-custodian. Weismann identified the determinants with the material in the nuclei of the cells, which material he (wrongly) supposed was divided unequally in the process of division or cleavage of the egg, so as to form a mosaic, the pieces (cells or regions) of which would then contain different determinants and would therefore be predetermined to develop in their respective different and definite directions.

According to the writer of the Hippocratic treatise *Peri Gones* and to Darwin, therefore, offspring resembles parent because the particles responsible for the development of the parts of the offspring come from the corresponding parts of the parent. According to Weismann, however, offspring resembles parent because both have derived similar particles (determinants) from a common source—the germ-plasm.

The question of the *origin* of the particles or hereditary factors and of their *distribution* from the parent to the offspring is one which principally concerns the science of genetics. The modern tendency is to accept the principle of a germ-plasm while recognising that it is not as inaccessible to the modifying action of external factors as Weismann contended. The question of the *function* of the particles or factors in converting the fertilised egg into the body of the adult is the concern of that modern and rather special branch of embryology usually called physiological genetics.

Before dealing with the conclusion derived directly from experimental work, a moment's attention may be turned to philosophical criticisms of the preformationist view that particles, determinants, or any hereditarily transmitted units or factors, can "explain" development. First of all, Aristotle pointed out that certain features in which offspring resembled parent could not be ascribed to the transmission of particles from corresponding parts, for the latter might be dead structures like nails or hair from which no particles could be expected to come, or again they might be such characters as timbre of voice or method of gait. He goes on to say, by way of illustration, that if a son resembles his father, the shoes he wears will be like his father's shoes, yet there can, of course, be no question of particles here. In other cases, resemblance may refer to structure, plan or configuration rather than to the material of which it is composed, and it is hard to see how particles can represent such structure, plan or configuration. Again, how is the eventual beard of a son to be explained if he was born to a beardless father? To these objections might be added the insuperable difficulty of accounting for the production of offspring structurally different from the parent, as when the egg laid by a queen bee develops into a worker, or, even more generally, when a mother bears a son or a man fathers a daughter.

If, then, particles coming from corresponding parts are not required in some cases and cannot be resorted to in others in order to explain development and hereditary resemblance, why should they be postulated in any case? This, of course, concerns genetics as much as embryology, but Aristotle came very close to the crucial problem of the latter when he wrote: "either all the parts, as heart, lung, liver, eye, and all the rest, come into being together, or in succession....That the former is not the fact is plain even to the senses, for some of the parts are clearly visible as already existing in the embryo while others are not; that it is not because of their being too small that they are not visible is clear, for the lung is of greater size than the heart, and yet appears later than the heart in the original development".[1]

Simple observation, therefore, had even in Aristotle's time given the lie direct to the view that the embryo is a spatially preformed miniature adult. Similar but more exhaustive and more crucial observational evidence against the preformationist view was supplied by William Harvey (who referred to development as "*epigenesin sive partium superadditionem*") and, notably, by Caspar Friedrich Wolff. The conclusion to which the latter came is the same as that of Aristotle. In the earliest stages of the development of the fowl, the microscope reveals the presence of little globules heaped together without coherence, and a miniature of the adult simply does not exist. Further, no refuge can be taken in the assumption that the miniature is too small to be seen, for its parts (globules) are clearly visible, and, *a fortiori*, therefore, the whole. The plain fact is that the miniature of the adult is not there.

The necessary epigenetic correlate of this fact has been admirably put by Delage in the following words: "latent or potential characters are absent characters.... The egg contains nothing beyond the special physico-chemical constitution that confers upon it its individual properties *qua* cell. It is evident that this constitution is the condition of future characters, but this condition is in the egg extremely incomplete, and to say that it is complete but latent is to falsify the state of affairs. What is lacking to complete the conditions does not exist in the egg in a state of inhibition, but outside the egg altogether, and can equally well occur or not occur at the required moment. Ontogeny is *not* completely determined in the egg".[1] We might sum up the position by saying that to maintain the full preformationist view would partake of the nature of fraudulent bookkeeping.

There is no way of saving the view that the adult is preformed in the egg as a diminutive replica. The more subtle idea of Bonnet's, of preformed "organic points", or of determinants unequally distributed between the cells into which the egg divides, also met its doom a century ago, when Etienne Geoffroy St Hilaire (1826) experi-

---

[1] Quoted from Russell, *loc. cit.*

mentally produced developmental monsters out of chick embryos, and rightly concluded that since there cannot have been any preformation of these experimentally induced monstrosities, normal embryos need not be preformed either. A better known death-knell for the preformationist hypothesis is Driesch's demonstration that in many forms, the parts (blastomeres or groups of blastomeres) of the dividing egg could, if separated, develop into complete little embryos. It is impossible to imagine any theory of preformation, however elastic, which will explain the fact that an egg normally develops into a single embryo, and yet can be made to give rise to two or four whole embryos.

§ 3

The inevitable conclusion is that development involves a true increase of diversity, a creation of differentiation where previously none existed, and that the interpretation of embryonic development must be sought along the lines of some epigenetic theory. The problem is narrowed down to a search for a principle on which it is possible to understand how the determinations of the future embryo can arise out of a non-diversified egg. It is the great merit of C. M. Child to have shown in theory how this is possible. Briefly, his view (which will be considered in detail later) is that certain external factors set up quantitative differentials in the egg and embryo, as a result of which qualitative differences of structure ultimately ensue. The egg contains a complex of inherent factors, notably the genes of Mendelian theory, which have been transmitted from its parents and ensure that it shall develop in a specific fashion, and that if the environment is normal it shall develop so as to resemble other members of its kind. However, these internal inherent and transmitted factors of the egg, though genetically preformed, cannot be regarded as a preformation in a spatial or embryological sense. What they do is to confer upon the developing organism the capacity to respond in a specific way to certain stimuli which in the first instance are external to the organism. It is, as Ray Lankester and Herbst first suggested, these responses of a specific hereditary outfit to stimuli outside themselves, which constitute development.

Differentiation is evoked out of the egg afresh in each and every generation: every individual organism is created by epigenesis during its own life-history. The environment is as important as are the internal and transmitted hereditary factors, and both must be normal for a normal embryo to be developed. If the environment is abnormal, there will either be no development at all, or an abnormal and abortive development, and the same fate befalls an abnormal hereditary constitution reacting with a normal environment. If both the environmental and hereditary factors are within the bounds of normality, then development will follow the lines which are characteristic for the particular species of organism in question.

The origin of differentiation and of the epigenetic process are therefore to be found in the processes by which in the first place quantitative differentials are induced in the egg by external factors, and in the second place qualitative structural diversities result from the interaction of the quantitative differentials with the inherited constitution. It is these problems which form the subject-matter of this book.

§ 4

Meanwhile, it is necessary to pause, and to consider for a moment how the causal postulate can be applied to development conceived as an epigenesis. On the preformationist view, the causes of development present no particular difficulty, for differentiation is then supposed to be there all the time and to require nothing but expansion or unrolling ("evolution" in the eighteenth-century sense) in order to become visible. Even after the discomfiture of the preformationist view at the hands of Wolff and others, and the acceptance in principle of an epigenetic theory of development, the need for an application of the causal postulate was cloaked by the unfortunate effects of Haeckel's theory of recapitulation. This view, pushed to its ultimate conclusion, maintained that ontogeny or embryonic development was inevitably a recapitulation of phylogeny or racial evolutionary history, and that phylogeny was the mechanical cause of ontogeny, whatever Haeckel may have meant by such a statement. If this was true, then clearly there was no need to look for other causes than the evolutionary history in order to explain development. But, as Wilhelm His saw, it was not true.

The Aristotelian view of the causes of epigenesis is complicated and somewhat grotesque from the modern point of view, but it introduces some notions which are very apposite in any discussion of this problem. First of all, Aristotle realised the principle of linked causes, which may be illustrated with reference to the interdependence of meshed cogwheels in machinery. He wrote: "that which made the semen sets up the movement in the embryo, and makes the parts of it by having first touched something, though not continuing to touch it".[1] This is the principle on which a clock works after it has been wound up, and many thinkers have imagined development as the working of machinery originally wound up and set going at conception, the continued working of which was due to the progressive assumption of causal activity by the results of previous causes.

But Aristotle did not regard this view as providing a sufficient explanation; in addition, he held that the "soul" was active in controlling the material forces and mechanical processes of development. Kindred views have been expressed by von Baer and by Driesch. The former held that each stage of development was a necessary *condition* for the production of the following stage, but was not in any full sense its *cause*, for in addition he regarded the "essential nature" of the parent as responsible for controlling the development of the offspring. Driesch has adapted Aristotle's view of the functions of the "soul" in his theory of entelechies.

On the other hand, Wilhelm His, having overthrown Haeckel's theory of recapitulation, regarded each stage of development as a sufficient cause of the following stage, and so paved the way for a new branch of science: Entwicklungsmechanik or causal embryology, the foundations of which were laid by Wilhelm Roux. In what may be regarded as the "charter" of the new science, Roux prescribes the analysis of development into so-called complex components, such as assimilation, growth, cell-division, etc. Ultimately he supposed these complex components to be reducible to simple components, which in turn would be capable of interpretation in terms of physics and chemistry (Roux, 1885).

Whether future research will succeed in so reducing the complex components of development as to render them susceptible of ex-

[1] Quoted from Russell, *loc. cit.*

pression in fundamental physico-chemical terms is a question of its own, and one which has been much obscured by the introduction of what are ambiguously called "mechanistic" explanations. As Woodger's (1928) analysis has shown, the term "mechanistic" as applied to biological phenomena may mean: either

1. That the structure and function of living organisms is to be completely explained in terms of "little bits of stuff pushing one another about" in accordance with the classical laws of mechanics; or

2. That all the phenomena presented by a living organism are ultimately capable of analysis in terms of the laws of physics and chemistry; or

3. That a living organism is in some sense analogous to a human-made machine and that its processes are explicable in terms of this analogy; or

4. That the causal postulate is perfectly applicable to living organisms and can be satisfactorily applied to the biological order of things, whether or no the phenomena of the biological order can ultimately be brought into line with physico-chemical phenomena and prove susceptible of analysis in physico-chemical terms.

The fourth of these alternatives is generally accepted, and, indeed, the whole science of causal embryology is based upon it. The second alternative is also widely accepted, and is the only fruitful working hypothesis for the biologist. It is clear, however, that it may require modification, for further study, notably of the phenomena of life, is likely to reveal new and hitherto unsuspected physico-chemical properties of matter. Accordingly, it is necessary to take physics and chemistry in the most extended sense. The advances made in physics itself have rendered the first alternative untenable, and the third cannot pretend to have more value than can ever be ascribed to processes of reasoning by analogy; thus, what may be called the cruder mechanistic view embodied in alternatives 1 and 3 may be excluded.

§ 5

We are not concerned, here, with the construction of a philosophical system, nor do we wish to prejudge the question of the relationship to one another of phenomena of the physico-chemical and of the biological order; the reader to whom these matters are of interest

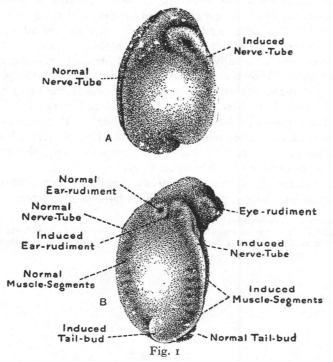

Fig. 1

Induction of secondary embryo by grafted organiser in *Triton*. A, 3 days after operation. B, Some days later. (From Wells, Huxley and Wells, *The Science of Life*, London, 1929; after Bautzmann.)

may most profitably be referred to the recent work of Drs von Bertalanffy and Woodger: *Modern Theories of Development*. But as biologists we do believe that the phenomena which we study in living organisms conform to a *biological order*, in which the causal postulate is strictly applicable. The great value of the new science of experimental embryology or developmental physiology (the term

"Entwicklungsmechanik" is hardly translatable, and, now that its birth has been described, may best be avoided in English writings) is that it is enabling biologists to discover the complex components of development, and so to explore new aspects of the biological order. The dorsal lip of the amphibian blastopore (the so-called "organiser") has been shown (see fig. 1) to be capable of inducing neighbouring tissues to give rise to all the essential structures of an embryo [1] (brain, spinal cord, eyes, ears, muscles, kidney tubes, etc.). The result of grafting an organiser into a suitable environment is just as definitely causally determined and predictable as the result of mixing two known reagents in a test-tube, although the phenomena are in the one case of the biological and in the other of the physico-chemical order. It may be confidently expected that in time the physiological basis of the organiser's action will be discovered and accurately analysed in physico-chemical terms. [2] Until then, however, it is both desirable and necessary to push the analysis as far as possible on the biological level.

It is as a contribution to the analysis of early development on the biological level that the following pages have been written.

[1] Spemann and Mangold, 1924.
[2] Already it is known that the organising action is due to a substance which is almost certainly lipoidal and probably a sterol (Waddington, Needham and Needham, 1933). See pp. 154 and 497.

Chapter II

EARLY AMPHIBIAN DEVELOPMENT: A DESCRIPTIVE SKETCH

§ 1

It will be best to base the analytical treatment of development upon a concrete example, and for this purpose the Amphibia are by far the most suitable material, as analysis is much more complete in them than in any other group of organisms. However, before embarking upon analysis, it will be desirable to give a brief descriptive sketch of amphibian development in so far as it is relevant to subsequent chapters; to do this is the purpose of the present chapter.

The chief stages of amphibian development are as follows: the changes associated with fertilisation; cleavage, leading to the blastula stage; gastrulation, leading to the gastrula; the elongation of the embryo and the formation of the neural folds and tube, constituting the neurula stage; the appearance of the tail, and of the remaining organ-rudiments, leading to the fully formed embryo, which then hatches as a young larva; and then the period of growth and of functional differentiation. These stages overlap somewhat, especially the last two, but they provide a useful broad classification.

The typical amphibian egg is a spherical object of which one hemisphere (known as the vegetative hemisphere) is loaded with yolk, while the other hemisphere (the animal hemisphere) is freer of yolk and contains the nucleus. There is, as a matter of fact, a graded distribution of yolk from the animal to the vegetative pole. In the Anura, the animal hemisphere is characterised by the possession of a layer of dark pigment at the surface, which distinguishes it at a glance from the lighter-coloured vegetative hemisphere. A similar distinction exists in the eggs of Urodela but is not so marked because the pigment is less dark. Yolk being of a higher specific gravity than the other constituents of the egg, it is found that after the egg has been laid and fertilised and is free to rotate within its membrane, the main egg-axis, or axis passing through the centres or poles of both animal and vegetative hemispheres, is practically vertical. (This is the rule in the majority of the Amphibia, but it should be mentioned that in *Rana esculenta* there are complications, into which there is no need to go here, as a result of which the egg-axis appears oblique. See Jenkinson, 1909 B.)

Even before development can be said to have begun, therefore, the egg possesses one mark of dissimilarity between its various regions, one mark of differentiation, which is expressed by saying

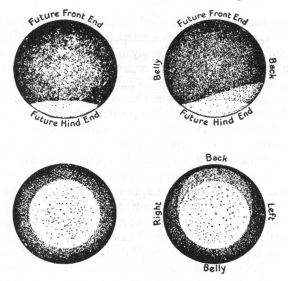

Fig. 2

Polarity and bilaterality in the frog's egg. Above, in equatorial view; below, seen from the vegetative pole: left, before fertilisation; right, after fertilisation. The unfertilised egg possesses a single main axis (polarity) at fertilisation, bilaterality is established through the formation of the grey crescent in or close to the future mid-dorsal line. (From Wells, Huxley and Wells, *The Science of Life*, London, 1929.)

that the egg has *polarity*. This polarity is of great importance for future development because the future front end of the animal will be formed in proximity to the animal pole of the egg, and the hind end of the animal close to the vegetative pole. Apart from this polarity, which concerns the differential distribution of pigment, yolk and cytoplasm, and the excentric position of the nucleus, the egg is undifferentiated.

As a rule among Amphibia, the 1st polar body is given off before fertilisation, and the 2nd polar body after that event. The fertilisation of the egg by the sperm has a threefold significance. In the first place it activates the egg to begin its development; secondly, it brings in to the resulting zygote its supply of paternal hereditary factors; and lastly, it is responsible for bringing about the next step in differentiation, which is the determination of a plane of bilateral symmetry.

In the frog it has been shown by experiment that the mid-ventral line of the embryo will be formed close to the meridian on which the sperm enters the egg.[1] The only visible differentiation at this stage, however, concerns the dorsal side, opposite the point of sperm entry. A region of this, rather below the equator of the egg, is marked soon after fertilisation by changes in the surface layer, leading in the case of the Anura to the formation of the so-called grey crescent, due to the retreat of pigment into the interior of the egg. Analogous, but less well-marked changes on the dorsal side of the recently fertilised egg are observable in the Urodela.[2]

After the entry of the sperm, therefore, the developing organism, although still a spherical object, has all three of its axes determined. The antero-posterior axis and the dorso-ventral axis of the future embryo lie in the plane of bilateral symmetry, which, in turn, passes through the original egg-axis of polarity. At the same time, the transverse, or left-right axis, is also necessarily fixed with the determination of the other two axes. The symmetry relations of the organism are thus completely and definitely fixed (fig. 2).

§ 2

The grey crescent of the Anuran egg (or its equivalent in the egg of Urodela) is the place at which the next marked step in differentiation appears. The egg has meanwhile undergone cleavage, and instead of being a single large cell, has come to consist of a large number (over a thousand) of smaller cells or blastomeres, which enclose a small cavity, the blastocoel. These blastomeres are smaller in the animal hemisphere than in the vegetative. This is a result of the prime differentiation of polarity, for yolk retards cell-division,

[1] Roux, 1903; Jenkinson, 1907, 1909 A.
[2] Vogt, 1926 B.

and the cells containing more yolk (those of the vegetative hemisphere) will necessarily divide less fast than the cells of the animal hemisphere which are relatively free from yolk. Consequently, the cells of the vegetative hemisphere will be larger than those of the animal hemisphere at any given time during cleavage.

There is also a slight difference in the size of the blastomeres at different positions on the same circle of latitude: a difference which is already shown by the animal hemisphere cells at the 8-cell stage.[1] Though the cause of this size difference at this early stage is obscure, at later stages of cleavage it is due to the fact that the cells on the dorsal side divide slightly faster and therefore become a little smaller than those on the ventral side.

The next stage in differentiation consists in the conversion of the ball of cells—the blastula—into a double-layered sac or gastrula, by means of the process of gastrulation. Owing to the large amount of yolk present in the amphibian egg, this process is not as simple as in other forms (such as *Amphioxus*) where gastrulation is a simple invagination of one side of the blastula into the other. In the amphibian, the same result is achieved by the spreading of the cells of the animal hemisphere and their downgrowth over those of the vegetative hemisphere, at the same time as they tuck in or invaginate and then extend forwards beneath the surface of the outer layer. This process of spreading and growing over (epiboly), and of tucking in (invagination), first takes place on the dorsal side of the embryo, in the region of the grey crescent, and gives rise to a lip known as the dorsal lip of the blastopore.

Eventually, this lip of overgrowth and tucking in forms a complete ring by extending laterally, until the two sides of the lip meet on the ventral side of the embryo. In this way the blastopore becomes a circular aperture leading into the cavity of the archenteron or future gut. This gut-cavity is a new formation and the direct result of gastrulation. Its lining is made up partly of the cells that have been tucked in round the rim of the lip of the blastopore, and partly of the yolk-laden cells which originally occupied the vegetative pole of the egg. The amount of these yolk-cells is too large for them to be completely accommodated in the newly formed gut-cavity, with the result that some of them protrude through the

[1] Morgan and Boring, 1903.

mouth of the blastopore forming the so-called yolk-plug. At the same time, the original cavity of the blastula, the blastocoel, has

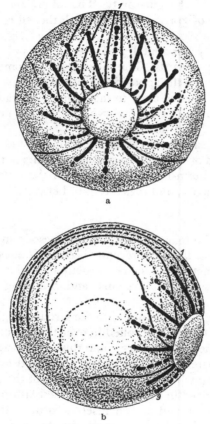

Fig. 3

Diagrams to show the directions of movement and displacement of the parts of the blastula in the process of gastrulation in Amphibia. *a*, Seen from the vegetative pole. *b*, From the left side. The thick lines show the tracks followed on the surface: the thin lines the tracks beneath the surface after invagination at the blastopore rim. (From Vogt, *Arch. Entwmech.* cxx, 1929.)

been more or less obliterated by the formation of the new cavity, the archenteron or gut (fig. 4).

Since it is on the dorsal side of the embryo that the overgrowth and ingrowth begins and continues with the greatest activity, the mass of heavy and inert yolk-cells becomes piled up on the opposite or ventral side of the gut-cavity. This alters the position of the embryo's centre of gravity, and as a result the entire embryo rotates ventralwards through about 100°, until the original egg-axis of polarity is nearly horizontal, and the animal pole faces forwards and a little downwards. The blastopore becomes smaller and eventually closes by the apposition of its lateral lips to one another. At closure, it is situated close to the original vegetative pole, which in its turn, as a result of the embryo's rotation, is now facing back-wards and slightly upwards.

Internally, meanwhile, the endoderm and the mesoderm are becoming sorted out, so that gastrulation results in the delimitation of the primary germ-layers, ectoderm on the outside, endoderm lining the gut-cavity, and mesoderm in between.

§ 3

The details of the manner in which the mesoderm and endoderm arise in Amphibia have only recently been made out and established, thanks to the method of marking definite regions of the living embryo with easily visible stains, and following them through development.[1]

The following account applies to the Urodele type. The material which becomes tucked and rolled in over the rim of the blastopore on the dorsal side of the embryo, and thus forms the primitive gut-roof, will ultimately give rise to the notochord and some of the mesoderm. Meanwhile, the yolk-laden cells of the original vegeta-tive pole are carried in under the lip of the blastopore by the pro-cess of invagination and find themselves forming the anterior end, floor, and sides of the gut-cavity. Later, these sides grow up be-neath the primitive gut-roof and meet one another in the mid-dorsal line, forming the definitive gut-roof. The remainder of the mesoderm is formed from the material rolled in at the lateral and ventral lips of the blastopore; though continuous dorsally with the primitive gut-roof, it is never in direct contact with the archenteric cavity (fig. 4).

The mesoderm thus forms paired sheets of tissue (right and left

[1] Vogt, 1929.

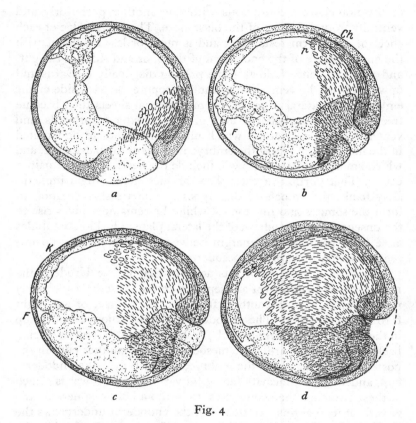

Fig. 4

Diagrams showing the process of invagination and mesoderm-formation in Urodeles. Each diagram is a median sagittal section on to which the mesoderm of one side has been projected. *a*, Early stage; the dorsal lip is well advanced, the ventral lip barely indicated, the sheet of mesoderm is beginning to spread forwards from the dorsal and lateral lips. *b*, The tip of the notochord (*Ch*) is growing forwards beneath the neural plate; the edge of the mesoderm sheet (*p*) has extended further forward; a small blastocoel (*F*) is still visible; *K*, the front of the neural plate. *c, d,* Further stages; mesoderm is growing in at the ventral lip; the mesodermal sheet has extended forwards and downwards, and leaves only a small area unoccupied. The paired rudiments of the heart are situated near the growing edge of the mesodermal sheet on each side. (From Vogt, *Arch. Entwmech.* cxx, 1929.)

of the notochord), continuous with one another posteriorly and ventrally, round the rim of the blastopore. The lateral edge of each sheet of mesoderm rests upon and is more or less confluent with the outer surface of the endoderm of the floor and sides of the gut, and this confluence follows a line passing diagonally forwards and upwards from the ventral lip of the blastopore, on each side of the embryo. The lateral edges of the mesodermal sheets then become free from the endoderm, and gradually extend forwards and ventrally. The mesoderm, which ultimately comes to be situated in the mid-ventral line of the embryo in front of the blastopore, and which among its derivatives will include the heart, is thus of paired origin. That portion of each sheet of mesoderm which immediately flanks the notochord undergoes metameric segmentation to form the somites and myotomes, while the remainder gives rise to the unsegmented mesoderm of the lateral plate. The kidney tubules arise from tissue on the margin between the segmented and unsegmented portions of the mesoderm.

It will be noticed from this account that in the Urodele, the mesoderm and endoderm are separate zones, more or less sharply marked off from one another, from the very outset of and right through gastrulation. The endoderm is soon fashioned into a cup in the antero-ventral region of the embryo, with its concavity facing backwards and upwards: the mesoderm forms another cup, in the postero-dorsal region of the embryo, inverted over the endoderm cup, and with its concavity facing forwards and downwards. Each of these two cups then completes itself into a hollow sphere by the growth of its margins. In this way, the endoderm undergrows the mesoderm and notochord to form the definitive gut-roof, while the mesoderm overgrows the endoderm until it eventually encircles it almost completely.[1]

---

[1] The detailed study of the processes of gastrulation and germ-layer formation in Urodela and in Anura throws an important light on the distinction (based on morphological considerations) between peristomial and gastral mesoderm. The former is regarded as derived from the active tissue round the rim of the blastopore, while the latter is supposed to be derived (by delamination or evagination) from the wall of the gut. In both Urodela and Anura the mesoderm is derived from a ring of tissue surrounding the blastopore, and is, strictly, peristomial. But in Anura the conditions of invagination are such that the mesoderm is rolled in as a mantle closely applied to the endoderm, and it is its subsequent delamination from the latter which gives the mesoderm the appearance of being of gastral origin.

The conditions in the Anuran type during gastrulation are in the main similar to those in the Urodele, except that, for reasons into which we need not here enter, the gut-cavity possesses its definitive endodermal gut-roof from the start. This definitive roof is complete except for a thin longitudinal strip corresponding to the notochord and to the cells immediately underlying it which will give rise to the hypochordal rod. When the notochord and hypochordal rod become lifted off from the gut-roof, a narrow gap is formed, but it soon becomes closed by the approximation of the free edges of the endoderm.[1]

§ 4

Since the cells that become tucked in during gastrulation were originally on the outer surface of the blastula before the process of gastrulation started, it is possible to outline on the surface of the blastula the various regions which will, in normal development, give rise to the various organs of the future embryo. By the method alluded to above, of making stains *intra vitam* in particular places on the surface of the blastula, and by following their changes of position during gastrulation and subsequent development, it is possible to discover the normal futures in store for all the regions of the blastula, and in this manner to ascertain their normal potencies. One may thus speak of the various regions of the blastula as *presumptive* organs: one region is presumptive notochord, another presumptive brain, and so forth.

By methods of this kind, and by making small injuries in definite places with the electric cautery, Vogt and his pupils have been able to map the amphibian blastula completely in terms of presumptive organ-rudiments. This has been accomplished both for a Urodele and an Anuran type.[2]

For purposes of description, a system of notation similar to that used in fixing the positions of places on the earth's surface will be found convenient. The dorsal meridian of the egg or blastula, which passes through the future dorsal lip of the blastopore, may be taken as a standard meridian, corresponding to the meridian of Greenwich in geography, and other meridians may be indicated by degrees of longitude, right or left, from the dorsal meridian.

[1] Mayer, 1931.   [2] Vogt, 1929; Suzuki, 1928.

In the same way, the great circle at right angles to the egg-axis is the equator of the egg. It coincides more or less with the line of demarcation between the pigmented cells of the animal hemisphere and the lighter-coloured cells of the vegetative hemisphere; frequently, however, the pigment extends well below the equator. Latitudinal position is not so easy to define as longitudinal, since the egg-equator is not clearly marked. In the meridian of symmetry, however, latitudinal position can be accurately defined as so many degrees above or below the dorsal lip of the blastopore.

With this in mind, it is now possible to pass to a description of the facts as found in the egg of the Urodele. Most of the cells of the vegetative hemisphere of the blastula eventually get tucked in or enclosed, and find themselves inside the embryo when gastrulation has been completed. A crescent-shaped region immediately above the position of the dorsal lip of the blastopore, and extending up some way above the equator, is presumptive notochord. On each side of this is a strip which will give rise to mesodermal somites and to the unsegmented mesoderm of the lateral plate. Below the latitudinal level of the dorsal lip is a region which includes the yolk-cells of the vegetative pole, and which will give rise to the front, ventral, and lateral walls of the gut-cavity and, eventually, to its definitive roof as well. Most of the ventral half (not to be confused with vegetative half) of the blastula, composed of portions of the vegetative as well as of the animal hemisphere, is presumptive epidermis. This leaves only one region unaccounted for; this, occupying most of the dorsal half of the animal hemisphere (minus the presumptive notochord and mesoderm regions mentioned above), is presumptive neural folds. This latter region may be described in the blastula as a crescent of which the horns extend down the sides of the embryo from the animal pole to the equator along meridians rather more than 90° right and left from the dorsal mid-line. The central part of the crescent extends from the animal pole to the point on the dorsal meridian to which the presumptive notochord region reaches, i.e. about 30° latitude above the equator. It is important to notice that at this early stage, in the blastula, the presumptive neural fold region occupies an elongated region which lies at right angles to the plane of bilateral symmetry.[1] While

[1] Goerttler, 1925; Vogt, 1926 A.

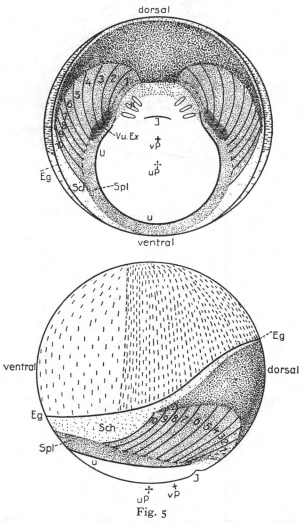

Fig. 5

Map of the presumptive regions of the Urodele embryo, projected on to the surface of the blastula, as seen from the vegetative pole and from the left side. Epidermis, sparse broken lines; neural plate, dense broken lines; notochord, dense dots; mesoderm, fine dots; endoderm, white. The future mesoderm segments are numbered. *Eg-Eg*, limit of invaginated region; *J*, site of formation of first invagination; *K*, gill-pouches; *Sch*, tail region; *Spl*, lateral plate mesoderm; *u*, position of future blastopore lip; *uP*, lowermost pole at this stage; *Vu.Ex*, forelimb; *vP*, vegetative pole. (From Vogt, *Arch. Entwmech.* CXX, 1929.)

24

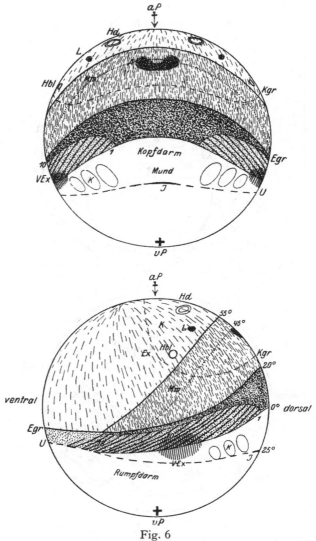

Fig. 6

Map of the presumptive regions of the Anuran embryo, projected on to the surface
of the blastula, as seen from the dorsal and left sides. Epidermis, sparse broken
lines; neural plate, dense broken lines; notochord, dense dots; mesoderm, fine
dots; endoderm, white. The future mesodermal segments are numbered.
A, eyes; *aP*, animal pole; *Egr*, limit of invaginated region; *Ex*, epidermis of limb
region; *Hbl*, ear vesicle; *Hd*, ventral sucker; *J*, site of formation of first invagination (dorsal lip); *K*, gill-pouches, and epidermis of gill region; *Kgr*, broken line
indicating limits of head; *L*, lens; *Mw*, neural fold; *U*, position of future blastopore lip; *VEx*, forelimb; *vP*, vegetative pole. *Kopfdarm*, foregut; *Mund*, buccal
cavity; *Rumpfdarm*, hindgut. (From Vogt, *Arch. Entwmech.* cxx, 1929.)

differing from the Urodele plan in certain details of relative life of regions, the Anuran plan is fundamentally similar (figs. 5 and 6).

The process of gastrulation entails remarkable streaming movements and displacements of the various regions of the embryo. Presumptive notochord, mesodermal somites, and mesoderm of the lateral plate become tucked in over the rim of the blastopore lip as already described. Their places on the surface of the embryo are taken by the presumptive neural fold region undergoing displacement, its original position being occupied by the expanding region of the presumptive epidermis. This displacement and expansion, however, takes place in a peculiar way. It must be remembered that at the start of gastrulation the lip of the blastopore is present only in the dorsal meridian, and the lateral lips are formed later. Consequently, the material on the dorsal meridian becomes tucked in first and reaches further forward on the under-side of the superficial layer of the embryo than does material which is situated more laterally. One result of this state of affairs has already been noted: the piling up of the yolk-cells on the ventral side of the gut-cavity with the resultant rotation of the whole embryo to conform to the new centre of gravity. There is another important result: since the disappearance of the cells from the surface of the embryo and their plunging in over the lip of the blastopore is more active on the dorsal side, there is a consequent stretching of the regions right and left of the dorsal meridian, and a movement towards that meridian to take the place of the invaginated material. In this way, the two horns of the crescent-shaped region of the presumptive neural folds, which at the start of gastrulation were situated at the sides of the embryo, now move nearer to the dorsal mid-line and to one another, so that they form parallel strips which eventually enclose the blastopore between their hindmost ends. Between these parallel strips, the central part of the presumptive neural fold region stretches backwards along the dorsal meridian to the dorsal lip of the blastopore, which it reaches. Thus, instead of lying as a transverse band across the embryo as at the blastula stage, the presumptive neural fold region after gastrulation occupies a position extending longitudinally along the dorsal side of the embryo, where the neural folds will in fact arise.[1] It may be referred to at this

[1] Goerttler, 1925.

stage as the neural plate. All the remainder of the surface of the embryo is now occupied by presumptive epidermis (figs. 3, 7 and 8).

The movements which have brought about gastrulation are therefore also responsible for bringing the presumptive neural fold material into place in preparation for the formation of the neurula, and this, in turn, as will shortly be seen, paves the way for the changes which result in the formation of the tail.

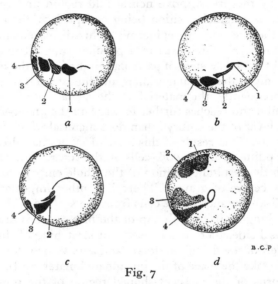

Fig. 7

The process of gastrulation in Urodeles, revealed by the movement of *intra vitam* stain marks placed on the surface of the blastula, as in *a*. The marks stretch and move towards the blastopore rim. In *b* mark 1 has become invaginated; in *d* only mark 4 is left on the surface; the others have become invaginated and passed forwards, forming the gut-wall, and can be seen by transparency through the epidermis. (After Goerttler, *Arch. Entwmech.* CVI, 1925, modified.)

Accompanying the processes of displacement and stretching which have just been described, growth also takes place, which process results in the elongation of the embryo along the line of the original egg-axis, now the antero-posterior axis—in other words, produces growth in length.

§ 5

The neural folds now rise up as a pair of parallel ridges along the dorsal side of the embryo, and come to enclose the blastopore, which is now reduced to a mere slit, between their hind ends. As soon as this has happened, the embryo may be termed a neurula. The groove between the neural folds becomes converted into a tube

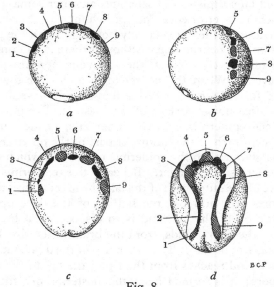

Fig. 8

The process of neurulation in Urodeles, revealed by the movement of *intra vitam* stain marks placed on the surface of the gastrula in a transverse line across the animal hemisphere. *a*, Seen from the dorsal side. *b*, From the right side. *c*, With the progress and completion of gastrulation, the band of stain marks becomes U-shaped, the arms parallel with one another along the dorsal side, and marking the site of formation of the neural folds (*d*). (After Goerttler, *Arch. Entwmech.* CVI, 1925, modified.)

as the neural folds arch over to join one another and fuse above it, and the blastopore is no longer at the surface of the embryo, but is covered over by these folds. In this manner, a neurenteric canal (actual or virtual according as to whether the blastopore is or is not still open) is formed, connecting the cavities of the neural tube and of the gut. After the fusion of the neural folds, epidermis covers the entire surface of the embryo, and the rudiments of all the other structures have come to lie beneath the surface (with the exception of a few sense-organs and placodes).

Meanwhile, inside the embryo, the notochord has become an elongated cylindrical rod above the roof of the gut in the mid-dorsal line. A split within the substance of the mesoderm gives rise to the coelomic cavity: this becomes restricted to the region of the unsegmented lateral plate, and separates an outer somatic from an inner splanchnic layer of coelomic epithelium.

The formation of the tail is closely bound up with the processes of gastrulation and neurulation. Although there is still uncertainty concerning one or two points, the following appears to be the course of events. When the neural folds arch over towards one another and fuse, there is formed a double arch or vault of tissue over the original dorsal surface of the blastula. The outer arch is the superficial epidermis, and the inner arch is the neural tube itself. A backgrowth of the hindmost part of the outer arch of the neural folds gives rise to the epidermis of the tail, which of course becomes progressively longer. Beneath this epidermis, and in consequence of the outgrowth of the tail, the inner arch of the neural folds becomes bent into a J, the bottom of the J occupying the region of the tip of the tail, and is so disposed that the anterior four-fifths of the neural folds, from the brain to the tip of the tail, form the long arm of the J. The other arm of the J is bent ventrally and forwards, and reaches from the tip of the tail to the region of the blastopore; it is formed from the posterior one-fifth of the neural folds. The notochord grows and stretches back between the arms of the J to the tip of the tail, and that part of the inner arch of the neural folds that lies dorsal to it (the anterior four-fifths) gives rise to the definitive neural tube; while that part of the inner arch of the neural folds that comes to lie ventral to the notochord gives rise to the myotomes or muscle-segments of the tail [1] (fig. 9).

There is therefore no undifferentiated tail-bud from which the structures of the tail arise: the neural tube and notochord are present in the neurula, and their hind ends simply grow and stretch backwards into the lengthening epidermal bag which forms the tail, and the material for the muscles of the tail is also present in the neurula in the hindmost part of the inner arch of the neural

[1] Bijtel and Woerdeman, 1928; Bijtel, 1931.

folds. But it is to be noticed that these caudal muscles arise from material that has never been invaginated.

This state of affairs need not give rise to undue astonishment, for the region from which this presumptive caudal muscle material

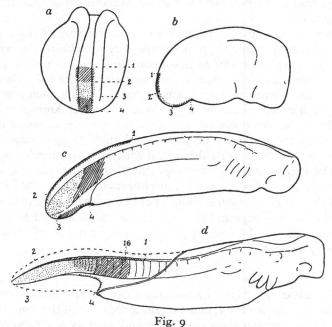

Fig. 9

Four stages of development of an embryo of *Amblystoma mexicanum* to which *intra vitam* stain marks were applied as shown in *a*; mark 1–2 is later found in the epidermis of the mid-dorsal line and in the hinder part of the neural tube; mark 2–3 in the epidermis of the tip of the tail and in the hinder muscles of the tail; mark 3–4 in the epidermis of the mid-ventral line and in the muscles of the base of the tail and the hinder part of the trunk. Mark 3–4 has been invaginated in part; the other marks have not been invaginated, but mesodermal muscles have nevertheless been formed from the median part of mark 2–3. (From Bijtel, *Arch. Entwmech.* cxxv, 1931.)

arises lies immediately to the side of the blastopore at the moment of the latter's closure, and at the blastula stage it lay touching the presumptive regions of the hindmost mesodermal somites of the trunk. It might be said that if the blastopore did not close so soon but remained open for a little time longer, it would tuck in and invaginate this material, which would then differ in no way from the presumptive mesoderm of the trunk. That the presumptive caudal muscle material does not get invaginated is probably due simply to the large amount of yolk present, which fills most of the interior of the embryo and decreases the space available for material to be invaginated.

However, the activities which lead to the uprising of the neural folds, and their fusion, appear of necessity to take in the whole region from anterior end to blastopore, and so this presumptive caudal muscle material, through the mere fact of its being left on the surface, is made to participate in this essentially alien process. Thus in the Amphibia the embryonic structures known as the neural folds do not represent a single ultimate morphological unit, but are composite and represent, in addition to epidermis, two distinct sets of structures, the neural tube and the muscles of the tail. The earliest stages of development of these sets of structures are merely bound up in a single morphogenetic process, the formation of the embryonic neural folds. The distinction between processes involving form-change and those involving chemical predetermination, which it will be necessary to discuss at more length later, is here very evident.

§ 6

The formation of the gut, the notochord, the neural tube, the mesoderm and coelomic cavity, and the tail, together with the elongation of the whole embryo along the antero-posterior axis, are examples of morphological differentiation, as a result of which the main organ-systems of the embryo become roughly blocked out as regards their position and their form. As development proceeds, the remaining organs become roughed out in the same way. Owing to the greater width of the groove between the neural folds in the anterior region, the neural tube is at its first formation already differentiated into regions of brain and spinal cord, the diameter of the cavity of the tube being greater in the region of the brain. From the brain the optic vesicles are pushed out on each side, and become converted into the optic cups by the invagination of their outer sides. Opposite the mouth of each optic cup, the lens is formed as a thickening of the overlying epidermis, and eventually becomes separated off from the epidermis to occupy a position in the mouth of the optic cup. The hypophysis grows in towards the ventral surface of the brain from the epidermis of the front of the head. On the under-side of the head, folds of epidermis give rise in Anura to the ventral sucker, while in many Urodela a finger-shaped outgrowth beneath the eye forms the so-called balancer.

On either side of the brain, behind the eyes, epidermal pits sink in to form the ear-vesicles. These pits arise from the deeper layers of the epidermis, and so the invagination may or may not be covered over by the superficial epidermal layer. At all events, the ear-vesicles soon become closed if they were open, and their original connexion with the epidermis and the exterior is reflected in the endolymphatic duct (open to the exterior throughout life in the Selachii). Another pair of pits, on the snout, gives rise to the nasal sacs and nostrils, and a larger median depression beneath them sinks in and breaks through into the anterior end of the endodermal gut. This anterior ectodermal portion of the gut is known as the stomodaeum, and its aperture of course constitutes the mouth-opening. A posterior ectodermal portion of the gut, or procto-daeum, is formed in a similar manner, close to the point at which the blastopore closed. Its aperture constitutes the anus, and in-ternally it fuses with and breaks through into the hinder end of the endodermal gut.

The fusion of the neural folds has not only resulted in the formation of the neural tube, but it has also led to the inclusion beneath the epidermis of narrow strips of cells, situated along the dorso-lateral sides of the neural tube, which constitute the neural crests. From the neural crests arise the nerve-cells or neurons which make up the ganglia or aggregations of nerve-cells situated on the dorsal roots of the segmented cranial and spinal nerves. Other cells derived from the neural crests give rise to the sheaths in which various nerves are enclosed. In the head region, it appears that the neural crests also give rise to parts of the visceral carti-laginous skeleton. In various places on the surface of the head, thickenings of the epidermis give rise to placodes, which form the sense-organs of the lateral-line system, and also contribute some nerve-cells to the ganglia of some of the cranial nerves. Outgrowths from the sides of the head form the rudiments of the external gills, while the limbs arise (early in Urodela, much later in Anura) as little thickenings which rapidly become conical and continue to elongate by growth.

As regards the internal development, the dorsal portions of the mesoderm of the trunk and the mesoderm of the tail become metamerically segmented, and give rise to the myotomes or muscle plates. These myotomes are at the outset connected with the meso-dermal lining of the general coelomic cavity by short stalks, called intermediate cell-masses or nephrotomes, which, like the myotomes, are segmental in arrangement. From some of these stalks, out-growths are formed, ultimately giving rise to the tubules of the kidney, and from these tubules a duct (the pronephric duct) grows back on each side into the proctodaeum, which from now on can be styled the cloaca.

The heart arises beneath the anterior part of the gut in the mid-ventral line, but the rudiments which form its muscular wall (parts of the splanchnic layer of coelomic epithelium) are at first paired. When they have fused together in the middle line, these rudiments roll up along the longitudinal axis of the embryo to form a tube, suspended by a mesentery (strictly, mesocardium) from the dorsal wall of the coelomic cavity, which in this region takes the name of pericardial cavity. Within the tube thus formed, some cells are en-closed which will give rise to the lining or endothelium of the heart. Originally these cells lay scattered irregularly between the floor of the gut and the splanchnic layer of coelomic epithelium, whence they arose.

The gut-cavity still contains a considerable quantity of yolk-cells, and these are heaped up and occupy most of the central and hinder parts of the gut, being piled up high on the floor, and reducing the actual free cavity to modest dimensions. Just behind the region of the heart, and in front of this mass of yolk-cells, a downgrowth is formed from the floor of the gut. This is the rudiment of the liver: its cavity will eventually develop into the lumina of the liver tubules and gall-bladder, while the connexion with the gut persists as the bile-duct. A lengthening of the gut takes place in the region immediately in front of the cloaca, and this gives rise to the in-testine, which later becomes coiled on itself like a watch spring.

By these processes of stretching, displacement, folding, and growth, morphological differentiation runs its course, and results in the placing of material in particular geometrical relations, roughly in the form and position of the various organs which are to arise. These simple rudiments then undergo growth at particular rates, which rates may be proportional to that of the whole embryo, or faster, or slower. It is obvious that the rate of growth of any particular rudiment relative to that of its neighbour, and any difference in the rates of growth of any one rudiment in the three dimensions of space, contribute essential factors in determining the final form of the organ and of the embryo as a whole (see pp. 225, 366).

§ 7

After the position and form of an organ has been roughly blocked out, there follows the process of elaboration of the cells of the organ for the function which they are to undertake in the organism. This is the process of histological differentiation, or *histo-differentiation*, as it may be more briefly styled. As a result of this process the cells of the neural tube, for instance, become diversified into supporting or ependyma cells and into neuroblasts, which latter produce axon-fibres and give rise to the tracts of the central nervous system and to the ventral nerve-roots. The dorsal nerve-roots are formed as a result of the production of fibres by the cells of the neural crests. In the eye, the various layers of the retina are very early differentiated from one another. Similarly, the cells of the myotomes become differentiated into fibres of striated muscle; mesenchyme cells in particular regions produce cartilage; others elsewhere produce connective tissue, and others again eventually give rise to bone. The cells of the hypophysis, which comes into relation with the floor of the fore-brain or infundibulum to form the pituitary body, become differentiated into the glandular elements characteristic of that body.

Thus, in every rudiment, the cells undergo specialisation to form characteristic tissues, differing from one another and from the simple undifferentiated blastomeres from which all the cells of the embryo arose. When histo-differentiation of an organ has approached completion, the organ is able to enter on a new phase of its development, viz. that of functional activity. Up to this point development has proceeded without function of the organs: indeed, they did not exist at the start and have had to be made. After this point (which does not occur at the same time for all the organs of an organism) development can only proceed with function. Function then perfects the results of the differentiation which has been achieved without it, and is necessary for full and final differentiation.[1]

The onset of function of the organs therefore marks an important epoch in development, and, following Roux, it is possible to distinguish a *prefunctional period* during which morphological and histological differentiation proceed to make the organs ready to enter upon their functions, from a *functional period* during which functional differentiation effects the final elaboration, interdependence, and control of the rudiments, and converts them into the perfected organs of the free-living organism.[2] It will be necessary to say more on this point in the final chapter.

This book concerns itself almost entirely with the prefunctional period. As has been shown, this period is characterised by certain remarkable sequences of morphological and histological processes of differentiation. Complications of structure and texture appear which had previously been absent. The next problem to be tackled, therefore, is the origin of differentiation. This concerns the question as to how developmental processes are causally related to one another in the sequence of events, i.e. whether the development of any given rudiment would take place as it normally does if it had not been for the previous development of some other rudiment, and also the question as to what are the factors, causes, or conditions which are responsible for initiating these sequences of processes of development and differentiation.

[1] The term function is here used to denote function in the ordinary physiological sense, as some specialised activity performed by the organ, normally for the physiological benefit of the organism as a whole. The tissues are always "functional" in the sense of being alive and working, and in addition they may be performing special developmental functions even in that period which is here denoted as the prefunctional period. Nevertheless, the distinction is an important and useful one.

[2] Roux, 1881.

Chapter III

EARLY AMPHIBIAN DEVELOPMENT:
A PRELIMINARY EXPERIMENTAL ANALYSIS

§ 1

It has been shown that even before the amphibian egg is fertilised it possesses one differentiation, in respect of its egg-axis, which determines the future positions of the anterior and posterior ends of the embryo. The factors determining this axis of polarity must be looked for at a stage before the egg is laid, for, while it is still in the ovary, the yolk is already concentrated into one hemisphere. It is possible that the orientation of the blood-vessels with regard to the follicles and developing oocytes in the ovary may be the determining factor. It has been asserted[1] that these blood-vessels are so distributed that the arterial blood reaches the oocyte from one side while the venous blood leaves it at the opposite side. This would cause a gradient in oxidation, and this in its turn would produce a gradient in the relative amounts of cytoplasm and yolk, more yolk being deposited in the regions of low oxidation.

In this particular case, the matter cannot be regarded as certain, since the same author has later qualified his assertion.[2] In other organisms, however, it appears assured that the regions of the oocyte where the rate of oxidation is highest will become the animal pole of the egg and the anterior end of the embryo (see Chap. IV). In the absence of evidence to the contrary, we are justified in assuming that some causal agency of this type is operative in producing the primary polarity of the amphibian egg.

Once the amphibian egg is fully formed, however, gravity will determine that the vegetative hemisphere (containing the relatively heavy yolk) shall be undermost. This is normally brought about by rotation of the egg within its membranes after being laid and fertilised. But if the egg is forcibly inverted and maintained in that position, gravity will determine that the yolk shall flow down to the new lower surface. It does this by means of streaming movements,

---

[1] Bellamy, 1919. [2] Bellamy, 1921.

and except in a few cases, where the vegetative pole is almost exactly uppermost, which condition must be expected to lead to special difficulties in the way of rearrangement of the yolk, such inverted eggs give rise to normal embryos. The cells at what is now the upper pole divide more rapidly than those at the lower pole, regardless of whether they are pigmented or unpigmented, and the dorsal lip of the blastopore appears at the proper level with regard to the vertical axis.[1]

Gravity is therefore responsible for the fact that in many forms the primary egg-axis is brought into a vertical position in normal development, but it is not responsible for the initial formation of the axis; nor is gravity an essential factor in normal development, for eggs withdrawn from the directive action of gravity by being forced to roll about continually in a clinostat,[2] or by being constantly disturbed by a stream of air bubbles,[3] nevertheless develop into normal embryos.

The original determination of the egg-axis, therefore, appears to be due to the development of a primary physiological gradient within the oocyte, which finds visible expression in the graded distribution of cytoplasm and yolk. And this in turn appears to be brought about by factors operative in the ovary which are external as regards the oocyte or egg itself. This point is of considerable importance, for it shows that even this first step in differentiation is externally determined, and is not due to an internal factor or factors.[4] Cases will be met with where the main axis of the future organism is normally not determined until after the egg is laid, and where its direction can be experimentally controlled (p. 60).

§ 2

The next step in differentiation is the acquisition of bilateral symmetry. Localisation of the future median plane of the organism has been shown to depend mainly upon the point of entry of the sperm. This has been demonstrated experimentally in the frog by making the sperm enter the egg on a selected meridian, either by means of a fine pipette, or by laying a thread against one side of the egg and allowing a drop of liquid containing sperm to creep along

---

[1] Pflüger, 1883; Born, 1885.   [2] Roux, 1884.
[3] Kathariner, 1901.   [4] Child, 1924, p. 133.

the thread. The result of the experiment can be checked by cutting the egg into sections, for the path of entry of the sperm is indicated by a trail of pigment leading into the interior of the egg, and the grey crescent which indicates the dorsal meridian can alsò be identified by the retreat of pigment from the surface By this means it can be proved that the grey crescent and therefore the mid-dorsal line is normally opposite or nearly opposite to the point of entry of the sperm. If, as sometimes happens, two sperms enter an egg simultaneously, the grey crescent is determined relatively

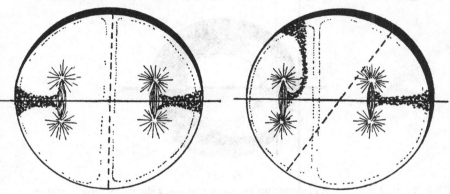

Fig. 10

Diagrammatic equatorial sections through dispermic frogs' eggs, showing that the grey crescent (position of which is indicated by thin outline) is formed opposite the midpoint between the two points of sperm-entry. The plane of symmetry is indicated by a broken line. (From Herlant, *Arch. de Biol.* xxvi, 1911, figs. ix, x, p. 250.)

to them both, and arises antipodally to the meridian half-way between their two points of entry.[1] The second step in differentiation, the acquisition of bilateral symmetry, is therefore also determined mainly in relation to a factor external to the egg (fig. 10).

But, as is very often found in the study of development, the main determining factor is not the sole one capable of exerting an effect. This conclusion is necessitated in this case by studying parthenogenetic eggs. Artificial parthenogenesis may be induced in the egg of the frog by pricking it with a needle dipped in blood or lymph. There is then no point of sperm-entry, and yet the eggs develop

[1] Roux, 1887; Jenkinson, 1909 A; Herlant, 1911.

bilateral symmetry. Furthermore, the plane of symmetry bears no relation to the point of pricking.[1] It is necessary, therefore, to assume that even in the unfertilised egg all the meridians are not perfectly equivalent, and that one of them has some slight differential in respect of the others. This meridional differential, however, must also be supposed to be due to some unequal incidence of external factors operating in the ovary. However this may be, the egg must acquire and possess some feeble determination of a plane of bilateral symmetry which becomes realised in the absence

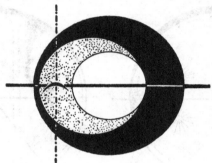

Fig. 11

Cortical localisation of dorsal lip region in frog's egg shown by forced rotation of the egg. Thick line, original plane of symmetry; chain line, new plane of symmetry, passing through centre of grey crescent region (stippled) and mass of yolk which has streamed down to lower pole by gravity. (From Weigmann, *Zeitschr. f. Wiss. Zool.* CXXIX, 1927.)

of any more powerful stimulus, as in the case of artificial parthenogenesis, but which may be overridden by such stimuli as the point of entry of the sperm, or the direction of incident light,[2] or the direction in which the yolk streams down when the egg has been forcibly inverted. In the latter case, the plane of symmetry is determined in such a way as to include the centre of the original grey crescent and the centre of the mass of yolk which has streamed down under the effect of gravity: the dorsal lip of the blastopore therefore arises in the normal position, but the lateral lips form a crescent the concave side of which is always turned towards the mass of yolk, wherever it may be (fig. 11).[3]

---

[1] Bataillon, 1910; Brachet, 1911.   [2] Jenkinson, 1909 A.
[3] Weigmann, 1927.

It appears that once a differential is established, the plane of symmetry will thereby be determined, and that it will be determined just as efficiently by a feeble differential as by a strong one. The possibilities of realising normal bilaterality are thus inherent in the egg; but the factors which determine the fact of its realisation and decide its localisation are external[1].

§ 3

The next step in development is cleavage, the splitting up of the egg by cell-division into a number of smaller cells, the blastomeres. Here, one of the effects of the axes already determined (the antero-posterior, and the dorso-ventral) manifests itself in a differential rate of activity and cell-division, and therefore a gradient in cell size, from the animal pole with its small, actively dividing cells, to the vegetative pole with its more sluggish yolk-containing cells; and, at any given circle of latitude, the cells on the dorsal side divide faster and are therefore smaller than those on the ventral side, at any given time. As will be pointed out in Chap. IX, the main organisation of the developing egg at this stage consists of these quantitative gradients, or, as we shall call them, *gradient-fields*.

The rate of cleavage and subsequent differentiation can be locally altered by subjecting the egg to differential temperature-exposure: one pole or side hot, the other cold.[2]

The amount of yolk present in the vegetative hemisphere of the amphibian egg, while responsible for the larger size of the vegetative blastomeres, is not too great to prevent holoblastic cleavage of the egg. It is possible, however, to make the cleavage of the frog's egg conform to the meroblastic type characteristic of Selachians and

----

[1] It might be supposed that the bilateral symmetry of the egg, once established, is necessarily identical with that of the resultant embryo. However, Jenkinson (1907, 1909 A) by means of an elaborate biometrical study has shown that the correlation between the two, though high, is not perfect: in other words, the grey crescent does not always lie exactly in the future mid-dorsal line. Thus both the determination of the grey crescent in the meridian of sperm-entry, and that of the axis of bilateral symmetry of the embryo in the meridian of the grey crescent are imperfect. In spite, however, of the slight elasticity of the determination at these two links in the causal chain, it is clear that in normal development the symmetry of the embryo is mainly determined by the point of sperm-entry. See also Tung, 1933.

[2] Huxley, 1927; Gilchrist, 1928, 1929; Vogt, 1928 B.

Sauropsida, by means of centrifugalisation. The eggs orientate themselves in the centrifuge tube in such a way that the animal pole is directed centripetally, and the yolk is concentrated into an abnormally dense mass at the vegetative pole. Cleavage then results in the formation of a disc of cells or blastoderm resting upon an undivided mass of yolk. The nuclei of some of the blastomeres migrate into the yolk and become enlarged, irregular and chromatic, and thus resemble the "yolk-nuclei" (bodies responsible for the precocious digestion of the yolk) characteristic of selachian development (fig. 12).[1]

The causes of cleavage concern the problem of cell-division, which, as such, lies outside the scope of this book.

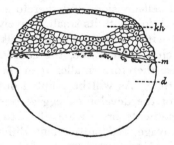

Fig. 12

Modified cleavage of frog's egg, under the influence of centrifugal force. The yolk (d) is concentrated in the vegetative hemisphere, and cleavage results in the formation of a blastoderm. m, yolk-nuclei; kh, blastocoel. (After Hertwig, from Jenkinson, *Experimental Embryology*, 1909.)

§ 4

Following upon cleavage, the next step is gastrulation. This process, which, of course, results in the conversion of a single-layered hollow ball (the blastula) into a double-layered sac (the gastrula), is heralded in Amphibia by the appearance of the dorsal lip of the blastopore at a particular latitudinal level on the blastula, in the dorsal meridian. The level at which the lip appears is under the control of the primary physiological gradient along the egg-axis,

[1] Hertwig, 1897, 1904; Jenkinson, 1915.

and can be altered by experimental means (see fig. 149 and p. 320).

In Amphibia, it has been found that the act of gastrulation can be analysed into a number of component processes. First, there is the tendency on the part of the cells of the animal hemisphere to expand and cover a larger surface. Next, the cells which constitute the *marginal zone* between the animal and vegetative hemispheres tend to stretch downwards towards the vegetative pole. This is accomplished by rearrangement of the cells, with the result that the ring-shaped band, increasing in depth, attempts to decrease in

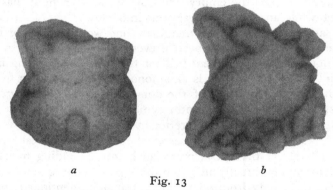

a   b

Fig. 13

The expanding growth-tendency of the presumptive epidermis of the Urodele embryo. *a*, Two ventral gastrula-halves grafted together, the epidermis of each of which is thrown into ridges and folds in vainly trying to overgrow the other. *b*, The same, 16 hours later, showing intensification of ridges and folds. (From Spemann, *Arch. Entwmech.* CXXIII, 1931.)

diameter. Thirdly, the cells just beneath the marginal zone in the dorsal meridian have the tendency to invaginate and form a pit-like depression. Normally, of course, all these processes take place together, with the result that the excess of material obtained by the stretching of the marginal zone becomes tucked into the invagination round the rim of what may now be called the blastopore. New material, as it arrives at the rim, becomes tucked in, and this tendency to roll or tuck in is also an independent process. Meanwhile, the space vacated on the surface by the material which has thus been invaginated, is occupied by the shifting and expanding regions of the animal hemisphere.

By a simple operation, the constituent processes of gastrulation can be dissociated from one another. Removal of a portion of tissue at the animal pole of a blastula leads to closure of the wound by approximation of the cut edges. This results in raising the marginal zone above the equator of the egg. Nevertheless, this zone soon shows its characteristic stretching movements, and decreases its diameter. Normally, of course, this decrease in diameter corresponds to the curvature of the egg from the equator to the vegetative pole. But as the marginal zone is now above the level of the equator, it cannot simply grow down over the vegetative hemisphere: instead, it constricts the embryo into the form of an hour-glass.

Meanwhile, an invagination appears in the lower half of the hour-glass, at a place which the marginal zone would normally have reached, but which it has been prevented from reaching by the conditions of the experiment.[1] That the process of rolling in or diving beneath the surface is an autonomous one is shown by the fact that isolated portions of the dorsal lip region, when grafted into strange situations in another embryo, promptly proceed to transfer themselves into the interior by this means.

The fact that all these processes should begin and take place more actively at the dorsal meridian before extending to lateral meridians and eventually all round the egg, is a consequence of the gradient of activity from dorsal to ventral side, mentioned above. While the marginal zone is stretching, overgrowing the vegetative hemisphere, and being invaginated and tucked in round the lip of the blastopore, and the presumptive neural fold region is being stretched and displaced, thus taking the place of the presumptive primitive gut-roof which is being invaginated, the presumptive epidermis region expands and extends by growth so as to cover the area vacated by the presumptive neural folds. This growth-tendency on the part of presumptive epidermis is also shown by isolated pieces when grafted,[2] and by two ventral half-gastrulae grafted together: the epidermis of each half tries in vain to overgrow the other (fig. 13), with the result that it is thrown into numerous folds.[3]

The harmonious co-operation of all these processes, which normally result in gastrulation, can be thrown out of gear by interference with the gradients, and alteration of the relative rates of activity in

[1] Vogt, 1922.   [2] Mangold, 1924.   [3] Spemann, 1931.

different parts of the embryo,[1] or by changes of shape, such as those which are consequent on the release of the embryo from its vitelline membrane[2] (see Appendix, p. 481).

With regard to the actual paths of displacement followed by the invaginated tissues during these "mass movements" which bring about gastrulation, it may be said that the nearer any given piece of tissue is to the dorsal lip of the blastopore at the outset of gastrulation, the farther forward in the embryo will it find itself when that process is completed. So, those cells which occupy the place where the invagination first forms become the front wall of the fore-gut; those cells of the marginal zone in the mid-dorsal line which are the first to be tucked in form the tip of the notochord (figs. 3 and 4).

§ 5

Attention may now be turned to the presumptive regions of the future organs. As has been shown in Chap. II, these regions can be mapped out on the blastula, although there are no visible limits to distinguish them. The question arises as to how these various regions have their respective fates allotted to them.

The first point to make clear in any discussion of the origin of differentiation is the fact that it is impossible to appeal to differences between the nuclei of the cells of the blastula in order to account for the eventual differentiation of those cells. By making eggs undergo cleavage under compression between glass plates, the normal regular sequence of directions of cleavage can be disturbed, so that the nuclei come to be situated in cells other than those in which they would find themselves in normal unhindered cleavage. Nevertheless, the development of embryos so treated and then released from pressure is normal, and it is therefore clear that it is quite immaterial whether any given nucleus finds itself in one particular cell or in another.[3] This is confirmed in other ways and on other forms (see p. 85 and fig. 36).

This means that there is no inequality in nuclear division during early cleavage, and it is therefore impossible to attribute any deter-

---

[1] Huxley, 1927; Vogt, 1928 B; Gilchrist, 1928, 1929; Dean, Shaw, and Tazelaar, 1928; Tazelaar, Huxley, and de Beer, 1930; Castelnuovo, 1932.
[2] Spemann, 1931.
[3] Hertwig, 1893; Spemann, 1914, 1928.

minative effect to differences between the nuclei. The position of
any given nucleus in one or another presumptive region is without

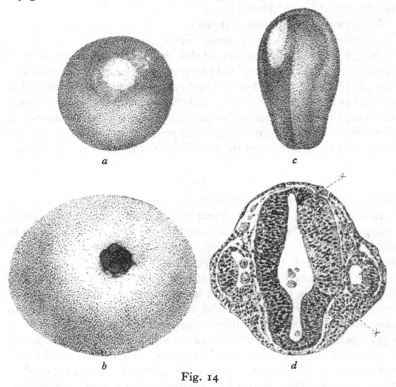

Fig. 14

The development of regions by dependent differentiation during the stage of
plasticity, before the onset of irreversible determination. At the early gastrula
stage, a piece of presumptive epidermis (gill region) of *Triton cristatus* is
exchanged for a piece of presumptive brain region of *T. taeniatus*. *a*, The dark
taeniatus embryo with the light *cristatus* graft. *b*, The light *cristatus* embryo
with the dark *taeniatus* graft. *c*, The *taeniatus* embryo at a later stage, with the
cristatus graft in the region of the left side of the brain. *d*, Transverse section
through the *taeniatus* embryo, showing part of the wall of the forebrain (between
X–X) formed from the grafted light-coloured *cristatus* tissue, which has undergone dependent differentiation according to its surroundings. (From Spemann,
Arch. Entwmech. XLVIII, 1921.)

effect on the subsequent normal differentiation of that region. The
key to the origin of the differentiation of the various regions must

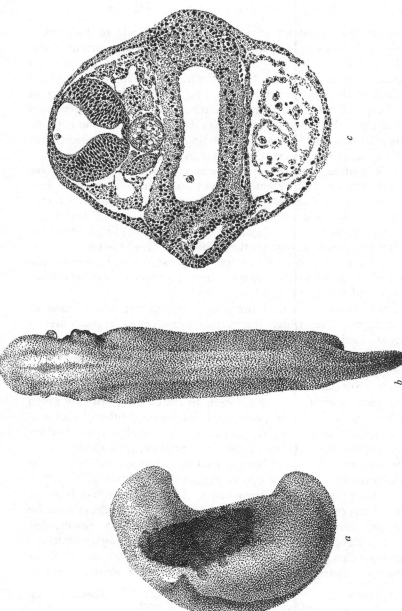

Fig. 15

a, The *cristatus* embryo, shown in fig 14 *b*, with the *taematus* graft in the region of the gill epidermis on the right side. *b*, Dorsal view of the same embryo at a later stage, showing the more advanced development of the gills formed from the grafted *taematus* tissue. *c*, Transverse section through *b* in the gill region. (From Spemann. *Arch. Entwmech.* XLVIII, 1921.)

therefore be looked for in some factor which affects the various regions of the cytoplasm in general, and not of the nucleus alone.

By the method of transplantation it can be shown that, up to a certain stage in the gastrulation of the newt, the fates of most of the regions of the embryo are not irrevocably determined. A piece of presumptive neural tube material removed from its embryo and grafted into the side of another, may differentiate into the external gills of its new host if it happened to be grafted into the presumptive gill region of the latter. Conversely, a piece of presumptive epidermis grafted into the appropriate region of the presumptive neural tube of another embryo, will undergo differentiation into part of the brain and the eye. Up to this stage of gastrulation, therefore, the regions develop according to their actual surroundings, and regardless of their origin and former surroundings:[1] they are in fact still plastic as regards their final fate. Even the future germ-layers are plastic up to this stage, for presumptive epidermis can be made to differentiate into mesodermal structures such as muscle fibres, and *vice versa* (figs. 14, 15 and 16).[2]

There comes a critical time, however, during the process of gastrulation, after which the various presumptive regions are no longer plastic. Their fates are then irrevocably determined, and, whatever the position into which they may be grafted, pieces of any given presumptive region will then undergo the differentiation which is typical of that particular region in normal development. *Pari passu* with the determination to differentiate in any given direction goes the loss of power to differentiate in other directions. In other words, the regions can then only develop towards their presumptive fates. One can then, for instance, graft the presumptive eye region from one late gastrula into another, and obtain the differentiation of a typical eye, facing into the body cavity (fig. 17)[3]. Something invisible has happened to fix the prospective fates on the various presumptive regions, and since this something must be due, presumably, to chemical changes in the various regions, this phase of development may be referred to as *chemo-differentiation*.[4] Through this process the organism has become a patchwork or mosaic of separately determined regions. It is of some interest to

---

[1] Spemann, 1918.   [2] Mangold, 1924.   [3] Spemann, 1919.
[4] Huxley, 1924; Goldschmidt, 1927; Bertalanffy, 1928.

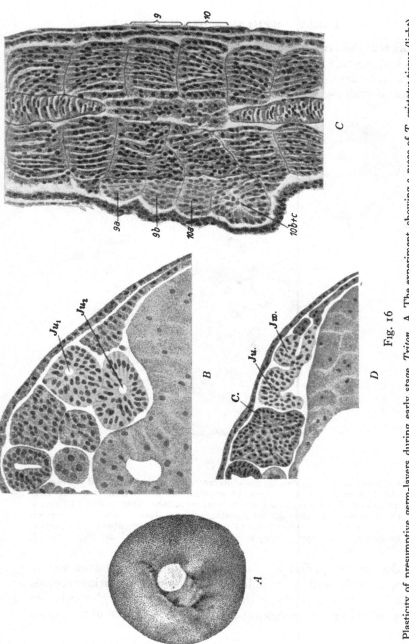

Fig. 16

Plasticity of presumptive germ-layers during early stage, *Triton*. A, The experiment, showing a piece of *T. cristatus* tissue (light) grafted into the dorsal lip of a gastrula of *T. alpestrs*. B, Mesodermal somites (*Ju₁, Ju₂*) formed from presumptive epidermis C, Myotomes formed from presumptive epidermis (*9a–10c*); 9, 10, normal myotomes of other side. D, Pronephric tubules formed from presumptive epidermis (*Jw.*); C. cutis. (From Mangold, *Arch. Mikr. Anat. u. Entomech.* C, 1924.)

note that the existence of this mosaic phase, so different in its total lack of plasticity and of power of regulation from anything known in other stages of development, was only detected through experimental analysis.

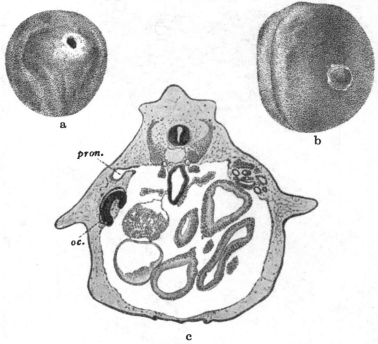

Fig. 17

The development of regions by self-differentiation after the stage of irreversible chemo-differentiation. *a*, A piece of presumptive eye region from an early neurula of *Bombinator* is grafted into the flank of another embryo of similar age, *b*. *c*, Transverse section through the resulting embryo showing the eye-cup which has developed by self-differentiation from the graft in its abnormal position; *oc.* grafted eye-cup; *pron.* pronephros. (From Mangold, *Ergeb. der Biol.* III, 1928, after Spemann.)

During the period of plasticity, chemo-differentiation sets in progressively until irrevocable determination of the various regions is achieved. But although the regions may still be plastic before this critical time, in the sense that they can be made to undergo a

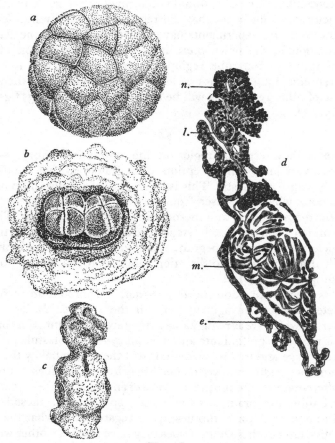

Fig. 18

The labile determination of regions. *a*, Morula of *Triton*, showing the cells subsequently isolated indicated by lines. The remaining cells were destroyed with needles, their contents forming a supporting and nutritive medium surrounding the four living cells. *b*, 1 hour after operation, the four living cells have divided to form eight. *c*, 9 days after operation, the cells have produced a complicated structure. *d*, Section through *c*; nervous tissue (*n*.), a lens (*l*.), muscle segments (*m*.) and epidermis (*e*.) have been differentiated. It is to be noted that no organiser was present in the explanted cells and that no gastrulation took place; the differentiations are therefore the effects of labile determinations of the cells themselves. (From Holtfreter, *Arch. Entwmech.* CXXIV, 1931.)

differentiation which they would not normally have carried out, this plasticity does not mean that the regions are entirely indifferent. On the contrary, experiments have shown that even at the start of gastrulation in the newt, there is a feeble determination of the presumptive neural tube region, in virtue of which it tends to differentiate along the lines of its prospective fate,[1] and the same is true of other regions, as will be seen later (pp. 136, 203, figs. 18, 62, 63, 64).

§ 6

There is, however, one region of the amphibian embryo which makes a very important exception to the statement that the tissues at early stages are plastic. This is the region of the dorsal lip of the blastopore, which has arisen from the grey crescent and is destined to form the notochord and mesoderm (chorda-mesoderm). This is determined from very early stages (possibly even in the fertilised egg before cleavage has begun). When grafted into other embryos it will differentiate in no direction other than that of its normal presumptive fate.[2]

This presumptive notochord, gut-roof, and mesoderm region is predetermined to invaginate beneath the surface. It has other properties which are as remarkable as they are important. If a portion of this region be grafted into another embryo in the blastula or early gastrula stage and in any position, it will there pass below the surface and proceed to induce the neighbouring host-tissues to undergo differentiation into the main organs of an embryo, often including neural tube and brain, eyes and ears, spinal cord, mesodermal somites and pronephric tubules, quite regardless of what the presumptive fates of these host-tissues may have been. In other words, the dorsal lip of the amphibian blastopore has the property of being able to force other tissues (during their state of plasticity) to undergo the organised differentiations and developments which lead to the production of an embryo. For this reason, the dorsal lip of the blastopore has received the name of *organiser*, as the German term *Organisator* coined by Spemann may be translated[3] (figs. 1, 19, 65).

[1] Goerttler, 1926; Holtfreter, 1931 A.
[2] In certain conditions, as when cultivated *in vitro*, etc., it may give rise to other organs, such as nervous system and gut (Holtfreter, 1931 A).
[3] Spemann and Mangold, H., 1924.

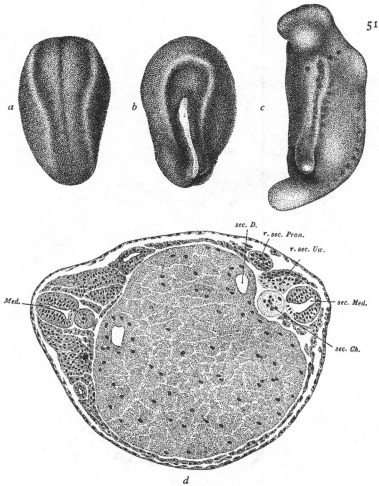

Fig. 19

The induction of secondary embryos by organiser grafts. *a*, Dorsal view showing primary embryo of *Triton taeniatus*. *b*, Side view showing secondary embryo induced by grafting an organiser of *T. cristatus* (distinguishable by lack of pigment) into the flank of *a*. *c*, Dorsal view of secondary embryo, and left side view of primary embryo, at later stage; note ear-vesicles of secondary embryo in line with that of primary. *d*, Transverse section through *c*. *Med.* neural tube of primary embryo; *r. sec. Pron.* pronephric duct; *r. sec. Uw.* mesodermal somite; *sec. Ch.* grafted notochord; *sec. D.* gut; *sec. Med.* neural tube; of secondary embryo. Note that most of the structures of the secondary embryo have been induced from host tissues, but that the graft has contributed to some (distinguishable by lack of pigment). (From Spemann and Mangold, *Arch. Mikr. Anat. u. Entwmech.* C, 1924.)

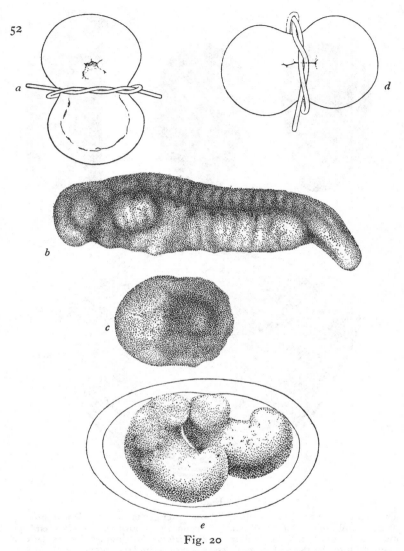

Fig. 20

The presence of the organiser region is essential for development. *a*, A newt's egg constricted into two in the transverse plane, thus separating dorsal and ventral halves. *b*, The result of isolation of a dorsal half (containing the organiser region): a perfect embryo. *c*, The result of isolating a ventral half (lacking the organiser region): a blastula-like ball of cells which develops no further. *d*, A newt's egg constricted in the plane of symmetry, thus separating lateral halves, each of which contains a portion of the organiser region, and, *e*, develops into a perfect embryo. (From Spemann, *Naturwiss.* IV, 1924.)

The vital importance of the organiser for development is shown by the classical experiment of separating the first two blastomeres of the newt's egg. If the plane of the first cleavage separates the future right and left halves of the body, both blastomeres will receive a portion of the organiser region, and both will organise themselves and produce miniature but otherwise normal embryos.[1] But if the first cleavage separates future dorsal and ventral halves, only the dorsal half will produce an embryo; the ventral half undergoes cleavage and makes an abortive attempt to produce germ-layers, but develops no further[1] (fig. 20). The same is true in the case of the frog.[2]

The action of the organiser raises a number of important problems which will receive more detailed consideration in a subsequent chapter. For the moment, attention may be focussed on the light which these phenomena throw on the analytical study of development.

§ 7

It has been seen that the newt's egg when fertilised has already had two determinations imposed upon it: that of polarity and that of bilateral symmetry. As a result of these determinations, one region, the future organiser, is localised and apparently fully determined at very early stages. Until a certain time, which is roughly half-way through the process of gastrulation, the various other regions of the embryo are still plastic, although they are presumably passing through the preliminary stages of chemo-differentiation. But the time comes when they, too, are irreversibly determined to follow the course of differentiation which characterises each part in normal development.

The terms *independent* or *self-differentiation* and *dependent differentiation* were introduced by Roux to characterise these two types or phases of differentiation. In Amphibia before gastrulation, all regions save that of the organiser show dependent differentiation: their developmental fate is dependent upon and conditioned by factors external to themselves—in this case the presence of an organiser in a particular spatial relation with them. This is proved

[1] Herlitzka, 1896; Spemann, 1903.　　　　[2] Schmidt, 1930, 1933.

by the two types of experiment we have mentioned; the grafting of tissues into abnormal positions relative to an intact organiser, and the grafting of an organiser in abnormal positions relative to an otherwise intact host embryo.

However, after a certain critical time during gastrulation, the various main regions develop, in respect of the type of tissue they produce, by self-differentiation. A piece of tissue grafted into an abnormal situation no longer has its fate determined by its position in relation to other tissues; the factors controlling its development are now situated within itself.

Of course, all differentiation is in certain respects dependent, in others independent. When grafts are made from one species to another before gastrulation, the grafted piece shows dependent differentiation as regards the organs and tissues which it forms, but self-differentiation as regards various fundamental characters such as cell-size and pigmentation (see p. 142). Conversely, in certain respects the fate of a piece of tissue in the self-differentiating phase is dependent on external conditions, for, as we shall see (p. 249), the development of its shape is dependent on mechanical factors in its new situation, whereas the type of tissue which it produces is not.

In experimental embryology, the terms are generally used in respect of dependence of *type of tissue* produced upon the activities of other parts of the embryo. Dependence upon external agencies is not usually discussed in this connexion (although some differentiations such as that of polarity are dependent upon them), but these are assumed to remain more or less constant, within the range permitting of normal development; and form-differences due to purely mechanical distortion are also usually omitted from consideration. Within these limits, the terms will be found very useful.

Other examples of self-differentiation are to be found in the development of the organiser region, of the eye-cup and of many other organs mentioned in Chap. VII, and of particular types of tumours and cancers irrespective of their site. Other examples of dependent differentiation which will be met with are the dependence of the lens and conjunctiva upon the eye-cup (pp. 178, 183), of the ear-capsule upon the ear-vesicle (p. 175), the dependence of amphibian metamorphosis upon a certain concentration of

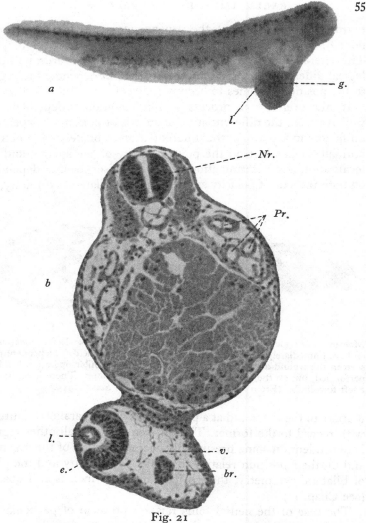

Fig. 21

The dependence of lens-differentiation on the optic-cup in *Triton*. *a*, Larva into which at the mid-gastrula stage a piece of presumptive brain region from another embryo of the same age was grafted. The graft (*g.*) developed by self-differentiation into parts of the brain, and an eye-cup which induced the formation of a lens (*l.*) from the ventral trunk epidermis of the host. *b*, Section through the same larva showing (*v.*) the vesicle formed from host tissue and containing the graft; *br.* portion of grafted brain; *e.* grafted tissue differentiated into eye-cup; *Nr.* spinal cord; *Pr.* pronephric tubules, of host embryo. (From Mangold, *Arch. Entwmech.* CXVII, 1929.)

thyroid hormone (p. 427), the dependence of the fine structure of bone upon the functional stresses to which it is exposed (p. 434). The lateral line in tadpoles is independent as regards its histological differentiation and increase in size, but dependent in regard to the position it comes to occupy (p. 355).

In first origin, each process of differentiation is dependent. As we have seen, the differentiation of an axis of polarity is dependent on factors in the ovary; the differentiation of bilateral symmetry is normally dependent on the point of entry of the sperm; and the localisation and determination of the organiser itself is dependent on both the axis of polarity and the plane of bilateral symmetry, for

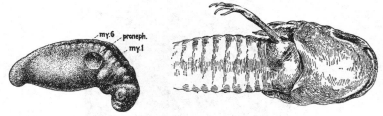

Fig. 22

Mosaic stage: localised determination of limb-potencies. Left: *Amblystoma* embryo immediately after removal of the right fore-limb field. The pronephros is seen in the wound-area. Right: a larva on which a similar operation has been performed, but on the left fore-limb field, 3 months later. There is no trace of a left fore-limb. (From Harrison, *Proc. Nat. Acad. Sci.* 1, 1915.)

it arises in the latter and at a particular level (or parallel of latitude) with regard to the former. The differentiation of all other regions is dependent on some presumably chemical action of the organiser and on their position relative to the axis of polarity and the plane of bilateral symmetry, though in a manner which is still obscure (see Chap. ix).

The case of the neural folds raises a problem of particular interest, for, as has been mentioned, not only can neural folds arise by chemo-differentiation *in situ* even if the organiser is removed or prevented from invaginating, but also the organiser is capable of inducing the formation of neural folds wherever it is grafted. In normal development, that tissue which the organiser normally induced to become neural folds is also that which in the absence of the organiser can become neural folds by self-differentiation. There seem, therefore, to be two methods by means of which neural folds can arise; such a phenomenon is referred to as "*double assurance*". Further discussion of this question is given in Chap. VI (pp. 139, 187).

As soon as some organs have reached the stage of full self-differentiation, they become able to induce other organs to arise by dependent differentiation. In many forms, for instance, the eye-cup induces the formation of a lens from the overlying epidermis (see p. 183), in a manner analogous to that by means of which the organiser induces the formation of neural folds (fig. 21). How general such secondary induction may prove to be in development is not as yet known.

However, we do know that in many cases what is first determined is a large region or field, and that later this region becomes split up into a further mosaic of independently determined subregions. For instance, as set forth more in detail in Chap. VII, the limb area is early determined as a region in the flank (fig. 22): only later are the various subregions, such as hand, forearm, upper arm, determined within the main region.

§ 8

During the period of self-differentiation, the embryo is thus a patchwork or mosaic of developing regions, the differentiation or localisation of all of them being originally dependent on something else, ultimately on the axis of polarity and plane of bilateral symmetry. The differentiation, however, is progressive, the mosaic coming to consist of more and smaller pieces, each of which eventually undergoes independent differentiation.

At this stage, almost the only integrating influences acting upon the embryo appear to be the simple ones of mechanical construction. Biological integration is almost absent: neither neural nor humoral correlation is yet possible, and little trace has been detected of influences analogous to that of the organiser or the optic-cup, or of chemical influence by contact. The chief exception appears to be that the polarity of the egg may persist to cause the polarisation of some on all of the separate organ-rudiments (see Chaps. VII and X). The embryo at this stage is like a multiple tissue-culture, the parts of which happen to cohere mechanically in a particular form: the only correlations are mechanical ones.

This lack of co-ordination accounts for the fact that, whether by regulation or regeneration, the making good of material or of parts that have been lost appears to be impossible during this stage of regional self-differentiation of the various organs,[1] although regulation was possible at the stage of the egg, blastula, and early gastrula, and regeneration will become possible in the larva. The loss of the earlier power of regulation seems to be due to the superposition upon the original unitary gradient-field system of a patchwork of independent chemo-differentiated regions (pp. 221, 350); while the later appearance of the power of regeneration is in the main due to the onset of growth, which in turn depends upon the acquisition of function by the nervous and vascular systems. The latter introduce the possibility of nervous and humoral correlation, and further make possible the mutual interplay of the functions of the various organs as soon as their histological differentiation has proceeded far enough to enable their tissues to function and so permit them to perfect their final development by functional differentiation (see Chap. XIII).

§ 9

From the foregoing sketch it will be obvious that development in Amphibia is epigenetic, and involves the creation of differentiation afresh in each and every generation. There can be no question of preformation, for the structures of the future organism are not there, nor are their positions localised or determined in the unfertilised egg. This epigenetic character of development is based on the capacity of the protoplasm of the egg to react in a particular way to certain stimuli which in the first instance are external, as when the egg-axis and plane of bilateral symmetry are induced, and then later internal, as when the tissues are induced to differentiate under the influence of an organiser. The whole of development is a series of such reactions or responses to stimuli. It therefore follows that no development can be normal in an abnormal environment, and, also, that the hereditary endowment of an organ-

---

[1] Harrison, 1915; Spurling, 1923. (See also figs. 22, 94.)

ism, represented by the inherited factors transmitted to it by its parents, is by itself insufficient to account for development.

Development is always the product of an interaction between a specific protoplasm and hereditary outfit on the one hand, and a particular complex of environmental factors on the other. The environmental factors operative with regard to any part of the organism are partly those of the external world, partly those of the internal environment provided by the rest of the organism.

Chapter IV

THE ORIGIN OF POLARITY, SYMMETRY, AND ASYMMETRY

§ 1

It has been seen that when the amphibian egg is laid, all that can be said about its future development is that the anterior end of the animal will be formed near the animal pole, and the posterior end near the vegetative pole. In all animals above the Protozoa, the primary differentiation during their development is this *axiation*, as Child calls the determination of the axis of polarity.

It is of great importance to realise that the factors invoked in order to explain the determination of polarity are external to the egg. In the sea-weed *Fucus*, it is found that the determination of polarity is normally due to the direction of incident light.[1] But it has been shown experimentally that the application of an electric current is also capable of inducing the determination of the axis of polarity in the egg of *Fucus*, in any direction, at will.[2] Further, it is found that when *Fucus* eggs are placed in groups very close to one another, each egg develops a polarity in such a way that its apical point faces away from the group.[1] Here it seems that a chemical factor is responsible, for the CO_2-tension will be higher and the oxgen-tension lower in the midst of the eggs in the group than in the surrounding fluid (fig. 23).

One of the agencies capable of inducing polarity in the egg of *Fucus* thus appears to be differential exposure to oxygen, and the same is true of many animals. In the sea-urchin the oocyte develops with one pole attached to the wall of the ovary, and the other pole projecting freely into the cavity, and exposed to the ovarian fluid and nutritive wandering cells. It appears that the centre of this portion, where physiological exchange with the immediate environment is most active, will become the animal pole of the egg.[3] Similar cases, where the attached and free surfaces of the

---

[1] Hurd, 1920; Whitaker, 1931.   [2] Lund, 1923 B.
[3] Jenkinson, 1911; Lindahl, 1932.

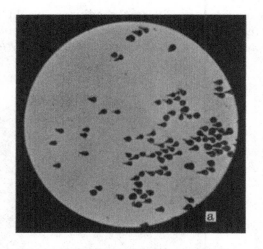

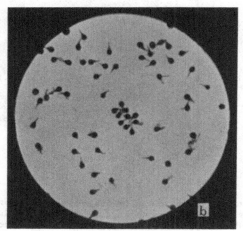

Fig. 23

Electrical control of primary polarity in the eggs of the brown alga *Fucus*. Two-cell stages. *a*, Eggs subjected to appropriate current density. They are practically all oriented with the smaller end towards the anode. *b*, Eggs serving as control, subjected to current density below the threshold requisite to secure orientation (equivalent to a fall of 0·025 volt across the egg). The eggs point at random. (From Lund, *Bot. Gaz.* LXXVI, 1923.)

oocytes are exposed to different conditions, would appear to account for the polarity of the unfertilised egg in many other forms, e.g. *Chaetopterus*,[1] *Sternaspis*,[2] *Cerebratulus*,[3] and *Cyclas*,[4] where

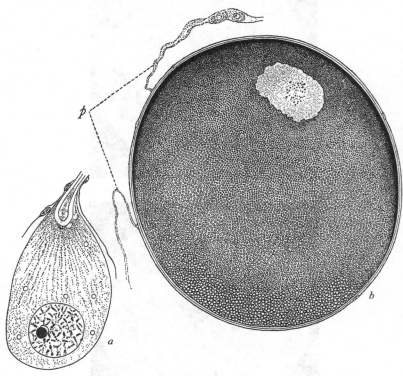

Fig. 24

The primary gradient in oocytes. *a*, In the Annelid *Sternaspis* the oocyte is attached by a narrow peduncle containing a vascular loop, and the nucleus is at the opposite end, which protrudes into the ovarian cavity. The attached end becomes the vegetative pole, the free end the animal pole. *b*, Gradient in amount of yolk and size of yolk-spheres in the oocyte of the frog; *p*, pedicle of attachment. (From Child, *Physiological Foundations of Behavior*, New York, 1924.)

the exposure is to the ovarian fluid, and the Coelenterate *Phialidium*,[5] where the exposure is to the surrounding sea-water. In all the above-mentioned cases the surrounding fluid, be it ovarian

[1] Lillie, 1906.   [2] Child, 1915 B, p. 341.   [3] C. B. Wilson, 1900.
[4] Stauffacher, 1894.   [5] Child, 1921 B, p. 54.

or sea-water, may be regarded as containing more oxygen than the tissues of the ovary. But in other forms, such as vertebrates, not only does the coelomic fluid in such small enclosed spaces as the ovarian cavity lose its respiratory function, but the ovary itself is well supplied with blood-vessels, and there is therefore reason to believe that the oxygen-supply of the tissues of the ovary is greater than that of the fluid surrounding the ovary. It is consequently of great interest to find that in birds the exposed side of the oocyte becomes the vegetative pole of the egg, while the attached side becomes the animal pole.[1] The same is true in *Amphioxus*, but here the attached side of the egg is turned towards the secondary ovarian cavity which is close to the atrial cavity, from which oxygen is probably derived. (For the frog, see p. 35.)

Asexual reproduction and regeneration phenomena also provide a number of examples in which polarity is induced from the outside, and such cases are, from the standpoint of general theory, as important as those concerning development from an egg. An axis of polarity can be experimentally induced in regenerating fragments of the Hydroid polyps, *Obelia* and *Corymorpha*. These organisms are built on a radially symmetrical plan, with an axis passing down from the oral end of the polyp along the stem. If a piece of *Obelia* stem be isolated, it normally retains its polarity, as shown by its regenerating a polyp at the original distal end earlier than at the proximal end. But if such cut pieces are subjected to the passage of an electric current of a certain strength through the water in which they are lying, it is found that regardless of the original polarity of the pieces, polyps are regenerated only at that end which points towards the anode: while stems (or stolons) may be formed from the end which is directed towards the kathode. This shows that the original polarity can be overridden by external stimuli such as an experimentally controlled electric current[2] (fig. 25).

A piece of the stem of *Corymorpha* regenerating in normal sea-water likewise retains its polarity, and regenerates a polyp at its originally distal end. But if such a piece is placed in water containing weak poisons, it dedifferentiates and loses its form, becoming converted into a banana-shaped mass lying on the bottom of the vessel. Replacement in clean water will lead to regeneration of a

---

[1] Conklin, 1932.   [2] Lund, 1921, 1923 A, 1924.

polyp, not, however, from either of the original ends, but from the central portion of the piece which is farthest from the glass bottom of the vessel and most freely exposed to the water and therefore to oxygen.[1] In this case, an original polarity-gradient has not merely

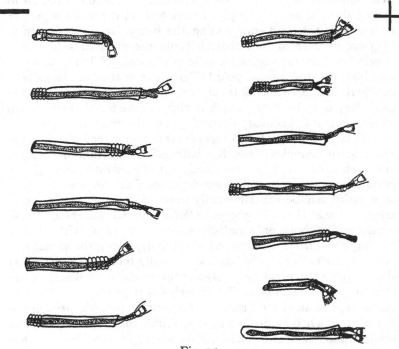

Fig. 25

Control of polarity by external agencies in Hydroids. A series of internodes of *Obelia* regenerating towards the anode when exposed to the passage of a weak electric current. The control series at the same stage had all regenerated hydranths at both cut ends. (From Lund, *Journ. Exp. Zool.* XXXIV, 1921.)

been reversed, but the original polarity has been obliterated, and a wholly new polarity induced (fig. 26).

The winter-buds of the social Ascidian *Clavellina* appear to be irregular aggregations of cells with no relation to the polarity of their parent. The polarity of the zooids which later arise from these

[1] Child, 1925 B, 1927.

buds must therefore be imposed on them from without.[1] An even more striking example is the masses produced by the joining up of cells and cell-groups after the tissues of sponges and hydroids have been strained through fine gauze, ground up with sand, or otherwise dissociated. Here, clearly, all traces of the original polarity must have been lost. However, the masses may later develop into miniature sponges with polarity of their own. This

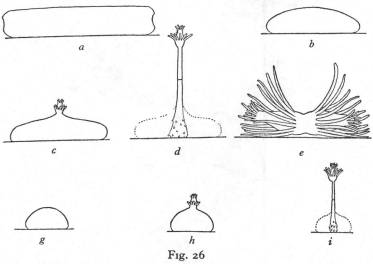

Fig. 26

Experimental imposition of a new primary axis in *Corymorpha*. In fragments of stem immediately after cutting (*a*) or after regeneration to form biaxial (*c*) or single hydranths, immersion in dilute alcohol causes dedifferentiation (*b*, *g*). On replacement in sea-water, redifferentiation occurs with a new axis at right angles to the old, with apical region at the centre of the free surface (*c*, *d*, *h*, *i*). (From Child, *Physiological Foundations of Behavior*, New York, 1924.)

polarity must have been induced by external factors.[2] Similar results have been obtained with hydroids (figs. 27, 132; see also p. 281).

Thus, apart from the cogent theoretical reasons advanced by Child, there is abundant evidence, experimental and circumstantial, for the view that the initial determination of an axis of polarity, or axiation, is due to the action of factors external to the developing organism.

[1] Huxley, 1926; Brien, 1930.
[2] H. V. Wilson, 1907, 1911; Child, 1928 B; Huxley, 1911, 1921 A.

Once the axis of polarity has been determined in an egg, it often becomes manifested by a stratified and graded distribution of egg-contents, some of which may be visibly distinct, such as pigment, fat, yolk, etc. (e.g. *Arbacia*). It is to be noted, however, that this stratification of materials is only an effect and not a cause of polarity.

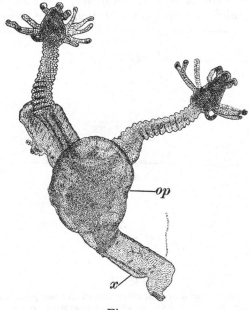

Fig. 27

Differentiation after dissociation in the hydroid *Pennaria*. The dissociated cells united to form rounded reconstitution-masses which surround themselves with perisarcs (*op*) and later form outgrowths which give rise to stolon-like structures (*x*) and normal hydranths. (After H. V. Wilson, from Gray, *Experimental Cytology*, Cambridge, 1931.)

A completely new restratification can be induced in any direction by means of the centrifuge, but development continues to be governed by the original axis of polarity.[1] It is probable in these cases (*Arbacia*) that the polar organisation of the egg, once it is determined, resides in the cortex and an invisible internal framework of more viscous cytoplasm which resists the displacing action

[1] Morgan and Spooner, 1909.

of centrifugalisation. In eggs which contain a large quantity of yolk (*Crepidula, Styela, Rana*) on the other hand, it seems that the viscous cytoplasmic framework can be permanently distorted and changed by displacement of the egg-contents as a result of prolonged centrifugalisation or inversion[1] (see also pp. 94, 218, 313).

§ 2

After the fixation of the axis of polarity, the most important determination in animals with bilateral symmetry is the determination of the plane of the latter. In the frog, this is normally due to the point of entry of the sperm. Before fertilisation, the egg is capable of forming its plane of symmetry in any one of an infinite number of possible planes passing through the egg-axis of polarity; the actual determination of a particular plane is fixed from the outside. The matter has been considered in detail in Chap. III (p. 36). We may sum up our conclusions as follows. The machinery for realising full normal bilateral symmetry is inherent in the egg; even very slight differential action of various external agencies can act as a trigger permitting a particular plane of symmetry to realise itself: normally, the entry of the sperm provides a strong differential which readily overrides the influence of other agencies.

The formation of the grey crescent in the amphibian egg appears to be bound up with the establishment of an activity-gradient of some sort extending dorso-ventrally across the equator of the egg. The existence of this gradient is revealed by various facts. In the first place, cleavage in the animal hemisphere proceeds more rapidly in the dorsal meridian, so that at the close of segmentation there is, a slight gradient in cell-size from dorsal to ventral along each circle of latitude. In the second place, there is the fact that gastrulation and invagination is initiated in the dorsal lip region, and then spreads progressively round each side until it reaches the ventral meridian, and the blastopore lip becomes circular.

Thirdly, there are the results of susceptibility experiments. These show that in anuran eggs exposed to lethal low temperatures or lethal concentrations of KCN, NH_4OH, $HgCl_2$, and other toxic agents, disintegration at any level begins at or near the dorsal meridian, and extends thence round the egg towards the ventral

[1] Conklin, 1924.

side. Further, in sub-lethal concentrations, the dorsal regions are the most inhibited in their differentiation.[1]

This last method allows us to make a further statement, namely that the dorso-ventral activity-gradient becomes progressively more intense (steeper) between fertilisation and gastrulation. In just-fertilised eggs, disintegration in lethal concentrations begins at the animal pole and then spreads along the dorsal side: in some cases a second centre of disintegration appears in the region of the grey

|←——mm——|

Fig. 28

Differential susceptibility in a frog's egg exposed to KCN from the 2-cell stage: disintegrated cells are shown light. The animal pole area (central) has disintegrated; also an area of cells, near the equator on one side, in the future organiser region. (After Bellamy, *Biol. Bull.* xxxvii, 1919, modified.)

crescent before the primary disintegration has spread to this area. During cleavage, the susceptibility of the dorsal lip region increases, until in late blastulae this region begins to disintegrate before or at the same time as the apical pole. In gastrula stages, the dorsal lip region is always the first to disintegrate.[1] It is probable that this process is correlated with the acquisition of organiser properties by the dorsal (grey crescent or dorsal lip) region (fig. 28).

The method by which bilateral symmetry is determined in the egg of Echinoderms is still problematical[2]; but the localisation of the plane can be revealed by susceptibility experiments at a stage before any bilateral symmetry is visible in the embryo.[3]

Further, a labile determination of bilateral symmetry has been

[1] Bellamy, 1919; Bellamy and Child, 1924.
[2] Hörstadius, 1928.   [3] Child, 1916 A.

discovered even in the egg of sea-urchins. If the egg is exposed to certain anaerobic conditions, a pit is formed in a particular place, but disappears on returning the egg to normal conditions. By means of *intra vitam* stains, it has been shown that the site of the pit coincides with the ventral side of the future larva.[1]

At the same time, as in the case of the amphibian egg, this labile determination of the plane of symmetry can be overriden by a variety of factors, of which, however, the point of sperm-entry is not one.[2] Artificial stretching and deformation of the egg (in a direction making some fairly large angle with the axis of polarity) leads to the determination of the dorso-ventral axis along that of tension. The primary axis of polarity is unaffected.[3]

Artificial rearrangement of the egg-contents has also been shown to influence the localisation of the plane of bilateral symmetry. In the sea-urchin, *Psammechinus miliaris*, the presence in the ripe egg of a subcortical layer of lipoid granules has been observed[4] and they may be displaced by means of the centrifuge. Neglecting those cases in which the granules are heaped up at either the animal or vegetative poles, it is found that the meridian of the egg on which the granules are accumulated becomes the ventral side of the larva.[5, 6] Similarly, the visible granules of the egg of *Arbacia* can be concentrated on any meridian, which then becomes the ventral side of the larva[6] (see also p. 218). The dorso-ventral axis, it seems, is determined as a gradient with high point ventrally. The curious fact that the ventral surface is associated in *Psammechinus* with centripetal lipoid granules, but in *Arbacia* with centrifugal yolk-particles, can be explained if their concentration leads to relatively higher metabolism. Similarly the ends of the stretched egg appear to be in a peculiar labile active condition. Interesting possibilities of analysis are here opened up.

In some Echinoderms, the dorso-ventral axis is visible in the unfertilised egg (*Psolus phantappus*)[7] and marked by a particular distribution of yolk, or (*Asterina gibbosa*)[8] by an elongation of the egg. The latter state of affairs is also found in some insects and

---

[1] Örström, in the press.
[2] Hörstadius, 1928.
[3] Boveri, 1901; Lindahl, 1933 A.
[4] Runnström, 1924.
[5] Runnström, 1925.
[6] Lindahl, 1933 B.
[7] Runnström and Runnström, 1921.
[8] Hörstadius, 1925.

Cephalopod Molluscs, where the egg is not only polarised, but bilaterally symmetrical in shape while still in the ovary.

It is of considerable theoretical importance to find that one and the same determination can occur either before or after fertilisation in different forms. In *Fucus*, no axes of symmetry at all are determined until after fertilisation. In most animals, radial symmetry and the primary axis are determined before fertilisation, bilateral symmetry at or after fertilisation. In some insects and Cephalopods, the determination of bilateral symmetry too has been shifted back to the period before fertilisation, and takes place under the influence of ovarian factors. It will be seen later that a similar shift in time-relations has occurred as regards the processes of chemo-differentiation in a number of forms, and that this shift contributes to the difference, which long puzzled experimental embryologists, between so-called "regulation-eggs" and "mosaic-eggs" (see Chap. v).[1]

§ 3

A further problem of symmetry is the determination of bilateral asymmetry. There are certain animals in which nothing is known as to the embryological determination of asymmetry, e.g. the skulls of certain whales and owls with asymmetrical formation of some of the bones; the bill of the wrybill plover; the various insects with spiral torsion of the genitalia;[2] the fish *Anableps* in which the copulatory tube points either left or right in males, and the genital aperture faces right or left in females;[3] the flatfish, in which either the left or the right side becomes uppermost when the fish is lying on the sea bottom, and the structure of the head is modified accordingly; or *Amphioxus*, in which the larval stages are markedly asymmetrical.

[1] In his large book on experimental embryology, Schleip (p. 842) argues at some length against the idea that the primary axes of the egg are imposed from without, and supposes that they arise by self-differentiation, though they may be modified by external agencies. It is logically almost impossible to conceive how a non-polarised fragment of living matter can acquire polarity by self-differentiation; and further, the experimental evidence in certain cases strongly supports the view that external differentials are responsible (*Fucus*, egg; *Corymorpha*, redifferentiation). The concept advanced above, that very slight external differentials may serve to release the capacity of the egg to develop polarity, reconciles both views. The *type* of the polarity is predetermined and therefore "self-differentiating"; the *direction* of the polarity is determined from without.
[2] Richards, 1927.　　　　　　　　　　[3] Garman, 1895, 1896.

It is clear that a fundamental difference must exist between the eggs of bilaterally symmetrical and those of radially symmetrical forms. In the former case, however, what is given by genetic constitution cannot be bilateral symmetry *per se*, but the capacity of developing such symmetry in relation to various external agencies. Harrison (1921 A, 1925 A) has suggested that the ultimate capacity for developing symmetry-relations is linked up with the intimate properties of the protoplasm and the "space-lattice" formed by its constituent parts.

The asymmetry of the large chelae found in many Crustacea either in one or both sexes, and also that of the opercula in certain

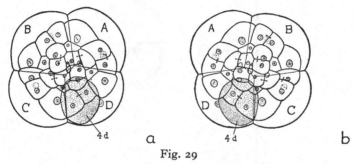

Fig. 29

Cleavage asymmetry in Molluscs. The position of the large mesoderm cell ($4d$) is reversed in laeotropic and dexiotropic cleavage. (From Morgan, *Experimental Embryology*, Columbia University Press, 1927.)

tubicolous Annelid worms, involves special problems of relative growth-rate, which are discussed by Przibram (1931 A).

The most marked asymmetry known is that of Gastropod Mollusca, most of which manifest a marked torsion of the internal anatomy together with unequal development of many paired organs. In addition, a large number of forms have their shells twisted into a spiral, which is usually dextral. Here it has been shown that the dextral or sinistral type of structure is under the control of Mendelian factors, whose action, however, is delayed for a generation[1] (see Chap. XII). The asymmetry of the adult is determined not by its own genetic constitution, but by that of the oocyte from which it arose, before it underwent the reduction divisions. The cleavage

[1] Boycott, Diver, Garstang, and Turner, 1930.

of the Gastropod egg is of the spiral type, and it has been shown that the direction of the initial spiral cleavage division is opposite in dextral and sinistral races of snails.[1] It is therefore probable that the orientation of the spindles of the first spiral cleavage division is responsible for the determination of the type of asymmetry which

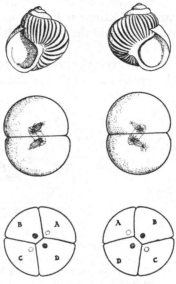

Fig. 30

Above, a sinistral (left) and a dextral (right) Gastropod shell. Below, corresponding asymmetry of cleavage. The obliquity of the spindles in the 2-cell stage (centre) and the portion of the cross-furrows in the 4-cell stage (bottom) are reversed in eggs with the left-handed (laeotropic) and the right-handed (dexiotropic) type of cleavage. (After Morgan, *Experimental Embryology* (Figs. 80, 79, 78 c and c', pp. 256–7), Columbia University Press, 1927, modified.)

the adult will exhibit,[2] and that the orientation of the spindles is, in turn, controlled by the Mendelian factors present in the oocyte (figs. 29, 30).

Reversed spiral cleavage has been observed exceptionally in the development of the leech *Clepsine*,[3] but as the adult is apparently perfectly symmetrical, no subsequent effects of the reversed cleavage can be detected. Occasionally, the leech egg gives rise to a

[1] Crampton, 1894.   [2] Conklin, 1897.   [3] Müller, 1932.

double monster, apparently by the production of two D-cells in place of one (see p. 108). In such cases the direction of spiral cleavage is reversed in the right-hand D-cell and all other cells on the right side of the plane of bilateral symmetry.

In the Echinoderms, most of the larval forms are asymmetrical, in that the left, but not the right, coelomic pouch acquires a water-pore placing it in communication with the exterior. Further, the fates of the various right and left coelomic pouches are very different. As a result, the hydrocoel and the rudiment of the body of the adult Echinoderm are formed on the left side of the body of the larva. It will be best to postpone the analysis of conditions in this group until the state of affairs in Vertebrates has been considered.

All Vertebrates are in reality asymmetrical. The stomach projects to the left of the middle line, while the heart and intestine show spiral twisting and are asymmetrical in other ways. The asymmetry of the gut and heart of the newt and frog has been experimentally shown to be dependent on a factor situated in the gut-roof. At the stage when the neural folds are still open, a square piece of presumptive neural tube material, together with the underlying gut-roof, is removed from the dorsal side, about half-way down the length of the embryo. The square piece is rotated through 180° and grafted back into place again so that the antero-posterior axis of the piece is reversed. The result of such an experiment is a normal embryo, except that it shows *situs inversus* of the asymmetrical organs, i.e. the stomach is on the right and the intestine and heart are twisted in the direction opposite to the normal. Rotation of the presumptive neural tube material alone, without the underlying gut-roof, does not interfere with the development of the normal asymmetry. The ventral regions of the embryo are not touched by the operation, and therefore the asymmetry of the heart and gut must be determined by some factor or agency differentially distributed across the gut-roof[1] (fig. 31).

If, however, the square piece which in the previous experiments was rotated, is simply removed, the embryo will show normal asymmetry. This may mean either that the differential factor extends, though with diminished intensity, on either side of the gut-

[1] Pressler, 1911; Meyer, 1913; Spemann, 1918.

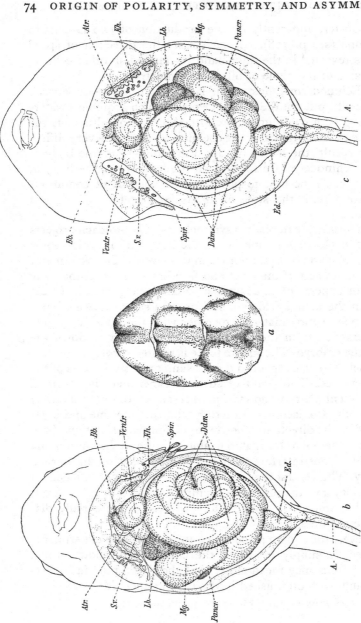

Fig. 31

The dependence of *situs viscerum* on the gut-roof. *a*, A neurula of *Bombinator* in which a rectangular piece of the dorsal surface together with underlying gut-roof has been cut out, rotated back to front, and grafted back again. *b*, Normal embryo seen as reconstructed from the ventral side. *c*, Embryo operated upon as in *a*, showing *situs inversus viscerum et cordis*, and the spiracle on the right side. *A.* anus; *Atr.* atrium; *Bb.* bulbus; *Ddm.* intestine; *Ed.* rectum; *Kh.* branchial chamber; *Lb.* liver; *Mg.* stomach; *Pancr.* pancreas; *Spir.* spiracle; *S v.* sinus venosus; *Ventr.* ventricle. (From Meyer, *Arch. Entwmech.* XXXVII, 1913.)

roof, or that other factors exist capable of controlling asymmetry, but normally overruled by the gut-roof factor.

Another line of attack on this problem is provided by those experiments in which a blastula of a newt is partially constricted by tying a fine hair round it in the plane of bilateral symmetry. The result is the production of double-headed monsters, and, while the

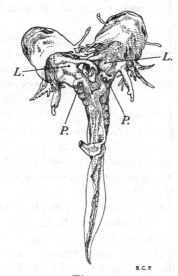

Fig. 32

Anterior doubling producing *situs inversus viscerum et cordis* in the right-hand member. The doubling was produced by partial constriction in the plane of symmetry of an early cleavage stage of *Triton*. The heart, gut, and position of liver (*L.*) and pancreas (*P.*) of the right-hand member (seen on the left in this ventral view) are reversed. (After Spemann and Falkenberg, *Arch. Entwmech.* XLV, 1919, simplified.)

left-hand member of such a pair always shows the normal asymmetry, the right-hand member nearly always shows *situs inversus*.[1] Double-headed monsters also occur in trout, in wild conditions and in hatcheries. When the two members are joined together only by the hinder region of the trunk (behind the abdominal cavity), both members have the usual vertebrate asymmetry. But when the join between the two members is farther forward, so that the alimentary

[1] Spemann and Falkenberg, 1919.

tract forks at a point between the stomach and the cloaca, the right-hand member frequently shows *situs inversus*, while the left-hand member is normal[1] (fig. 32).

The rudiment of the heart can be divided in amphibian embryos (at the tail-bud stage) by means of a longitudinal cut in the middle line; each half rudiment will give rise to a heart, and while the asymmetry of the left one is normal, that of the right one is reversed[2] (fig. 115).

The remarkable point about these experiments and observations is the constancy of normal asymmetry in the left-hand member, and the restriction of *situs inversus* to the right-hand member. This fact emerges still more clearly from those experiments in which the blastula of the newt is constricted by a hair in the plane of bilateral symmetry, and the hair is pulled tight, thus resulting in the complete separation of two half-blastulae, of which one represents the right and the other the left half of the original embryo. The left halves develop into perfect little newts with normal asymmetry; of the right halves, about equal numbers show normal asymmetry and *situs inversus* respectively.[3]

Whatever the asymmetry factor may be, it cannot be regarded as an absolute and localised producer of one specific type of asymmetrical structure—at least, not during the earliest stages of development—and for the following reasons. It is true that when newt embryos are divided into left and right halves at the blastula stage, about half of the right-hand portions show reversed asymmetry. But if the left and right blastomeres are separated from one another (likewise by constricting in the plane of bilateral symmetry with a hair) at the 2-cell stage, the right-hand blastomeres do not show any greater tendency to production of reversed asymmetry than is found to be the case in normal development of newts' eggs —2 to 3 per cent.[4] At the 2-cell stage, therefore, the asymmetry factor has not become predominant on the left side at the expense of the right. The same conclusion emerges from the simple experiment of reversing an egg and forcing it to continue its development in that position. If the prepotent normal-asymmetry factor were definitely located on the left side at this stage, then since the

---

[1] Stockard, 1921; Morrill, 1919; Swett, 1921.   [2] Ekman, 1924, 1925.
[3] Ruud and Spemann, 1923.   [4] Mangold, 1921 B.

plane of bilateral symmetry is already determined at fertilisation, rotation of the egg so that the vegetative pole becomes uppermost, while the dorsal meridian remains unchanged, would cause the original left side of the egg to become the right side of the embryo. But as a matter of fact, the embryos arising from such reversed eggs do not show *situs inversus*.[1]

The asymmetry factor must therefore be regarded as a factor which results in a greater activity of the tissues on one side (the left) of the body as compared with the other: such greater activity developing progressively. It is interesting to note that a similar progressive accentuation of a differential or gradient is to be observed in the dorso-ventral axis of the amphibian egg, between the time when the grey crescent is first formed and the establishment of full organiser capacity in the dorsal lip (see p. 68). Further research should be directed to discovering whether such accentuation or steepening of activity-gradients is a regular feature in their development.

The result of the action of this asymmetry factor is seen in the more rapid growth and differentiation of the left side, as regards certain organs. Experimental proof of this is provided by the mesodermal rudiments from which the muscular wall of the heart is formed. These rudiments are at first situated on each side of the body, and later on move towards the middle line. But if the rudiments are removed from the embryo while they are still lateral in position, and are made to develop in isolation (explantation in ectodermal jackets, in suitable culture media), the remarkable thing is that the rudiments from the left side show pulsations while those from the right do not. Further, the histological differentiation of which the rudiments from the right side are capable is inferior in degree to that of the rudiments from the left.[2] The difference between the left and right rudiments of the heart is a physiological one, and appears to be quantitative rather than qualitative, and in every case the left side is prepotent (fig. 33).

If the asymmetry factor is, as suggested, concerned with the relatively greater activity of one side, it should be susceptible of experimental control. One method of affecting its action is the simple mechanical one of removing some of the material from the

[1] Hämmerling, 1927.   [2] Goerttler, 1928.

left side of an otherwise normal developing embryo. This experiment has been performed on the blastula of the newt, and resulted in the production of *situs inversus*.[1] Another method is to subject the developing embryo to physical factors which are calculated to affect the rate of activities of the tissues, and to direct these physical

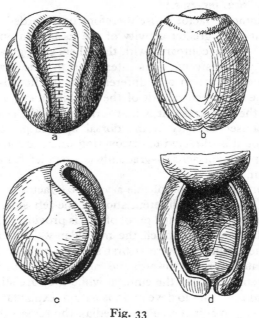

Fig. 33

a, Dorsal; *b*, Ventral; *c*, Left side views of an embryo of Urodele to show the position of the paired rudiments of the heart in the mesoderm beneath the surface. *d*, Embryo from which the dorsal surface has been cut off and the entire gut peeled out, thus revealing the ventral mesoderm, with the position of the paired heart rudiments indicated by circles. The left rudiment, when isolated, develops further than the right. (From Goerttler, *Verh. Anat. Ges.* XXXVII, 1928.)

factors in such a way that one side of the embryo is affected more than the other. This would appear to be why chick embryos show *situs inversus* when they are locally damaged by overheating on the left side during incubation.[2] Here, the intensity of action of the tissues on the left side has been decreased, while in the previously described experiment it is their amount which has been reduced.

[1] Wilhelmi, 1920, 1921.   [2] Warynsky and Fol, 1884.

Now, since the embryos that develop from eggs in which the plane of bilateral symmetry has been artificially selected (by controlling the point of sperm-entry; see p. 36) show the normal asymmetry, it follows that the determination of the left as the ultimately prepotent side must be made at the same time as that of the plane of bilateral symmetry. And if this determination were due to an external factor, it would be impossible to understand why it invariably acts so as to produce its effect on a meridian 90° right of the meridian of sperm-entry, and convey a greater power of activity to this eventual left side of the embryo. The conclusion is therefore enforced that the determination of the left-right axis is in some way connected with that of the plane of bilateral symmetry, and is the result of some factor acting within the egg.

There is as yet no indication of how this factor acts, but it may be pointed out that the determination of a third axis (in the case of the egg, the left-right axis) as a consequence of the determination of the other two axes of space (in the egg, the primary egg-axis, i.e. antero-posterior, and the dorso-ventral), is a phenomenon not without an analogy in the inanimate world. It is well known that if a conductor carries an electric current through an independent magnetic field of force which is orientated at right angles to the conductor, then the conductor will be subjected to a force acting at right angles both to the magnetic field and to the conductor. If it be imagined that the magnetic field is vertical with the North Pole uppermost, and a horizontal conductor carries an electric current away from an observer, the force acting on the conductor will tend to displace it to the observer's left. It is not pretended that the egg-axis is the site of a magnetic field, nor that the dorso-ventral axis is a simple conductor. But the physical analogy described above does show how it is possible to obtain a determination of a third axis, and a polarisation in it, as a consequence of the determination and polarisation of axes in the other two planes of space.

In the larva of *Amphioxus*, asymmetry is very marked. In this form[1] double monsters can be artificially produced by disarranging the 4-cell stage. In such cases, both components always show the normal asymmetry: symmetry is never reversed. The difference

[1] Conklin, 1933.

between these results and those on the newt is doubtless to be explained as a result of the precocious appearance of bilaterality in the egg of *Amphioxus*, extending to chemo-differentiated substances, which is established immediately after fertilisation. As suggested

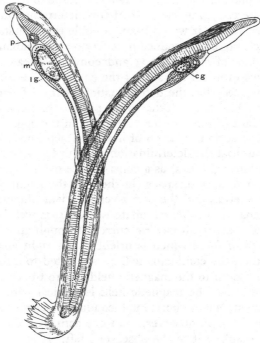

Fig. 34

Double monster of *Amphioxus*, produced by mechanical disarrangement and partial separation of the blastomeres in the 2-cell stage. Note that both components show normal asymmetry. *1g*, 1st gill-cleft; *cg*, club-shaped gland; *m*, mouth; *p*, preoral pit. (From Conklin, *Journ. Exp. Zool.* LXIV, 1933.)

in the preceding paragraph, this rigid bilaterality might establish an equally rigid asymmetry-gradient (fig. 34).

§ 4

The conclusions arrived at from a consideration of the results obtained from experiments on amphibian development are supported and extended by experiments on Echinoderm larvae. In these, also, it is the left side of the body which is prepotent as compared with the right, and this prepotency manifests itself in the fact that the hydrocoel, the water-pore, and the rudiments of the adult animal are formed on the left side of the body of the larva. In *Asterina*, the gastrula can be divided into left and right halves by section in the plane of bilateral symmetry. The left halves develop into larvae with normal asymmetry: they have a hydrocoel on the left. The right halves can do one of three things; they may have a hydrocoel on the left side only; they may have a hydrocoel on the right side only; or they may have a hydrocoel both on the left and on the right side.[1] This last condition is sometimes found in otherwise normal echinopluteus larvae,[2] and can be experimentally produced by subjecting the larvae to hypertonic sea-water,[3] while it is the rule in ophioplutei.

The occurrence of halves produced from the right side of the original larva, which develop a hydrocoel both on the left and on the right, is of great interest, for it provides a situation which could scarcely be realised in the amphibian embryo. There, the gut and heart must be twisted either one way or the other, but cannot be twisted in both ways at once in the same embryo.[4]

Both this result on Echinoderm larvae, however, and that obtained by dividing newt blastulae into right and left halves (p. 76), can be plausibly explained along similar lines. It has already been found necessary to postulate a main activity-gradient, concerned with asymmetry, and extending transversely across the body from left to right. This is presumably superposed on minor activity-gradients extending inwards from the surface towards the centre of the embryo. In any case, when the developing egg is cut in half, the inner surfaces of each half are damaged or interfered with, and their activity reduced. In the left-hand halves, the effect of this will merely be to steepen the existing asymmetry-gradient; all resulting organisms will therefore be of normal asymmetry. In the right-hand halves, however, the effect will be in the contrary direction to that

---

[1] Hörstadius, 1928.   [2] MacBride, 1911.   [3] MacBride, 1918.
[4] In the Gastropod *Limnæa* (see p. 411) occasional specimens have been bred in which the dextral and sinistral forces are so delicately balanced that the result is an animal with a flat shell coiled in one plane, like that of *Planorbis*. Most of these specimens are abnormal in their anatomy and die early (Boycott, Diver, Garstang and Turner, 1930).

of the main asymmetry-gradient. If the result is merely to flatten this gradient, animals of normal asymmetry will still result. But if the effect of the cut is strong enough to reverse the existing gradient, animals of reversed asymmetry will arise. This applies both to Echinoderms and to Amphibia: in addition, in Echinoderms the almost complete balancing of the two lateral halves of the gradient will give rise to bilaterally symmetrical forms (with both left and right hydrocoels), whereas this result is impossible in Amphibia, where the normal and reversed asymmetries are mutually exclusive alternatives.

The main points of this chapter may be briefly summarised as follows. In amphibian development, polarity or axiation and bilateral symmetry are both established as the result of agencies external to the egg. In both cases, an important effect of these agencies is the production of activity-gradients extending through the whole egg. In the production of bilateral asymmetry, an activity-gradient is also involved. At the moment it is not possible to state what is the originating cause of this asymmetry-gradient; we do know, however, that its establishment is in some way dependent upon the establishment of the dorso-ventral gradient which determines bilateral symmetry. Both these latter gradients appear to become progressively accentuated during the period of cleavage.

Chapter V

CLEAVAGE AND DIFFERENTIATION

§ 1

The most obvious visible change during the first phase of development of the fertilised egg is its cleavage into a number of separate cells. We must now ask whether other equally important but less obvious changes may not be taking place at the same time, and enquire into the relation between cleavage and the processes leading to morphological differentiation.

The pattern of cleavage is normally oriented in relation to the existing major axis of the egg, e.g. the first two cleavage planes are, in all known cases except one, meridional: the exception is provided by the Nematodes, where the plane of the first cleavage is still oriented with reference to the axis, but at right angles to it, and therefore latitudinally. In Cephalopods and Ascidians, the cleavage pattern is oriented with reference to the secondary axis of bilateral symmetry as well.

The orientation of cleavage-pattern can, however, be modified. It may be modified in relation to a new, induced, axis of polarity. For instance, in the sea-urchin *Lytechinus* and the star-fish *Patiria*, cut fragments of the unfertilised egg, subsequently fertilised, always have the first two cleavage planes at right angles to the cut surface, which, as we shall see later (p. 313), has established a new polarity.[1]

The cleavage-pattern may also be modified by mechanical means, e.g. by a restratification of the egg-contents by the use of the centrifuge (see p. 218). In the sea-urchin *Arbacia*, for instance, the first two cleavages are perpendicular to the stratification, whatever its relation to the original axis.[2] Or the cleavage-pattern may be altered by forcing eggs to undergo cleavage while compressed between glass plates. The orientation of the division spindles in these cases is governed by the principle known as "Hertwig's rule", which lays down that at mitosis the spindle will form with its long axis in the

---

[1] Taylor and Tennent, 1924; Taylor, Tennent and Whitaker, 1925; Taylor and Whitaker, 1926.   [2] Morgan and Lyon, 1907.

direction of the longest axis of the cytoplasm of the cell.[1] The distortion occasioned by the glass plates causes the third cleavage plane to form meridionally instead of latitudinally[2] (figs. 35, 36).

One of the consequences of the experiments of forcing eggs to cleave under compression is that the normal distribution of the

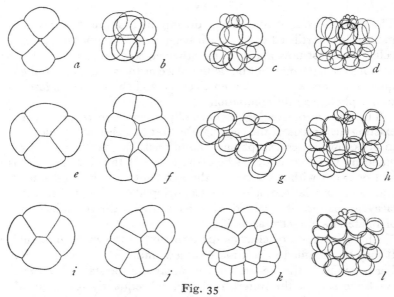

Fig. 35

Disarrangement of cleavage by pressure in sea-urchin eggs. *a–d*, Normal cleavage. *e, f* and *i–k*, Flat plates arising from cleavage under pressure. *g, h*, Subsequent cleavage of *f* when released from pressure. *l*, The same for *k*. The late cleavage stages are drawn with the vegetative pole uppermost. The disarrangement of the nuclei does not prevent the development of normal plutei. (After Driesch, from Morgan, *Experimental Embryology*, Columbia University Press, 1927.)

cleavage nuclei is altered, but subsequent development is normal in spite of the fact that a number of nuclei find themselves in blastomeres other than those in which they would be situated in

[1] Although this rule is of very general application, there are some notable exceptions to it. For instance, cleavage in the star-fish *Patiria* occurs in relation to the polarity of the egg, whether original or induced by operation (see p. 313), even when the egg is deformed by pressure (Taylor and Whitaker, 1926). Other exceptions are found in the first cleavage of *Ascaris* eggs, and in the divisions of the cells forming the germ-bands of Crustacea (see Jenkinson, 1909 B, p. 34).

[2] Frog, Hertwig, 1893; sea-urchin, Driesch, 1893.

normal cleavage. As already mentioned (p. 43), these facts prove
that during cleavage the nuclei divide in such a way that their
daughter-nuclei are quantitatively and qualitatively equal.

A more recent and very elegant demonstration of the equivalence
of the nuclei of the blastomeres was carried out as follows. By

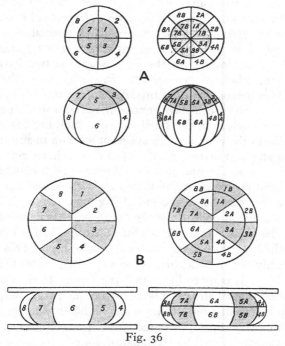

Fig. 36

Diagram to show the altered distribution of nuclei in frogs' eggs made to segment
under pressure. A, Normal eggs. B, Eggs subjected to pressure. Left, 8-cell
stage; right, 16-cell stage. In each case a polar view is shown above, a side view
below. Cells produced by the division of corresponding cells are numbered alike.
(From Wells, Huxley and Wells, *The Science of Life*, London, 1929.)

means of a fine hair, the fertilised egg of the newt can be constricted into the shape of a dumb-bell, in such a way that the zygote
nucleus is confined to one side. This side will then undergo cleavage
as the nucleus divides, while the other side of the dumb-bell will
remain uncleaved. By releasing the ligature, the constriction can
be relaxed, and one nucleus—any one, at random—may be allowed

to pass across from one side to the other. This may be done at the 2-cell, 4-cell, 8-cell, 16-cell, or 32-cell stage of the cleaved side. If the constriction lay in the plane of bilateral symmetry of the original fertilised egg, and if after the passage of one nucleus from the cleaved side to the other the ligature is then drawn tight again so as to separate the two halves completely, each half will develop into a normal little newt. One of these little embryos will contain only the nuclear material of one blastomere of the normal 2-, 4-, 8-, or 16-cell stage, or as we may for brevity write it, a 1/2, 1/4, 1/8 or 1/16 nucleus, depending on the time when the two halves were separated; the other embryo will contain all the rest of the nuclear material. This means that in normal development, the nuclei of the blastomeres of the 16-cell stage contain material which is equivalent to that of the nucleus of the fertilised egg. Nothing has been lost by the nuclei in the process of cleavage, at least up to and including the 16-cell stage. Further, all the 1/16 nuclei have retained this equivalence, for in the numerous experiments performed it would not have been possible for the nucleus which passed across from one side to the other to be the same[1] (fig. 37).

When cleavage in one-half of the dumb-bell has reached the 32-cell stage, the passage of a nucleus into the other half is insufficient to enable the latter to undergo normal development. This is, however, probably not to be attributed to a qualitative insufficiency of a 1/32 nucleus. It is more likely that the failure to develop is due to some alteration of the cytoplasm of the uncleaved half, in turn due to the length of time during which it has been deprived of a nucleus, and therefore prevented from prosecuting its normal physiological activities. This explanation follows from the fact that a 1/16 nucleus is incapable of ensuring development beyond the late gastrula, or, rarely, early neurula stage, if the constriction had been placed in such a way as to separate dorsal and ventral halves of the future embryo, and the zygote nucleus had been restricted to the ventral half. A 1/16 nucleus is therefore unable to do in a previously enucleate dorsal half what it can do perfectly well in a previously enucleate lateral half. It would appear that the failure in this case lies with the cytoplasm. The susceptibility of the cytoplasm of a dorsal half is greater than that of a lateral half (see

[1] Spemann, 1928.

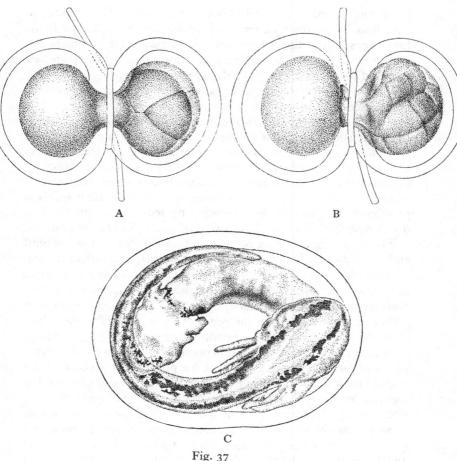

A   B

C

Fig. 37

The equality of nuclear division during cleavage. A fertilised egg of *Triton taeniatus* was constricted by a ligature, restricting the nucleus to the right-hand half, in which cleavage has reached the 8-cell stage, while the left-hand half is still undivided. A, At the 16-cell stage one of the cleavage-planes coincides with that of the ligature, and 1/16 nucleus has passed across into the as yet undivided left half. B, The ligature was then drawn tight so as to effect complete separation between the two (lateral) halves. C, Each developed into a perfect embryo (one slightly further advanced than the other); 140 days after the operation, they were identical. A nucleus of the 1/16 stage is therefore equivalent to that of the whole egg. (From Spemann, *Zeitschr. Wiss. Zool.* CXXXII, 1928.)

p. 68, and Chap. IX); accordingly the latter can survive absence of a nucleus during the time required for the zygote nucleus to divide four times, while the former loses its capacity for complete development if it has remained enucleate for a longer time than that required for three divisions of the zygote nucleus.[1]

The equivalence of nuclei at later stages of cleavage has been established from experiments conducted on the eggs of insects. The egg of the dragon-fly *Platycnemis* is an elongated structure in which the nucleus is central and divides several times before its products of division reach the surface of the egg and the cytoplasm is partitioned off into separate blastomeres. By focussing a pencil of ultra-violet rays on a nucleus at the 2-nucleus stage (corresponding, of course, to the 2-cell stage of forms with ordinary cleavage) it is possible to kill it. But the remaining nucleus and its products of division are sufficient to allow a normal embryo to be formed.

The insect egg is further peculiar in that it possesses near its hind end a region which is essential for the subsequent differentiation of the embryo (see Chap. VI). But the activities of this region are not manifested unless some of the nuclei which have resulted from cleavage migrate into it. This "population" of the hinder end, and indeed of all the surface of the egg, by nuclei, normally takes place after the 5th cleavage, corresponding to the 32-cell stage. Again, by means of ultra-violet rays, it is possible to affect a zone of cytoplasm of the egg in such a way that the products of division of the nuclei are delayed in passing through it, and instead of receiving nuclei after the fifth cleavage, the hinder end only receives them after the eighth cleavage, i.e. at the 256-cell stage. Nevertheless, these nuclei are adequate to activate the region in question, and normal embryos are produced. Here, then, is evidence that the division of the nuclei is qualitatively equal as far as the 256-cell stage.[2]

---

[1] We have already noted that an isolated ventral half, since it does not contain any of the organiser-region, is incapable of development beyond a stage roughly equivalent to the late blastula. It might be supposed therefore that nuclei which had been restricted to a ventral half had been in some way affected so as to be unable to promote full development on passage into a dorsal half. There is, however, no positive evidence for such a possibility, while the greater susceptibility of the dorsal half of the egg is an established fact (Spemann, 1901 B, 1902 1903, 1914, 1928; Ruud and Spemann, 1923).

[2] Seidel, 1932.

§ 2

It is clear from these experiments that whatever the first manifestation of differentiation in the embryo may be, it is not to be found in the division of the nuclei of the blastomeres during cleavage. Attention must therefore be turned to the cytoplasm, in order to see whether it, too, is equivalent in the different blastomeres.

Considering first the case of the newt: the fact that a lateral or a dorsal half of an egg at the 2-cell stage, a blastula or an early gastrula will develop, but that a ventral half will not (p. 53), shows that all the regions are not equivalent; and since this non-equivalence cannot reside in the nuclei, it must concern the cytoplasm. Actually, the importance of the orientation of the constriction separating the halves in the experiments described above, has been shown to lie solely in the fact that the presence of some of the organiser area (grey crescent, dorsal lip region) is essential if development beyond the blastula stage is to take place. When the constriction coincides with the plane of bilateral symmetry, the halves will be lateral and each will possess a portion of the region of the grey crescent. But if the constriction is at right angles to the plane of bilateral symmetry and separates a dorsal half from a ventral half, the former will contain the whole of the region of the grey crescent and will develop, while the latter will not contain any portion of the region in question and will not develop.

A ventral half of an embryo (blastula or early gastrula) can be made to develop if the dorsal lip of the blastopore of another embryo is grafted into it,[1] and this proves conclusively that the inability of a ventral half to develop is due, not to lack of any nuclear

Fig. 38

The formation of an embryo with neural tube, somites, and notochord, out of a ventral half (see fig. 20, c) of a *Triton* embryo, by grafting an organiser into it. (From Bautzmann, H., *Arch. Entwmech.* CX, 1927.)

material or factors, but to lack of a definite portion of cytoplasm—the organiser (fig. 38).

In the newt, therefore, there is already a differentiation of the cytoplasm just after fertilisation and before the first cleavage, and

[1] Bautzmann, 1927.

this differentiation is of such a kind that a certain cytoplasmic region is essential for development. There is no other qualitative difference between the blastomeres of the 2- or 4-cell stage, as is shown by the fact that a single blastomere of the 4-cell stage will—provided that it contains a portion of the region of the grey crescent—develop into a complete but miniature embryo, although in normal development this blastomere would have furnished material for only one-quarter of an embryo.[1] Further, two embryos of the

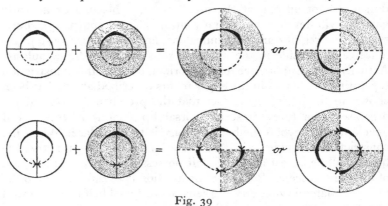

Fig. 39

Diagram showing results of uniting pairs of *Triton* eggs in the 2-cell stage. The future organiser-region (dorsal lip) is represented as a black crescent. Above, condition when the first cleavage of both divides dorsal from ventral halves. Below, condition when the first cleavage of both is median and divides right and left halves. The result expected is a multiple monstrous form with three components. (After Seidel, from Morgan, *Experimental Embryology*, Columbia University Press, 1927, modified.)

newt placed together cross-wise over one another at the 2-cell stage can undergo development to form one single large embryo, provided that the grey crescent regions of both are adjacent. (If these regions are not adjacent, a double monster is produced[2] (figs. 39, 40).)

The findings in the newt can be extended to other Urodela, and in them it can be said that except for the determination of the region of the grey crescent which will give rise to the dorsal lip of the blastopore, or organiser, the cytoplasm of the egg is not unequally distributed between the blastomeres up to the end of the 4-cell stage inclusive. When animal and vegetative regions are divided, as

[1] Ruud, 1925.   [2] Mangold, 1921 A; Mangold and Seidel, 1927.

happens at the 8-cell stage, it might be expected that this separation of different portions of the primary egg-axis would mean an unequal distribution of potencies, and, as will be seen later, this expectation is in fact realised.

The case of the frog is in principle similar to that of the newt. But the experiments conducted from time to time on the frog have suffered so much from unforeseen complications, that the conclusions drawn from them were for a long time misleading. The chief difficulty arises from the fact that the eggs of Anura have long defied attempts to secure the constriction and separation of blastomeres. Experimental technique has therefore been largely

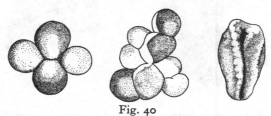

Fig. 40

One embryo from two eggs. Left, two *Triton* eggs in the 2-cell stage are laid across each other, so that their blastomeres alternate. Centre, each blastomere has divided once. Right, a giant neurula resulting from such a fusion. (After Seidel, from Morgan, *Experimental Embryology*, Columbia University Press, 1927, modified.)

restricted to injuring one of the blastomeres: this is usually accomplished with a hot needle. The result of such an experiment at the 2-cell stage is that the uninjured blastomere develops into a half-embryo, and does not produce much more than it would have done if its sister-blastomere had developed normally alongside it, for it is a condition of the experiment that the injured blastomere remains attached to the uninjured one.[1] (For the present purpose, the subsequent attempt of the half-embryo to complete itself by "post-generation"[2] may be passed over here as irrelevant (fig. 41).)

[1] Roux, 1888.

[2] In some of the cases originally described by Roux, the half-embryo obtained by injuring one blastomere with a hot needle appeared to be subsequently converted into a whole embryo, by the utilisation of the materials of the injured blastomere. To this restorative process, the name "post-generation" was given. It was imagined that the reorganisation of the injured blastomere was brought about either by belated cleavage of its nucleus, or by invasion of cells from the uninjured half, or by overgrowth of the injured half by layers of tissue from the

The result of this experiment was at first interpreted to mean that the two blastomeres were already differentiated at the first cleavage and were determined to give rise, each of them, to one-half of the future embryo. But this conclusion was later shown to be erroneous in a number of ways. In the first place, it was noticed that the half-embryo which developed might be a lateral half, or a dorsal half, or an oblique half, according as to whether the plane of the

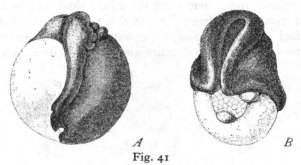

A   *B*

Fig. 41

A, Lateral, and B, anterior, partial embryos of the frog produced from eggs in which one of the first two blastomeres have been killed but allowed to remain in place. (After Roux, from Morgan, *Experimental Embryology*, Columbia University Press, 1927.)

first cleavage coincided with, or was perpendicular, or oblique, to the plane of bilateral symmetry. The alleged determination of the blastomeres at the 2-cell stage was therefore not constant.[1] Then it was found that if a normal embryo at the 2-cell stage is inverted, each of the two blastomeres will then develop into as much as it can of a complete embryo. The limitations on completeness are due

uninjured half, or by a combination of these methods. Morgan (1895) was unable to confirm these findings, and the position is still obscure. Discrepancies between various results seem to be due to the relative degrees of injury inflicted by the hot needle. Where the coagulation of the protoplasm is extensive and cleavage of the injured blastomere cannot proceed, it is unlikely that the half-embryo ever becomes complete, although it may appear to be more complete than it really is, as a result of the spreading of the epidermis from the uninjured half and consequent concealment of the underlying defects. If, on the other hand, the cleavage of the injured blastomere is only delayed but it nevertheless reaches the blastula stage by the time that the uninjured half is ready to gastrulate, the rapid restoration of the missing half would be possible. At all events, the theoretical arguments originally based on the alleged phenomenon of post-generation have long ceased to be important.

[1] Hertwig, 1893; Brachet, 1903, 1905.

to purely mechanical reasons; and the result is the formation of a double monster. This means that each blastomere at the 2-cell stage of the frog is capable of giving rise to more than it would produce in normal development, and therefore the various regions of the egg cannot all be determined at this stage.[1]

It is clear, therefore, that it is not the mere presence of the injured blastomere, when the latter is pricked with a hot needle at the 2-cell stage, which prevents the other blastomere from developing into a complete embryo. This is still more evident from the experiment in which one blastomere of the 2-cell stage is injured as before, and then the embryo, injured and uninjured blastomeres together, is inverted and maintained in that position. The uninjured blastomere will then develop into a more or less complete embryo.[2] The inversion results in a streaming of the contents of the uninjured blastomere so that the yolk again becomes undermost, and it is to this rearrangement that the power of developing into a whole embryo on the part of a single blastomere must be ascribed. It must be because there is no such rearrangement in the case where a blastomere is injured and the embryo is not inverted, that the uninjured blastomere in such an experiment develops into a half. The presence of the injured blastomere necessitates the retention of the hemispherical shape on the part of the uninjured blastomere, and no possibility is provided for the rearrangement of its contents, which appears to be necessary if the half is to regulate into a whole. Indeed, it is difficult to see what kind of stimulus other than inversion could upset what in the uninjured blastomere is merely the continuation of normal development. In one case, two frog's eggs were found enclosed within one membrane, which deformed both of them into a hemispherical shape. In the subsequent development each embryo was deficient on the flattened side.[3]

In another anuran, *Chorophilus*, it has been found possible to remove the injured blastomere altogether by sucking it out with a fine pipette, and the uninjured blastomere then develops into a whole embryo, presumably as a result of the rearrangement of its contents, for after removal of its injured sister the uninjured blastomere becomes spherical.[4] Lastly, improved technique has made

[1] Schultze, 1894; G. Wetzel, 1895.   [2] Morgan, 1895.
[3] Witschi, 1927.   [4] McClendon, 1910.

94   CLEAVAGE AND DIFFERENTIATION

it possible to separate the blastomeres of the frog at the 2-cell stage, and it has been found that each blastomere thus isolated can (provided that it contains a portion of the grey crescent region) develop into a whole embryo.[1]

It will be remembered, as explained in Chap. III, that grafting experiments have shown in the newt that the various regions (except that of the organiser) are plastic up to a certain stage in gastrulation, and that tissue which was presumptive epidermis can differentiate into part of the brain and eye. Recent improvements in technique have permitted of analogous experiments on anuran material, and it has been found that the tissues of the frog (again with the exception of the organiser) are plastic up to a similar stage.[2]

Cleavage of the egg of Anura, then, does not result in the separation of qualitatively unequal cytoplasm between the blastomeres, certainly of the 2-cell stage, and presumably of the 4-cell stage, with the exception of the specialised region of the grey crescent. In this respect, therefore, the anuran egg does not differ from that of the Urodele.

A few more words may be added concerning the cause of the production of the double monsters from embryos of the frog which have been inverted at the 2-cell stage. It has been found possible to obtain such monsters by inverting the undivided egg, and therefore the duplicity of the monsters is not due to the number of blastomeres into which the egg has cleaved when it is inverted. Triple monsters may also arise from inversion. These anomalies have been shown to be due to the fact that when the streaming of the yolk takes place, consequent on the inversion, a streak of inert

[1] Schmidt, 1930, 1933.
[2] Schotté, 1930; Schmidt, 1930.

The question of the existence at early stages (fertilised but unsegmented egg) of anuran development of cytoplasmic regions possessing a determination has been attacked by the method of making small injuries with a heated or unheated needle. Loss of tissue (by damage *in situ* or by extra-ovation) at this early stage leads to the development of imperfect larvae, and it has been held that even the unsegmented egg possesses (labile) determinations (Brachet, 1905, 1906, 1911, 1923, 1927; Pasteels, 1932). But it is necessary to point out that the eventual malformation or non-appearance of an organ after injury to the egg is, by itself, no logical justification for the view that the rudiment of the organ in question was determined at the stage operated upon: the injury done to a particular part of an egg persists, and may exert an inhibiting influence on the *subsequent* determination and chemo-differentiation of whatever rudiment comes to occupy its site. On the other hand, clean removal of pieces of *blastulae* (other than the organiser-region) in *Bombinator* and *Triton* allows normal development to occur (Bruns, 1931).

yolk is left near the surface, and this interferes with the process of gastrulation. The lip of the blastopore becomes as it were split on this obstacle, and invagination takes place in opposite directions, away from the streak of yolk. In other cases, the blastocoel is displaced, and it seems that the pressure within it causes the cells which form its wall to present an obstacle on which the blastopore lip

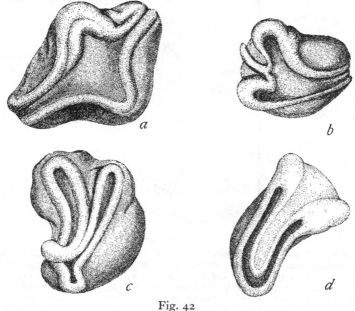

Fig. 42

Double monsters of *duplicitas cruciata* type, produced by inverting the 2-cell stage of frogs' eggs. (After Schleip and Penners, from Morgan, *Experimental Embryology*, Columbia University Press, 1927, fig. 157, p. 393.)

becomes split, and likewise forks. Each portion of the blastopore lip then invaginates on its own, and gives rise to the essential features of an embryo, in so far as this is mechanically possible[1] (fig. 42).

Double monsters have also been obtained in the frog simply by fertilising over-ripe eggs. The cleavage of such eggs is abnormal in that the blastomeres of the vegetative hemisphere are relatively much too large. Presumably, the physiological condition of such

[1] Penners and Schleip, 1928.

eggs involves a decrease in the activities of the cytoplasm, or, in other words, a relative increase in the inertia of the mass of yolk. At all events, the splitting into two of the blastopore lip has been observed in such eggs, at the onset of gastrulation.[1] An additional result obtained in these experiments is cases of disorganised growth, leading to tumour-like proliferations, which increase at the expense of the embryo itself, may give rise to metastases, and can be propagated by grafting (fig. 43).

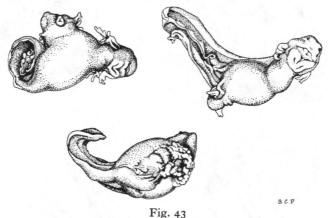

Fig. 43

The effect of delayed fertilisation in frogs' eggs. Duplication, teratological monstrosities, and tumour-like growths in tadpoles derived from late-fertilised eggs (over 3 days over-ripe). Top left, anterior duplication, the upper head imperfect, with single sucker. Top right, tadpole with irregular tail and rudimentary secondary ("parasite") head. Below, larva with much reduced head and tumorous growths ventrally. (Redrawn after Witschi, *Verh. Naturforsch. Ges. Basel*, XXXIV, 1922.)

§ 3

Turning now to the experiments of isolating blastomeres in other groups of animals, it was found that the results differ considerably in the various groups. In some forms, the isolated blastomere develops into a complete and normally proportioned larva, differing from a normal larva merely in its small absolute size. In other forms, the isolated blastomere is incapable of doing this, and gives rise to a partial structure only. As extreme examples of these two types we may take the Hydrozoa and the Ascidians, respectively.

[1] Witschi, 1922, 1930.

In the Hydrozoan *Clytia*, for example, if the blastomeres are isolated at the 4-cell stage, all four of them can give rise to complete little planula larvae which then settle down and develop into hydroid polyps.[1] To a certain extent, this totipotence of Hydroids continues up to the 16-cell stage, at which isolated blastomeres can still produce larvae, though apparently not polyps: whether this

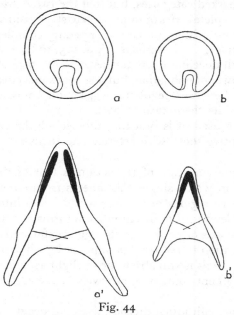

Fig. 44

Sea-urchin gastrulae and plutei from a whole egg (left) and a 1/2 blastomere (right). The latter are normal except in size. (From T. H. Morgan, *Sci. Monthly*, XVIII, 1924, p. 532.)

is due to lack of material or to a real restriction of potency is obscure. In the Ascidian *Styela*, on the other hand, even the first two blastomeres, if isolated, will produce only half-embryos.[2] It is true that the ectoderm grows over the whole surface of the half-embryo, that its notochord develops to form a normally shaped but half-sized rod, and that there is some rounding off of the general form. But in its essentials, the organisation is that of a left or right

---

[1] Zoja, 1895, 1896; Maas, 1905.
[2] Conklin, 1905, 1906.

half. The same mosaic development is seen in anterior and posterior 2/4 halves (fig. 45).

In the early period of study of experimental embryology, these two types were sharply distinguished from one another as "regulation-eggs" and "mosaic-eggs" respectively. Later work has, however, shown first, that all forms do not fall into one or the other of two sharply marked categories, but that the two extremes are connected by a complete series of intermediate steps; and secondly, that at least two very distinct processes impeding complete regulation may be operative in "mosaic-eggs" (pp. 105, 108). Furthermore, it appears that all developing organisms at some stage of their career possess the power of regulation, but lose it at some later stage. Thus the distinction between "regulation-eggs" and "mosaic-eggs" loses a great deal of its theoretical importance, and if the terms are to continue being used, it is best that they should be employed in a purely descriptive sense with reference to their behaviour during cleavage.

The most extreme case of regulation is that of the Hydrozoa, already cited, in which single blastomeres from either the animal or the vegetative regions of the egg will develop into larvae as if they were whole eggs. But in a number of forms, the differentiation along the main axis of polarity of the egg is sufficiently fixed by the time of fertilisation to render this impossible, while differentiation round the main axis is still absent or so slight as to permit of regulation in a fragment containing all levels of the egg along its main axis.

In most eggs, latitudinal division does not occur until the third cleavage (leading from the 4- to the 8-cell stage), and this means that isolated 1/8 blastomeres, or isolated animal or vegetative halves, will be unable to give rise to whole larvae, whereas 1/2 or 1/4 blastomeres, or isolated lateral halves, will be capable of complete regulation. This is the case, for instance, in Echinoderms[1] and to a certain extent Nemertines[2] (fig. 44).

The eggs of Amphibia approach this last type, but the capacity of their blastomeres to achieve complete development is limited by the restriction of organiser capacity to the dorsal side. The organiser-region is determined at fertilisation, and therefore 1/2 or

[1] Driesch, 1900.   [2] E. B. Wilson, 1903; Zeleny, 1904.

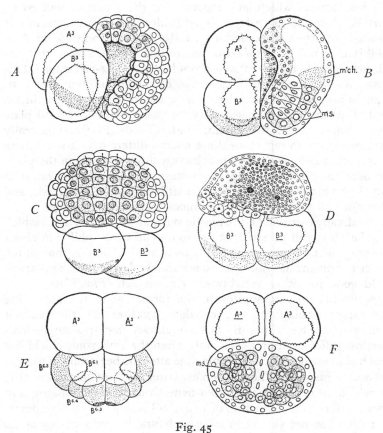

Fig. 45

Mosaic development in the Ascidian *Styela*. (Compare fig. 59.) Two blasto-meres of the 4-cell stage have been killed in each case. In A and B, these are the two left blastomeres, resulting in the formation of (A) a right half-gastrula, (B) a right half-larva with one muscle rudiment (*ms.*) and mesenchyme-rudiment (*m'ch.*). In C and D, the two posterior blastomeres have been killed, leading to the formation of anterior half-embryos with complete neural plate but no muscles. In E and F, the two anterior blastomeres have been killed. E, The segmentation is typical of a posterior half. F, A posterior half-embryo is pro-duced with complete muscle-rudiments but no neural plate or notochord. The only regulation is the overgrowth of the ectoderm and the form-regulation of the notochord. (From Conklin, Chap. IX of Cowdry, *General Cytology*, Chicago, 1924.)

1/4 blastomeres which lack a portion of the organiser will go no further than the germ-layer stage (unless a foreign organiser is grafted into them). Something similar to the conditions in Amphibia is found in *Amphioxus*, where it has been shown that the blastomeres are totipotent at but not beyond the 2-cell stage.[1] This restriction is due to the localisation of chemo-differentiated substances necessary for mesoderm formation in the ventral meridian, and of other substances necessary for notochord and neural plate formation in the dorsal meridian. The fertilised egg is thus bilaterally symmetrical with regard to those chemo-differentiated substances it contains; and, since the first cleavage always occurs in the plane of bilateral symmetry, the 2-cell stage is therefore the latest at which the blastomeres can contain all levels of the main axis, and therefore all these various substances (see below, p. 123).

As already mentioned, the plastic stage of development, in which regulation is still possible, comes to an end in Amphibia at about the stage of mid-gastrulation. A similar state of affairs, though the precise moment has not been so accurately determined, appears to hold good for other vertebrates; e.g. in fish (*Fundulus*) defect-experiments on stages prior to the formation of the germ-ring (i.e. early gastrulation) give rise to defects in the size of the resultant embryos.[2] Other experiments have shown that qualitative irreversible differentiation begins only when the embryonic shield has reached a distinct size—i.e. some time after the beginning of gastrulation. In birds, it is known from experiments (see Chap. VI, p. 161) in which an organiser is grafted beneath another blastoderm and there induces the formation of neural folds that irreversible determination has not yet set in after 22 hours' incubation, but as no interchange experiments have been performed, it is not known at what stage determination of the various regions is definitely fixed. In this connexion it should be mentioned that isolation experiments demonstrate the "competence" (Waddington, 1932) to differentiate into various structures, but they give no information as to whether the power to differentiate into any other structures has been lost. In mammals, nothing has as yet been experimentally determined with regard to these points.

[1] Conklin, 1924, 1933.
[2] Lewis, 1912.

A special case is found in the Nematoda (*Ascaris*). Here the first cleavage division is latitudinal, at right angles to the main axis, and separates animal and vegetative portions. The developmental potencies of the blastomeres have been tested by killing unwanted ones with ultra-violet light. It is then found that the surviving blastomeres develop just as they would have done under normal conditions,[1] and produce anterior or posterior half-embryos. However, by means of the centrifuge, the first cleavage division in these eggs may be made to pass meridionally, and then both of the first two blastomeres will develop a set of reproductive organs, i.e. will produce more than they would normally have produced.[2] The regulative capacity of the Nematode egg before cleavage is shown in the fact that fusion may occur between two eggs, which can then regulate to form a single giant embryo of normal proportions.[3] The inclusion of the Nematode egg among "mosaic-eggs" is therefore merely a consequence of the fact that in this group the first cleavage division is latitudinal (see p. 398, and figs. 192, 193).

The Echinoderms are of further interest in this respect. In some forms, such as the star-fish *Patiria* and the sea-urchin *Lytechinus*, at the earliest stage the apico-basal differentiation is absent, or at least ineffective (see p. 313); both animal and vegetative halves of unfertilised eggs, subsequently fertilised, are capable of giving rise to normal miniature larvae.[4] This is, however, not the case in another sea-urchin, *Paracentrotus lividus*, for here, the apical organ and stomodaeum-forming potency is restricted to the animal half, and the gut-forming potency to the vegetative half of the unfertilised egg. The egg can be cut into two equatorially, and then both halves fertilised. The animal half will give rise to a blastula with long cilia, the ciliation covering an abnormally large area and thus forming a very diffuse apical organ, but such larvae have no gut and no mesenchyme. The vegetative half will produce a larva with a normally tripartite gut, mesenchyme and skeletal spicules (the latter without any regular arrangement or orientation), but without stomodaeum, cilia, apical organ, or arms. The same is true for 4/8 animal and vegetative fragments. Animal and vegetative

---

[1] Stevens, 1909.   [2] Boveri and Hogue, 1909.
[3] Zur Strassen, 1898.
[4] Taylor and Tennent, 1924; Taylor and Whitaker, 1926.

halves isolated at later (blastula) stages show the same developmental potencies, the only difference being[1] that a stomodaeum is formed in such animal halves (see p. 166, and fig. 46).

Already in the unfertilised egg of this sea-urchin (*Paracentrotus*), therefore, the cytoplasm of an animal half, which represents only presumptive epidermis and other epidermal structures, lacks the potency to form an enteron, while that of a vegetative half is incapable of forming an apical organ or stomodaeum.[1] There is therefore an important distribution of potencies along the primary egg-

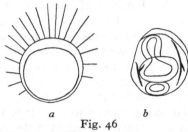

a   b

Fig. 46

Partial larvae from fragments of sea-urchin eggs. *a*, Blastula with abnormally extensive apical organ, derived from animal half of unfertilised egg, fertilised egg, or young blastula. Note absence of gut, mesenchyme, stomodaeum. *b*, Ovoid larva, without stomodaeum, apical organ, ciliated band or arms, derived from vegetative half of unfertilised egg, fertilised egg, or blastula. Note presence of spicules and tripartite gut. (From Hörstadius, *Acta Zool.* IX, 1928.)

axis, and it is because they include all the levels of this axis that the blastomeres of the 2- and 4-cell stages of *Paracentrotus* and meridional halves of gastrulae (see p. 81) are totipotent. This distribution of potencies along the egg-axis has been further analysed by studying the developmental potencies of pieces smaller than halves.

At the 32-cell stage, the cells of the animal half of *Paracentrotus* form two plates or discs of mesomeres, one above the other. They may be designated as *an. 1* and *an. 2* (presumptive ectoderm). The cells of the vegetative half (at the 64-cell stage) also form two discs of macromeres, which may be referred to as *veg. 1* (presumptive ectoderm) and *veg. 2* (presumptive endoderm). Lastly, at the extreme vegetative pole of the egg, there are the micromeres (presumptive primary mesenchyme). Accordingly, the egg of *Paracentrotus* can be divided latitudinally into five layers, each of which

[1] Hörstadius, 1928.

is capable of being isolated, at the 32- or 64-cell stage, and studied in respect of its developmental potency [1] (fig. 47).

An isolated *an. 1* disc develops into a blastula covered all over with the long stiff cilia characteristic of the apical organ. An apparent regulation later occurs in that these sensory cilia are lost and replaced by mobile short cilia, with which the larva swims about.

An isolated *an. 2* disc develops into a blastula, three-quarters of the surface of which are at first covered with the large stiff cilia. In both of these two cases, a true pluteus larva is never formed.

An isolated *veg. 1* disc develops into a larva which may or may not possess an apical organ. Ordinary cilia are present, and a small gut is invaginated.

An isolated *veg. 2* disc produces a larva without an apical organ but with cilia, and a gut is invaginated which may become tripartite in the normal manner.

The micromere group when isolated produces a ball of cells which soon falls apart. Disc *veg. 2* together with the micromeres produces a larva in which the gut is so disproportionately large that it fails entirely to invaginate: instead it protrudes outwards and forms a so-called exogastrula.

It is clear, therefore, that not only are the potencies of the animal half different from those of the vegetative half, but that these differences are graded along the main axis of the egg.

As a result of this differentiation, whereas two sea-urchin eggs or blastulae can give rise to a single double-sized larva when united with their primary axes parallel, union with divergent axes results in a double monster. The same applies to the results of uniting two previously separated 1/2 blastomeres [2] (fig. 48).

The Echinoderms present another curious phenomenon. Isolated 1/2 or 1/4 blastomeres, though they give rise to whole larvae, cleave as parts (see p. 128); e.g. a 1/2 blastomere will form four mesomeres, two macromeres, and two micromeres, just as it would have done if it had been left forming part of a whole egg: the early blastula too is clearly a half and not a whole. If the consistency of the cytoplasm in the developing Echinoderm egg were so stiff as to prevent a half or quarter blastula, produced in this way, from rounding up into a sphere, the fragment could not have formed a

---

[1] Hörstadius, 1931. [2] References in von Ubisch, 1925.

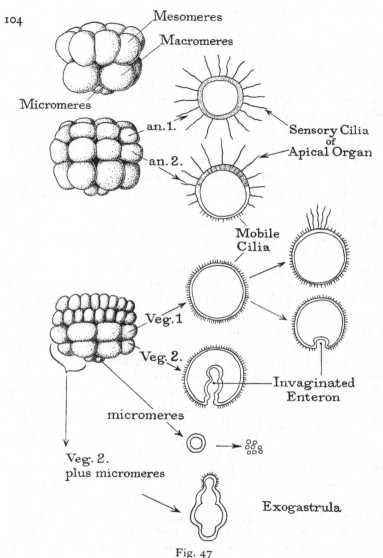

Mesomeres

Macromeres

Micromeres

an. 1.

an. 2.

Sensory Cilia
of
Apical Organ

Mobile
Cilia

Veg. 1

Veg. 2.

Invaginated
Enteron

micromeres

Veg. 2.
plus micromeres

Exogastrula

Fig. 47

Diagram illustrating the developmental potencies of isolated fragments of the sea-urchin embryo, representing different levels along the egg-axis. Note distribution from animal to vegetative pole of potencies for the formation of apical organ, ectoderm with cilia, and invaginated endoderm. The position of the third (equatorial) cleavage plane varies; when nearer to the animal pole, an isolated *veg. 1* disc has more animal and less vegetative potencies, which it shows by forming an apical organ and not forming a gut; when the cleavage plane is nearer to the vegetative pole, an isolated *veg. 1* disc invaginates a small gut and forms no apical organ. (Original, based on Hörstadius.)

whole larva, but would have been forced to continue development as a part.

It appears, in point of fact, that one of the reasons for mosaic development from egg or blastula fragments is extreme viscosity of the cytoplasm.

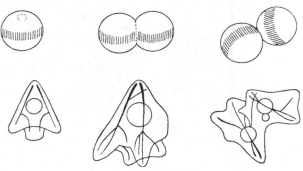

Fig. 48

Diagram to show the influence of the primary axial gradient in fusion-experiments with sea-urchin eggs. Left, single egg and resultant pluteus. Centre, two eggs united with their axes parallel produce a single pluteus. Right, two eggs united with their axes at an angle produce double monsters. (From Przibram, *Handb. norm. and pathol. Physiol.* XIV, 1925, fig. 411, p. 1099.)

A good example of this is found in Ctenophores. Here the adult has eight swimming plates or costae. But although in these forms the first two cleavage divisions are meridional, larvae developed from 1/2 blastomeres have only four costae: 1/8 and 1/4 blastomeres give larvae with one and two costae respectively.[1] In the uncleaved egg of *Beroë*, there is a complete and uniform peripheral layer of a clear substance which appears green by dark-field illumination. By an elaborate series of changes, due to streaming movements of the peripheral zone, and to alternation of more viscous and less viscous phases in the general cytoplasm, the end of cleavage sees this green substance lodged in the micromeres and forming their entire contents, while none of it remains in the macromeres. The micromeres give rise to the ectoderm, including the costae,[2] and contain some materials, precociously chemo-differentiated in the green substance, needed for costa-formation (figs. 49, 50).

At the beginning of each cleavage division during the early

[1] Fischel, 1898.   [2] Spek, 1926.

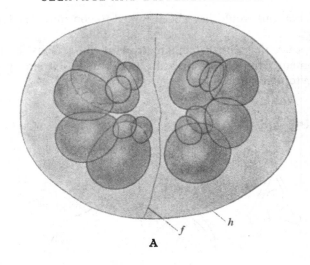

A

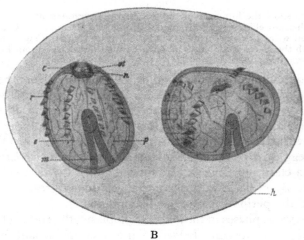

B

Fig. 49

Mosaic development of Ctenophores. A, 16-cell stage divided into two equal lateral halves. B, Partial larvae developed from these halves; each has four costae. *c*, cilia; *e*, endoderm; *f*, fold in egg-case showing line of division of the halves; *h*, egg-case; *m*, stomach; *n*, nerve centre; *ot*, otolith; *p*, pigment; *r*, costae. (After Fischel, from Schleip, *Determination der Primitiventwicklung*, 1929, fig. 18, p. 51.)

stages, the green substance is largely localised at one end of the blastomeres. During this period the cytoplasm is highly viscous: it then becomes more fluid, and the green substance is redistributed uniformly round each blastomere. After the 8-cell stage, however, it remains localised near one pole, and is progressively separated off by a series of unequal cell-divisions into the micromeres.

If the egg is cut at stages up to the 8-cell stage, the result will depend on two factors: first, whether the distribution of the green

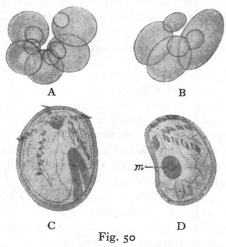

Fig. 50

Mosaic development of Ctenophores. A, B, Fragments of 16-cell stage, divided unequally, so that A has five macromeres and five micromeres; B, three macromeres and three micromeres. C, D, Partial larvae developed from A and B, respectively; C has five costae, D has three. (After Fischel, from Schleip, *Determination der Primitiventwicklung*, 1929, fig. 19, p. 52.)

substance at the moment is uniform over the surface of the blastomeres, or if it is asymmetrically localised; and secondly, whether the egg is in a more fluid state when redistribution of the green substance is easy, or in a very viscous state when redistribution may be impossible before the next cleavage. These facts account for the certain amount of regulation which has been obtained in some experiments on Ctenophore eggs. Immediately after being laid, the egg of *Beroë* is in a highly viscous state, but with the approach of the first cleavage division it becomes more fluid. If in the former period portions of cytoplasm are removed from the egg, some of the costae may be entirely absent; if, on the other hand, portions (even quite large) of cytoplasm are removed from the egg in the latter period (which, it may be noted, is also later in time), none of the costae are absent, although they may be small.[1]

Another interesting example in which viscosity plays an important part is provided by the Ascidian egg. The unfertilised egg is very fluid, and, indeed, as will be seen below (p. 119), extensive internal rearrangements of the contents take place at fertilisation. But 10 minutes after fertilisation, the cytoplasm takes on a high degree of viscosity; this is reduced for a short period at 40 minutes after fertilisation, and then rises again.[2]

§ 4

In *Beroë*, in addition to a variable high viscosity, we find, as mentioned above, the precocious formation, prior to fertilisation, of certain specific substances, which are apparently of an "organ-forming" nature. As we shall see, precocious chemo-differentiation of such substances is universal among so-called mosaic-eggs. As a result of this precocity in their formation, the specific organ-forming or morphogenetic substances are already formed in the just-fertilised egg, instead of being produced only after gastrulation as in Amphibia. If these morphogenetic substances are distributed unevenly during cleavage, mosaic development is the result. One of the classical illustrations of this is the Mollusc *Dentalium*.

Dentalium is an example of that group of animals which exhibit the remarkable form of determinate segmentation known as spiral cleavage, to be found in most Molluscs and many worms. It will be advisable to give a brief general description of this type of cleavage before continuing our discussion of *Dentalium*. The first two cleavages are meridional, and are often unequal, so that one of the cells at the 4-cell stage (blastomere *D*) is larger than the other three (*A*, *B*, *C*). The next cleavage is latitudinal but very unequal,[3]

[1] Yatsu, 1912 A, B; Spek, 1926.   [2] Dalcq, 1932.

[3] The inequality which characterises these cleavage divisions seems to depend on a gradient of permeability extending through the cytoplasm of the dividing cell. Ultra-violet rays and $MgCl_2$ render the permeability of the cytoplasm uniform throughout the cell, and after exposure to these agencies cleavage divisions (of the Lamellibranch Molluscs), which would normally be unequal, take place equally (Pasteels, 1931).

separating four micromeres ($1a$ to $1d$) from four macromeres ($1A$ to $1D$). At the next three cleavages, the micromeres divide sub-equally, but the macromeres bud off three further quartets of small cells or micromeres ($2a$ to $2d$, $3a$ to $3d$, $4a$ to $4d$). After a perfectly definite and fixed number of cleavage divisions, which differs for

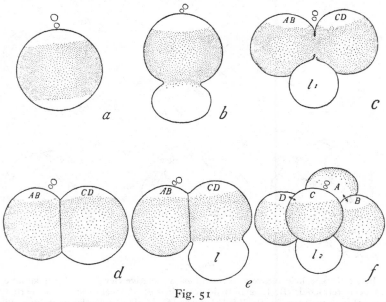

Fig. 51

The polar lobe in *Dentalium*. *a*, Fertilised egg with animal and vegetative clear zones (pole plasms). *b*, Protrusion of first polar lobe. *c*, It passes to one of the first two blastomeres. *d*, 2-cell stage, retraction of polar lobe. *e*, 2-cell stage, protrusion of second polar lobe. *f*, End of second cleavage, second polar lobe passes to the *D* blastomere. (After Wilson, from Morgan, *Experimental Embryology*, Columbia University Press, 1927.)

different blastomeres, a larva with a fixed number of cells is produced.

In *Dentalium*, at the approach of the first cleavage, a portion of the vegetative region is partially constricted off from the rest of the egg as the so-called *polar lobe*. This passes to one of the first two blastomeres and is then withdrawn into it. The blastomere with the polar lobe is destined to be posterior, and is called *CD* in contradistinction to the *AB* or anterior blastomere (figs. 51, 52).

The *AB* blastomere, if isolated, produces a larva which lacks the apical organ and the region of the body behind the main ring of cilia (the post-trochal region), including the coelo-mesoderm. But the apical organ and post-trochal region are present in larvae developed from isolated *CD* blastomeres: these structures are, how-

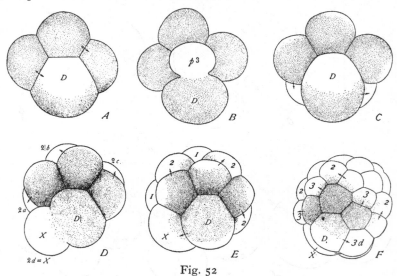

Fig. 52

A, B, 4-cell stage, later cleavage in *Dentalium,* vegetative views. A, The polar lobe has been retracted. B, It is protruded again but of smaller size than earlier (fig. 51) in preparation for the third cleavage. C, 8-cell stage: polar lobe retracted into 1*D*. D, fourth cleavage: 1*D* divides into 2*D* and a 2*d* cell (first somatoblast) containing polar lobe material. E, 16-cell stage. F, 32-cell stage, showing formation of third quartet of micromeres. (After Wilson, from Morgan, *Experimental Embryology*, Columbia University Press, 1927.)

ever, of normal full size, and therefore disproportionately large for the half-sized larva[1] (fig. 53).

At the approach of the second cleavage, the polar lobe is protruded again from blastomere *CD*, and becomes incorporated into *D*. If blastomeres *A*, *B*, or *C* are isolated, they resemble *AB* in that the larvae into which they develop lack the apical organ and the post-trochal region. These structures are present, but relatively much too large, in the miniature larvae developed from isolated *D* blastomeres.

[1] E. B. Wilson, 1904 A.

The third cleavage separates the first quartet of micromeres ($1a$, $1b$, $1c$, $1d$) from the four macromeres ($1A$, $1B$, $1C$, $1D$).

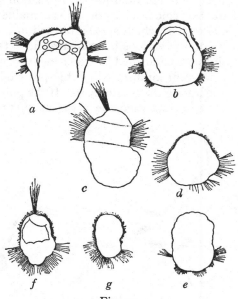

Fig. 53

Dentalium, development of isolated blastomeres. *a*, Larva resulting from an isolated *CD* blastomere (therefore containing the first polar lobe); this larva is of reduced size but normal in form except that the apical organ and post-trochal region are proportionately too large. *b*, Twin larva to *a*, resulting from an isolated *AB* blastomere from the same egg; this larva lacks the apical organ and post-trochal region. *c*, Larva resulting from an isolated *D* blastomere (therefore containing the second polar lobe); this larva is of reduced size but normal in form except that the apical organ and post-trochal region are proportionately much too large. *d*, Twin larva to *c*, resulting from an isolated *C* blastomere from the same egg; this larva lacks the apical organ and post-trochal region. *e*, Larva resulting from an isolated *A* or *B* blastomere; this larva lacks the apical organ and post-trochal region. *f*, Larva resulting from an isolated $1d$ blastomere; this larva possesses the apical organ but lacks the post-trochal region. *g*, Twin larva to *f*, resulting from an isolated $1c$ blastomere from the same egg; this larva lacks the apical organ and post-trochal region. (From Jenkinson, *Experimental Embryology*, Oxford, 1909, after Wilson.)

Isolated $1a$, $1b$, or $1c$ blastomeres give larvae which possess a ring of cilia, but lack gut, apical organ, and post-trochal region. An isolated $1d$ blastomere gives a similar larva, except that it possesses an apical organ. It is clear, therefore, that in the egg of *Dentalium* there is a particular portion of the cytoplasm which is precociously chemo-differentiated, and essential for the formation of the apical organ and post-trochal region. This portion is contained in the first polar lobe, the contents of which are distributed in a definite and unequal way between the various blastomeres (fig. 53).

These conclusions are confirmed by experiments (also on *Dentalium*) of a rather different kind, in which the polar lobe is simply cut off, without separating the blastomeres. If the polar lobe is cut off at the onset of the first cleavage, the larva (like that from isolated

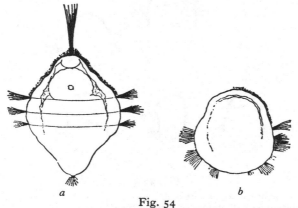

a   b

Fig. 54

Organ-forming substances in *Dentalium*. *a*, Normal larva, 24 hours old. *b*, Larva lacking apical organ and post-trochal region, obtained after removal of the first polar lobe. (From Jenkinson, *Experimental Embryology*, Oxford, 1909, after Wilson.)

AB, *A*, *B*, or *C* blastomeres) lacks the apical organ and post-trochal region. If the polar lobe is cut off at the onset of the second cleavage, the larva possesses the apical organ and lacks the post-trochal region. At the approach of the second cleavage, therefore, the organ-forming materials for the post-trochal region become separated from those for the apical organ, for the latter are not included in the polar lobe at its second extrusion; instead, they presumably migrate into the animal end of the *D* blastomere where they are in a position to become included in 1*d* at the next cleavage (fig. 54).

Similar occurrences whereby chemo-differentiated substances present in the uncleaved egg are restricted by specialisations of the cleavage mechanism to particular blastomeres, and therefore later distributed to particular and circumscribed regions of the embryo, are found in other Molluscs and Annelids. A polar lobe very similar to that of *Dentalium* is found in the Gastropod *Ilyanassa* and the Polychaetes *Chaetopterus* and *Myzostoma*. In *Ilyanassa*, if the polar lobe is removed, no mesoderm is produced and the larva is abnormal in form. Isolated blastomeres give rise to incomplete embryos which die before reaching the larval stage.[1] No experiments appear to have been performed on *Chaetopterus* and *Myzostoma*, but there is every reason to think that the polar lobe in them plays a similar part.

In the Oligochaete *Tubifex*, the egg possesses so-called pole-plasms—clear areas of cytoplasm at the animal and vegetative poles. By means of extremely unequal cleavage, these are entirely restricted, first to blastomere *CD*, and then to blastomere *D*. The two pole-plasms then unite near the centre of the cell. At the next cleavage the united pole-plasms remain entirely within the macromere ($1D$), but a portion of them passes to the first (ectodermal) somatoblast, $2d$. The remainder, which is left in $2D$, passes at the fifth cleavage entirely into $3D$, and then at the sixth into the second (mesodermal) somatoblast, $4d$. The course of events in the leech *Clepsine* appears to be the same.

In *Tubifex*, none of the blastomeres *AB*, *A*, *B*, or *C* is capable by itself (the unwanted blastomeres being killed by ultra-violet light) of developing into anything approaching a complete embryo; blastomere *D*, however, can develop into a complete and properly proportioned embryo.[2]

Other experiments on *Tubifex* have confirmed and extended these results. By certain methods (application of heat, or deprivation of oxygen) the first cleavage can be made to take place equally instead of unequally, and in this case both blastomeres of the 2-cell stage possess equal amounts of the pole-plasms. From such eggs, double monsters (of the *cruciata* type, see p. 156) are produced. It would appear that when the time comes for the formation of micromeres, each set of pole-plasms will give rise to a set of somatoblasts (one ectodermal and one mesodermal), and these will differentiate independently into the main organs of the trunk.[3]

[1] Crampton, 1896.   [2] Penners, 1925.   [3] Penners, 1924.

Similarly, in *Chaetopterus*, it has been found possible to make the first cleavage take place equally instead of unequally, by temporary compression exerted after extrusion of the second polar body and released when the first cleavage plane has cut half-way through the egg. In this case, both the blastomeres of the 2-cell stage receive a half of the polar lobe, and the result is the formation of double monsters (likewise of the *cruciata* type). If, however, the 1/2 blastomeres of such eggs are isolated, each can give rise to a single whole embryo.[1] Double monsters have also been found in the leech *Clepsine*,[2] where they are probably due to equality of cleavage divisions. It is interesting to note in this case that the spiral cleavage of the right-hand member is reversed.

It further appears from experiments on *Clepsine*, in which the method of damaging small areas of the unsegmented egg was employed, that the animal pole-plasm is necessary for cleavage to occur at all. When only the vegetative pole-plasm has been destroyed, cleavage is more or less normal except that it is delayed in the *D* quadrant. Both the somatoblasts ($2d$ and $4d$) are formed, though $4d$ contains more yolk than normal. However, although $4d$ produces the rudiments of mesodermal germ-bands, these are incapable of differentiation, and the embryo dies. It is not known whether the ectodermal germ-bands (derived from $2d$) could differentiate, as this only occurs at a later stage.

The cleavage-pattern, even in the absence of vegetative pole-plasm, is thus predetermined down to the formation of the rudiments of mesodermal germ-bands of typical appearance. But the chemo-differentiation of these to definitive mesoderm is dependent on the presence within them of an organ-forming substance derived from the vegetative pole-plasm.[3]

We may here draw attention to the work on the limpet *Patella*,[4] which demonstrates the extraordinary restriction of potency shown by the micromeres in forms with spiral cleavage. In *Patella*, no polar lobe is formed, but, as in *Tubifex*, the potentiality for producing mesoderm is restricted to the *D* quadrant. An isolated micromere of the first quartet (i.e. a 1/8 animal blastomere, $1a$ to $1d$) produces a purely ectoblastic structure with cells typical of

[1] Titlebaum, 1928.   [2] Müller, 1932.
[3] Leopoldseder, 1931.   [4] E. B. Wilson, 1904 B.

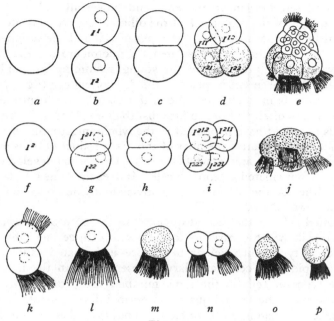

Fig. 55

Development of isolated blastomeres of *Patella*. *a*, An isolated 1/8 blastomere (micromere of the first quartet: 1). *b, c*, Its division into two cells: 1^1 and 1^2. *d*, Their division into four cells: 1^{11} and 1^{12}, 1^{21} and 1^{22}; one of these (1^{11}) is an apical rosette-cell, another (1^{12}) is a "Molluscan cross" cell, and the remaining two (1^{21} and 1^{22}) are trochoblasts (stippled). *e*, Resulting larva 24 hours old, containing an apical organ (formed from derivatives of 1^{11}), intermediate cells (derivatives of 1^{12}), and four cilia-bearing trochoblasts, the results of division of 1^{21} and 1^{22}. These cells are exactly those to which 1/8 micromere would give rise in normal development in a whole embryo. *f*, An isolated 1/16 blastomere (primary trochoblast, 1^2). *g, h*, Its division into two cells (1^{21} and 1^{22}). *i*, Their division into four cells (1^{211} and 1^{212}; 1^{221} and 1^{222}). *j*, The same, 24 hours later; each of the four cells (trochoblasts) has put out cilia, but divides no further. These cells are exactly those to which a 1/16 blastomere (primary trochoblast) would give rise in normal development in a whole embryo. *k*, A pair of trochoblast cells, the only product of an isolated 1/32 blastomere (trochoblast: 1^{21} or 1^{22}). *l, m*, Isolated 1/64 blastomere (trochoblast: 1^{211} or 1^{212}, 1^{221}, 1^{222}); such a cell puts out cilia but does not divide. *n*, A pair of "secondary trochoblast cells," the product of an isolated 1/32 blastomere (1^{12}, "Molluscan cross" cell). *o, p*, Isolated 1/64 blastomere ("secondary trochoblast," 1^{122}); such a cell puts out cilia but divides no further. (From Jenkinson, *Experimental Embryology*, Oxford 1909, after Wilson.)

the apical sense-organ at one end, and powerfully ciliated cells characteristic of the prototroch at the other: these are separated by non-ciliated epidermal cells. The types of cell and the number of each type produced by the isolated 1/8 micromere are the same as it would have produced in the young swimming trochophore larva if it had been left in place in the developing egg (fig. 55).

Descendants of a 1/8 micromere, if isolated later, continue to divide just as often and to produce just the same kind and number of cells as would have happened in the whole intact embryo. For instance, the vegetative member of the first product of division of a micromere of the first quartet ($1a_2$ to $1d_2$) in normal development produces four ciliated prototroch cells. It does the same if isolated; while either of the products of its division divides once only to produce two ciliated cells.

Isolated cells of the second quartet ($2a$ to $2d$) produce certain ciliated cells which contribute to the prototroch, certain others of a different type which belong to the pre-anal ciliated band, non-ciliated epidermal cells, and larval mesenchyme in the interior. These types of cells in the same numbers are produced by the micromeres of the second quartet in normal development.

The exact meaning of these facts has not been determined. Presumably, two agencies are at work. First, certain chemo-differentiated substances are probably restricted to particular micromeres; secondly, it appears that the number of cleavage divisions possible to any isolated blastomere is fixed. This may perhaps be correlated with the fact (described below, see p. 132) that nuclear synthesis during cleavage takes place at the expense of certain materials in the cytoplasm, present in finite amount. When these materials are exhausted in an isolated cell (deprived of contact with the yolk of its egg, or other food-supply), cleavage stops.

It is probable that in all Annelids and Molluscs (other than Cephalopods), even when no differentiated substances can be detected in the uncleaved egg, they do in fact exist, and are distributed during cleavage in a similar way. Only on these lines can such facts be explained as the almost universal restriction of the potentiality of forming mesoderm bands to $4d$, and of forming ectoderm bands to $2d$. The very general fact that the $D\ 1/4$ blastomere is larger than its three sisters is doubtless to be explained by the presence in it alone of such specific organ-forming stuffs.

Summing up the evidence, we may say that the animal pole-plasm, and its presumable homologue, the slightly thickened cap of cytoplasm at the animal pole of the egg of *Dentalium*, are in some way responsible for normal cleavage, though this has only been demonstrated for *Clepsine* (see above). On the other hand, organ-forming substances for apical organ (where present), and mesodermal and ectodermal germ-bands, appear to be located in the vegetative pole-plasm or polar lobe. This has been demonstrated for the apical organ in *Dentalium*, for the mesodermal germ-bands in all forms investigated, and for the ectodermal germ-bands in *Dentalium*. The migration in the direction of the animal pole of the material for the apical organ occurs before the second cleavage (*Dentalium*), for the ectodermal and mesodermal germ-bands only later, in some cases (Annelida) after a union of the two pole-plasms within the *D* macromere, to be segregated at the fourth and sixth cleavages respectively.

In general we may say that determinate spiral cleavage provides an effective method of distributing precociously differentiated substances to particular regions of the embryo, and that special advantage of this has been taken by the Annelids and Molluscs, though in varying degrees by different forms.

§ 5

The conditions found in *Beroë* and *Dentalium* introduce us to another principle of considerable importance. In *Beroe*, the formation by chemo-differentiation of the green ectoderm-producing substance and the uncoloured endoderm-producing material is effected prior to fertilisation, but the localisation of these substances in their definitive positions is only brought about during cleavage. In regard to the egg, these substances are *preformed*, but not *prelocalised*.

The same is true, though the details are even more elaborate, concerning the distribution of the materials contained in the pole-plasms and polar lobes of Mollusca and Annelida.

When the distinction between mosaic- and regulation-eggs was regarded as fundamental, this distinction between the preformation and the prelocalisation of organ-forming substances appeared to be of considerable theoretical importance. From the point of view here adopted, the absence of prelocalisation in such cases is seen to be a frequent (though not universal) consequence of precocious

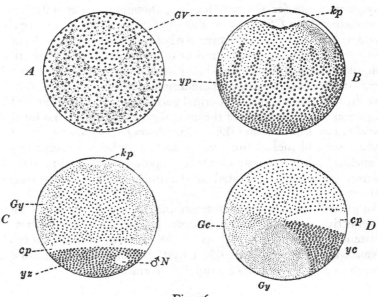

Fig. 56

Localisation of organ-forming substances in Ascidians. Views of eggs of *Styela*. Yellow cytoplasm containing mitochondria (*yp*) small circles. Yolk (*Gy*) stippled. Clear cytoplasm (*cp*) white. A, Before fertilisation, showing germinal vesicle (*GV*), yellow cytoplasm evenly distributed over the surface. B, Immediately after fertilisation, showing clear region (*kp*) derived from germinal vesicle at animal pole, yellow cytoplasm streaming down to vegetative pole. C, The yellow cytoplasm forms a cap at the vegetative pole (*yz*) containing the male pronucleus. The clear cytoplasm forms a layer just above it. D, Left side view of egg just before first cleavage, showing yellow crescent (*yc*) and clear crescent (*cp*) posterior, and grey crescent (*Gc*) anterior. (From Conklin, Chap. ix of Cowdry, *General Cytology*, Chicago, 1924.)

chemo-differentiation. When chemo-differentiation occurs prior to fertilisation, the differentiated substances thus produced are able to shift their relative positions, either in the uncleaved egg, or as a result of manœuvres effected during cleavage. If, on the other hand, it does not occur until after the end of the cleavage period (as in Amphibia), the substances are precluded from this type of movement through their being confined within cell-membranes, and redistribution can only be effected by the movements of cell-regions.

The examples so far given concern the migration of organ-forming substances during cleavage. Other forms show striking localisation phenomena in the uncleaved egg, usually initiated by polar body formation, as the result of fertilisation. The classical example of this is the Ascidian *Styela* (*Cynthia*). Before fertilisation, the egg contains a cap of clear cytoplasm at its animal pole, a central mass of yolk, and a superficial layer of yellow cytoplasm laden with mitochondria. The clear cytoplasm is chiefly derived from the breaking down of the large germinal vesicle. Almost immediately after the entry of the sperm, the polar bodies are given off, and the clear cytoplasm and the yellow cytoplasm flow down to the vegetative pole, leaving the animal pole occupied by the yolk, except for a very small cap of clear cytoplasm. Next, the sperm moves towards the centre of the egg, along an apparently predetermined path (indicating that a plane of bilateral symmetry already exists in the egg), and another rearrangement of the cytoplasmic regions ensues. The sperm appears to drag much of the yellow cytoplasm with it into the interior of the egg, and this yellow cytoplasm now forms a crescent on the surface, beneath the equator, with its horns extending a quarter of the way round the egg on each side. The clear cytoplasm forms a crescent immediately above the yellow cytoplasm, and the centre of these crescents marks the future ventro-posterior side of the embryo. After the first cleavage, another crescent, light grey in colour, is formed opposite the yellow crescent (fig. 56). A pattern almost precisely similar is found in *Amphioxus*.[1] Thus, in these forms, both radial and bilaterally symmetrical localisation are effected prior to cleavage.

Other examples of such rearrangements are afforded by other Ascidians (e.g. *Ciona*) and by *Myzostoma*, in which a green vegetative area is formed in the oocyte, while fertilisation results in the withdrawal of a red substance to the animal pole, leaving a clear equatorial zone.[2] In the leech *Clepsine*, the pole-plasms, or areas of clear cytoplasm at the two poles, only form after the polar bodies

[1] Conklin, 1933. [2] Driesch, 1897.

have been extruded. Here, and in various other Annelids, the material for the pole-plasms and polar lobes appears to have been previously distributed over the whole surface of the egg, and to some extent in the interior. This follows from the fact that all and sundry unfertilised egg-fragments of *Chaetopterus* subsequently fertilised are capable of developing into normally formed miniature larvae,[1] whereas the polar lobe in later stages is sharply localised.

Further evidence of a rearrangement of materials as a result of fertilisation is provided by other experiments in which the development of egg-fragments is studied. In the Nemertine *Cerebratulus*, for instance, such experiments show that there is a *progressive* increase, from before fertilisation, to the onset of the first cleavage, in the restriction of the potencies of animal and vegetative regions. The animal region becomes progressively less able to produce digestive tract and larval lappets, while the proportion of vegetative fragments which produce an apical organ becomes smaller, during the period in question.[2]

A remarkable fact, whose precise interpretation is not clear, is found in *Dentalium*. If the unfertilised egg is cut across, latitudinally or obliquely, and then fertilised, the vegetative portion segments as a whole, with a polar lobe usually of correct proportional size.[3] The resulting larva also has a correctly proportioned apical organ and trunk. It will be remembered that when a *CD* blastomere is isolated, it forms a polar lobe as large as in the whole egg, and the larva is disproportionate. A further remarkable fact is that when an already fertilised egg is cut so as to produce an enucleate vegetative fragment, though this does not cleave, it will protrude its polar lobe synchronously with the first division of the nucleated portion: the polar lobe is of the same size as in an intact egg. Some irreversible change concerning the quantity of material in the polar lobe must take place at fertilisation.

Further light on the mechanism of formation of polar lobes is thrown by experiments on the Mollusc *Ilyanassa*. Here, a polar lobe is protruded four times: when the first and second polar bodies are forming, and at the first and second cleavages. At its first appearance, the degree of protrusion is extremely slight; at its second, moderate; while at its last two appearances, it is very marked, and

---

[1] E. B. Wilson, 1929.   [2] Yatsu, 1910.   [3] E. B. Wilson, 1904 A.

the polar lobe is at one moment only connected with the rest of the egg by a narrow stalk (figs. 57, 58).

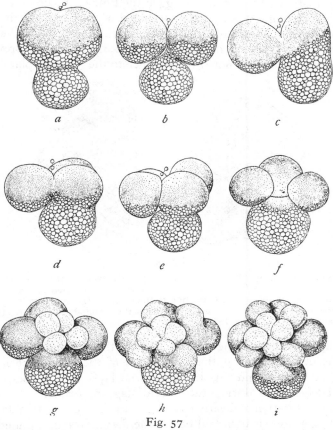

Fig. 57

Polar lobe formation in the normal cleavage of the Mollusc *Ilyanassa*. The small circles represent yolk-spheres. *a*, Protrusion of the lobe by the uncleaved egg. *b*, *c*, Its passage to the *CD* blastomere. *d*–*f*, Second cleavage; the polar lobe passes to the *D* blastomere. *g*, Formation of first quartet of micromeres by unequal dexiotropic division. *h*, *i*, Formation of second quartet of micromeres by an unequal laeotropic division. (From Morgan, *Experimental Embryology*, Columbia University Press, 1927, fig. 135, p. 360.)

Normally, the lobe is composed of very yolky material. Centrifuge experiments, however, show that its protrusion occurs irrespective of the materials which it contains, for it may contain only oil and cytoplasm, or these on one side and yolk on the other. It was further found that on centrifuging the egg at the time of the first cleavage, the nuclear spindle may be disarranged so as to divide the egg equatorially with reference to the original axis. In such cases the lobe appears with reference to the original axis, i.e. at the end opposite the polar bodies, and not with reference to the cleavage plane.[1]

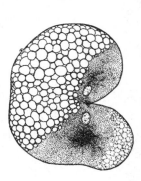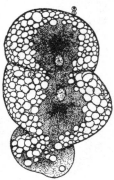

Fig. 58

Effects of centrifuging the uncleaved eggs of the Mollusc *Ilyanassa*. A (modified) polar lobe is formed at the first cleavage; it forms at the region opposite to the polar bodies—i.e. in its normal relation to the original egg-axis—irrespective of the direction of the first cleavage furrow or of the materials it contains. (From Morgan, *Journ. Exp. Zool.* LXIV, 1933.)

It appears that during mitosis, most of the egg undergoes some degree of gelation, but that the region of the polar lobe is not involved in this, and that the superficial layer of the lobe-region is predetermined during the oocyte stage to behave as it does during maturation and early cleavage. It is interesting to note that if the polar lobe is detached before the first cleavage, it undergoes spontaneous changes of form, apparently synchronised with the cleavages of the egg.[1]

§ 6

For mosaic development to occur, some degree of precocious chemo-differentiation must have been effected, prior to the onset of cleavage. The organ-forming materials thus available may be

[1] Morgan, 1933.

differentially distributed by a specialised cleavage, or regulation may be prevented by a high degree of viscosity in the egg, at any given stage. *Beroë* affords an example of the latter method, but the best illustration is provided by the Ascidians.

It will be remembered that the Ascidian egg is in a fluid state before fertilisation, but that after this event its viscosity is enormously increased. Further, we have already mentioned the localisation of different organ-forming substances which takes place at fertilisation. Elaborate experiments on the effects of killing blastomeres and of centrifugalisation have shown that the fertilised egg is already a mosaic of chemo-differentiated regions. 1/2, 1/4, and 3/4 blastomeres all develop into those parts of a larva to which they would have given rise in normal development. The partial embryos round themselves off, and this process in some forms (e.g. *Phallusia*) goes much further than in others (e.g. *Styela, Ciona*), so that the products of single blastomeres may appear superficially to be whole larvae, but sections invariably show that they are only parts, halves or quarters (e.g. with mesoderm only on one side; see p. 97 and figs. 45, 59). The egg of *Amphioxus* behaves in an extremely similar manner.[1] In the experiments on *Amphioxus*, the blastomeres of the 2- and 4-cell stages were frequently disarranged without being totally separated. In such cases, they always preserved their inherent polarity, though there was complete fusion between their products. The result was the formation of double monsters in various orientations (fig. 34).

In *Styela*, the fertilised egg contains yolk in the animal hemisphere, cytoplasm with yellow mitochondrial granules at the vegetative pole, and clear cytoplasm in between. In the centrifuge tube the eggs tend to orientate themselves in such a way that the animal pole with the relatively heavy yolk is centrifugal, and the stratification of the egg is then increased by the centrifugalisation. But if the eggs are slightly compressed, either by mutual pressure, or by being placed in fine tubes so that they cannot rotate, centrifugalisation can restratify the egg-contents in such a way that, for instance, all the yellow granular cytoplasm is confined to one of the blastomeres of the 2-cell stage. The resulting embryo then possesses muscle-fibres only on one side of the body.[2] The mosaic

---

[1] Conklin, 1933.   [2] Conklin, 1924, 1931.

has been forcibly disarranged by the centrifuge, taking advantage of the different specific gravities of the various egg-contents, but each part of the mosaic continues its predetermined course. In this way, organ-forming substances have been shown to be present in the fertilised egg, and respectively responsible for the formation of ectoderm ("ectoplasm"), endoderm ("entoplasm"), neural plate ("neuroplasm"), notochord ("chordoplasm"), muscle fibres ("myoplasm"), and mesenchyme ("chymoplasm")[1] (figs. 56, 59). It is thus clear that the fertilised egg of the Ascidian is already a highly complex mosaic of chemo-differentiated stuffs, and we may now turn to the experiments in which the developmental potencies of fragments of the unfertilised egg have been tested.

Latitudinal halves of unfertilised eggs of *Ascidiella*, subsequently fertilised, show that there is already at this stage a differential repartition of potencies along the egg-axis. The larvae obtained may be deficient in one or more kinds of tissue according to the level of the cut: myoplasm can be separated from chymoplasm, neuroplasm from chordoplasm, the former in each case being situated nearer to the animal pole. The various substances must, therefore, occupy different levels.[2]

In view of the rigid mosaic behaviour of isolated blastomeres, and of the definite localisation of substances at fertilisation (as tested by the centrifuge experiments), the further result may seem surprising that meridional halves of unfertilised eggs, subsequently fertilised, may give rise either to apparently normal and symmetrical larvae, or to lateral half-larvae. The former type appear to provide a case of regulation, which would be remarkable in such a form as an Ascidian. These results can, however, be explained on the view that the various organ-forming substances in the unfertilised egg occupy circular zones at particular levels surrounding the egg (or possibly crescentic zones, the horns of which quite or

---

[1] Careful analysis has shown that the *visible* prelocalised substances in *Styela*, such as the mitochondria, which impart the yellow colour to the region destined to give rise to muscles, are not themselves morphogenetic substances. Muscles can develop without mitochondria. The various regions differ in the consistency of their cytoplasm, and it is these sharply marked off differentiated regions which appear to constitute the true organ-forming substances. The mitochondria and other gross differences are symptoms, not causes (see Duesberg, 1928). This question of the relation of organ-forming substances to raw materials will be discussed in Chap. VII.   [2] Dalcq, 1932.

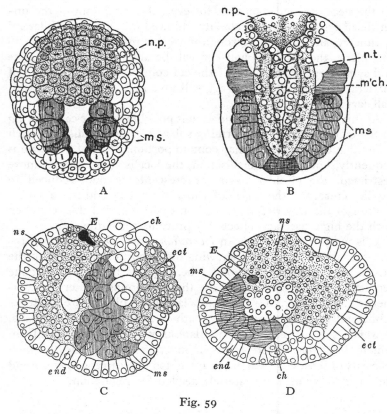

Fig. 59

Mosaic development and prelocalisation in the egg of Ascidians (*Styela*). A, B, Normal development, dorsal surface views: A, Late gastrula. The neural plate (*n.p.*) overlies the notochord rudiment (not seen); and the muscle-rudiments (*ms.*) border the blastopore laterally. B, Neurula. The neural tube (*n.t.*) has formed, and the mesenchyme (*m'ch.*) is visible. C, D, Sections of abnormal neurula stages derived from eggs centrifuged before the first cleavage. The disarrangement of organ-forming substances by centrifuging had led to the disarrangement of organ-rudiments. Endoderm (*end.*) and notochord (*ch.*) appear on the outside, ectoderm (*ect.*) and neural plate substance (*ns.*) on the inside. Eye spots (*E.*) and muscle-rudiments are also ectopic. (From Conklin, Chap. IX of Cowdry, *General Cytology*, Chicago, 1924.)

almost meet). Any meridional half will thus contain a portion of all the necessary substances. However, the cytoplasm of the unfertilised egg appears to be already endowed with a plane of bilateral symmetry, and if the cut through the egg is made at right angles to this plane, the resulting half-egg will be able to form a complete and symmetrical larva. But if the cut coincides with the plane of bilateral symmetry, the half-egg will give rise to an asymmetrical half-larva.[1]

At this stage, therefore, regulation is possible in some cases, owing to the fact that the organ-forming substances are localised in such a way that egg-fragments may contain portions of all of them. Subsequently, however, at fertilisation, the localisation becomes more restricted, the circular bands or crescents become reduced to smaller crescents, the horns of which do not extend more than a quarter of the way round the egg on each side, and this, together with the high viscosity, effectively prevents regulation.

It is worth stressing that in *Dentalium*, the *CD* blastomere produces a larva which, while showing disproportion in regard to the organs derived from the polar lobe, appears to have undergone regulation round the major axis, thus becoming bilaterally symmetrical. Similarly, in *Amphioxus* a lateral 1/2 blastomere produces a bilaterally symmetrical larva. Both in *Beroë* and the Ascidians, however, 1/2 larvae preserve the laterality of the blastomere from which they arose. Here again, it must presumably be the high viscosity of these eggs which has prevented the rearrangement round the main polar axis of materials needed for regulation.

§ 7

Returning to the question of the relation of cleavage to differentiation, it may then be said that the part which cleavage plays is only indirect. Cleavage is a process whereby the single-celled fertilised egg is split up into a number of separate cells whose differing qualities depend upon factors which are originally independent of cleavage, and concern the viscosity of the egg and the time of chemo-differentiation of its cytoplasm.

In this connexion, we may refer to the very interesting case of the insect egg. Here, cleavage of the nucleus begins and continues

[1] Dalcq, 1932.

for a long time in the interior of the egg, while the peripheral cytoplasm or blastema remains undivided. It is only later that the nuclei, now very numerous, migrate to the surface of the egg, and the cytoplasm becomes partitioned off into blastomeres, forming the blastoderm (see also p. 88).

Experiments on the regulatory capacity of the insect egg have given different results in the various groups. In the house-fly *Musca domestica*, the nuclei have already begun to divide when the egg is laid, but the cytoplasm is still quite undivided. Nevertheless, all the parts of the cytoplasm are already determined and chemo-differentiated; damage done to any part of the cytoplasm results in damage to or absence of some definite structure in the developed organism, and no regulation is possible. Here, then, is a clear case of precocious chemo-differentiation of the cytoplasm and mosaic development in which cleavage plays no part at all.[1]

In the ant *Camponotus ligniperda*, it has been possible to determine the time at which chemo-differentiation sets in. This is found to coincide with the start of the visible differentiation of the blastema into various regions, such as those of the future embryonic shield, extra-embryonic blastoderm, etc., which takes place before the nucleus has begun to divide at all. Prior to this time, the egg is undetermined and capable of regulation: after this time the cytoplasm is chemo-differentiated, and development strictly mosaic.[2]

In the dragon-fly *Platycnemis pennipes*, the time of onset of chemo-differentiation is relatively later, during the blastoderm stage, and the early egg is therefore capable of regulation. It has been possible to obtain a normally proportioned diminutive insect from one (posterior) half of an egg constricted transversely into two; duplications and triplications of structures after making longitudinal slits in the blastoderm; and two insects from one egg, the blastoderm of which was divided transversely.[3] Later on, however, constrictions and injuries result in the development of partial embryos only. In this case, as in that of *Camponotus*, it has been possible to establish the very interesting fact that the process of chemo-differentiation emanates as a stream from an activating centre, situated near the hinder end of the egg (figs. 60, 84 and 122; see also pp. 170, 252).

---

[1] Reith, 1925; Pauli, 1927.   [2] Reith, 1931.   [3] Seidel, 1926, 1928, 1929.

The relative unimportance from the point of view of differentiation of the way in which the egg cleaves is revealed by the following experiments.

When a blastomere of a sea-urchin is isolated at the 2- or 4-cell stage, it develops, as already mentioned, into a whole larva, but the

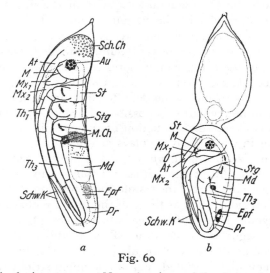

a   b

Fig. 60

Regulation in the insect egg. *a*, Normal embryo of the dragon-fly *Platycnemis pennipes*, seen from the left side. *b*, Dwarf embryo, obtained by partial constriction of the egg at the 4-nucleus stage; the dwarf is normally proportioned and developed and its organs have arisen from regions the presumptive fate of which was quite different; their fates were therefore not irreversibly determined at the stage operated upon, and regulation has been possible. *At*. antenna; *Au*. eye; *Epf*. hindgut; *M*. mandible; *M.Ch*. chitinous muscle-attachments; *Md*. midgut; *Mx*$_1$, first maxilla; *Mx*$_2$, second maxilla; *O*. labrum; *Pr*. proctodaeum; *Sch.Ch*. apical chitin; *St*. stomodaeum; *Stg*. spiracle; *Th*$_{1-3}$, first to third thoracic legs; *Schw.K*. gills. (From Seidel, *Biol. Zentralbl.* XLIX, 1929.)

cleavage which it undergoes is the same as that which it would have undergone if it had been left in contact with its sister-blastomeres. In normal development in these forms, the first and second cleavages are meridional and equal: the third cleavage is latitudinal and equal; the fourth cleavage in the animal hemisphere is meridional and equal, in the vegetative hemisphere it is latitudinal and unequal. Each cleavage division is therefore recognisably distinct. Now, in a blastomere isolated at the 2-cell stage, the first cleavage which it undergoes after isolation is meridional and equal (corresponding to the second normal cleavage), and its next cleavage is latitudinal and equal (corresponding to the third normal cleavage), and so forth. The first cleavage of a blastomere isolated at the 4-cell stage is latitudinal and equal (corresponding to the third normal cleavage). In other words, the isolated blastomeres cleave as if they were still parts of a whole, but they develop into whole larvae. Here, clearly, the method of cleavage is without effect on the subsequent development and differentiation.

The system of cleavage in the sea-urchin egg has been shown to depend on a number of factors. First, there is the control which the cytoplasm exerts on the orientation of the division spindles; this is of such a kind that for a certain period of time (normally corresponding to that between fertilisation and the attainment of the 4-cell stage) any nuclear spindles that there may happen to be are restricted to a latitudinal plane so that division will be meridional; after this period, the spindles are rotated into the longitudinal axis so that division will be latitudinal. From now onwards there will be two sets of division spindles; one in the animal and one in the vegetative half of the egg. Those in the former set revert to the latitudinal plane (meridional division of mesomeres), while those in the latter remain longitudinal (latitudinal division of macromeres from micromeres). Experiments of cutting eggs at varying times after fertilisation have shown that the fixation of a division spindle to a given axis is progressively determined: a 1/2 egg cut meridionally within a quarter of an hour of fertilisation can as it were start again with the determination of its spindle axis, and the 1/2 will cleave as a whole egg; a similar 1/2 egg cut meridionally three-quarters of an hour after fertilisation has its spindle axis set and fixed, and it cleaves as a 1/2 blastomere.

Secondly, there is localised at or near the vegetative pole a special region of cytoplasm which determines a marked inequality of cleavage, leading to the formation of tiny micromeres split off from the large macromeres. Thirdly, there is the fact that this special region of the cytoplasm at the vegetative pole does not acquire its property of causing unequal division until after a certain definite

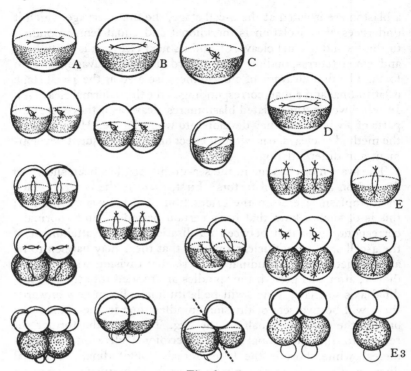

Fig. 61

Cleavage of the sea-urchin egg. Column A, normal cleavage as far as the 16-cell stage (eight mesomeres, four macromeres, four micromeres), serving as time-scale (read from top to bottom) for the other columns. By treatment with hypotonic sea-water or shaking, the formation of the mitotic spindles can be delayed: the other columns show the effects of increasing retardation of spindle-formation. Column B, the first two cleavage spindles latitudinal, the third vertical but so delayed that it falls within the period of micromere-formation: result, four micromeres at the 8-cell stage. Column C, first cleavage spindle latitudinal, the second fall within the period during which the spindles are rotated into the vertical position; they have not quite achieved it here and are oblique; the third cleavage spindles, at right angles to the second, are also oblique: result, two meso-meres and two micromeres at the 8-cell stage. Column D, the first cleavage spindle latitudinal, the second vertical, the third similar to the fourth of normal cleavage, i.e. latitudinal in animal, vertical in vegetative cells: result, four mesomeres, two macromeres, two micromeres at 8-cell stage. Blastomeres isolated at the 2-cell stage cleave according to this pattern. Column E, cleavage of blastomeres isolated at the 4-cell stage or of eggs cut into meridional halves (in which the mitotic apparatus is so delayed that the first cleavage spindle coincides with the third of normal cleavage and is vertical); the second (like the fourth normal) cleavage spindles are latitudinal in animal, vertical in vegetative cells. (From Hörstadius, *Acta Zool.* IX, 1928, slightly modified.)

period of time. In normal cleavage, this time corresponds to the attainment of the 8-cell stage.

By various methods (use of dilute sea-water, shaking, and cutting the egg into halves), it is possible to alter the time-relations of mitosis relatively to these three factors. By delaying the rate of cell-division, it is possible to make the second, or even the first cleavage of an egg fall into the period when the nuclear spindles are forced into the longitudinal axis. The result will be latitudinal division at the 2-cell and 1-cell stages respectively, whereas it normally happens at the 4-cell stage. Very instructive are the cases in which the cleavage division falls *during* the change of position of the nuclear spindles, i.e. when the latter are oblique. One more cleavage division in eggs whose mitoses are thus delayed will lead to formation of micromeres precociously (fig. 61).

It will thus be seen that it is possible to make a whole egg cleave as if it were an isolated blastomere of the 2-cell or 4-cell stage. When a blastomere is isolated from a normal egg, the mitotic speed of which has not been interfered with, the subsequent cleavage divisions continue to be governed by the same factors as in the normal egg, with the result, therefore, that the blastomere cleaves as a part.[1]

The second example of the relative unimportance of cleavage as regards differentiation is provided by those cases in which a frog's egg has been penetrated by several sperms. One sperm-nucleus fuses with the egg-nucleus, but the other sperm-nuclei remain isolated in the cytoplasm of the egg. When the egg begins to undergo cleavage, not only does the zygote-nucleus divide and induce the division of the cytoplasm into blastomeres, but each of the isolated sperm-nuclei has a portion of cytoplasm allotted to it, and this becomes separated off as a little blastomere and subsequently divides. Cleavage is therefore very irregular, and the embryo is composed of an indiscriminate mixture of blastomeres, some containing the products of division of the zygote-nucleus and representing the normal blastomeres of typical cleavage, and some representing blastomeres which would normally never have come into existence. The two kinds of blastomeres can be recognised without difficulty, for those derived from the zygote-nucleus are of course diploid,

[1] Driesch, 1900; Hörstadius, 1928.

while the others are haploid. Since the volume of the cell is proportional to the quantity of nuclear material which it contains, it is easy to recognise the descendants of the two kinds of blastomeres in the tissues to which they give rise. In spite of their abnormal cleavage, such polyspermic frogs' eggs can sometimes develop normally, the stage ultimately reached depending on the number of supernumerary sperms present. A pentaspermic egg can produce a free-swimming tadpole which lives for 10 days after hatching:[1] a dispermic egg can produce a tadpole which lives for three months.[2]

Lastly, it has been shown in the case of *Chaetopterus* and *Nereis* that a certain amount of differentiation can take place even if cleavage is totally suppressed, by treatment of the egg with KCl.[3] Cilia are put out and internal rearrangements occur, the most interesting of which is the assumption by certain granules of the position in the egg which corresponds to that of the cells of the prototroch, which cells in normal cleavage come to contain these granules.[4]

§ 8

But besides splitting up the cytoplasm of the egg into smaller units, cleavage has one very important effect, though its bearing on differentiation is indirect, and this concerns the adjustment of the ratio between amount of nuclear matter and amount of cytoplasm present in the cell.

In the oocyte of the sea-urchin (*Echinus microtuberculatus*) it has been shown that the ratio between the volume of the cytoplasm and that of the nucleus is 7 : 1. Maturation results in a certain increase in cytoplasmic volume and a reduction in nuclear volume, so that the ratio of cytoplasm to nucleus in the ripe egg is 400 : 1. But the volume of the cytoplasm has been only about doubled, so that the explanation of the high ratio in the ripe egg must be looked for to a small extent in the extrusion of nuclear material in the polar bodies, and to a large extent in the passage of nuclear material into the cytoplasm. Now the total amount of nucleic acid in the egg and in subsequent stages of cleavage up to the blastula is constant.[5]

---

[1] Brachet, 1910.   [2] Herlant, 1911.   [3] Lillie, 1902; Spek, 1930.
[4] What is in some ways a complementary experiment has been carried out by removing the zygote nucleus from uncleaved axolotl eggs by means of a micropipette. In spite of the absence of nuclei, the cytoplasm makes an attempt to carry out cleavage, though this is partial and irregular. Jollos and Peterfi, 1923.
[5] Masing, 1910.

But at the start of cleavage, most of this nucleic acid is in the cytoplasm. At each cleavage division, the nuclei of the daughter-blastomeres are slightly larger than half the nucleus of the blastomere that gave rise to them. There is consequently a gradual return of nuclear material from the cytoplasm into the nuclei of the blastomeres, and this is shown by the drop in the ratio of total volume of cytoplasm to total volume of nuclei at successive stages of cleavage. At the 4-cell stage the ratio is about 18 : 1, at the 64-cell stage it is 12 : 1, while in the blastula the ratio has returned to the original value of 7 : 1.[1]

These results are of considerable interest, and for two reasons. In the first place, the return of the cytoplasmo-nuclear ratio to the original value occurs in the blastula, when cleavage has ended, and when the hereditary effects of the nuclear material can begin to manifest themselves, as will be shown in Chap. XII. It is not improbable that these two sets of events are causally related. In the second place, the recognition of the existence in the cytoplasm of the ripe egg of a finite amount of nuclear material accounts for the termination of cleavage. It is well known that eggs which are made to develop in the haploid condition (as by artificial parthenogenesis) go on cleaving until their cells are half the volume of normal diploid cells.[2] The haploid nuclei of the blastomeres requiring only half the amount of nuclear material from the cytoplasm, the supply in the cytoplasm will last longer than is the case with diploid nuclei; cleavage will therefore go on for a longer time, and the cells will be smaller. Conversely, it is known that if half an egg, containing a nucleus, is fertilised (that is to say, diploid nuclei but only half the normal quantity of cytoplasm is present), the resulting larva has cells of normal (diploid) volume but is itself of half size. It follows that it has half the number of cells that the normal has, and this is what would be expected since it had only half the reserves of nuclear material in the cytoplasm. Lastly, it is possible in some cases to obtain fertilised eggs with tetraploid nuclei. The size of the embryos which these produce is normal, but their cells are twice as large and half as numerous as normal. The quantity of nuclear reserve materials in the cytoplasm has given out sooner than during normal cleavage, with the result that the division of the blastomeres has not proceeded so far.

[1] Godlewski, 1925.   [2] Boveri, 1905.

Chapter VI

ORGANISERS: INDUCERS OF DIFFERENTIATION

§ 1

The remarkable organising properties of the dorsal lip of the blastopore of amphibian embryos were discovered in the following manner. In the experiments with newts' eggs of grafting pieces of the presumptive neural fold region into other positions, in order to discover the time at which they became irrevocably determined to develop by self-differentiation, it was observed that the determination of the posterior part of the presumptive neural fold region (i.e. that portion which lies near the dorsal lip of the blastopore) was effected sooner than that of the anterior part (i.e. farther away from the dorsal lip). It looked as if some agency emanated from the dorsal lip of the blastopore like a "flow of determination", and either streamed or was carried forwards[1] (see also p. 173).

This suspicion was confirmed when it was found that if the animal hemisphere is cut off from an early gastrula of the newt, rotated through any angle about the egg-axis, and then stuck on to the vegetative hemisphere again, the neural folds arise in line with the dorsal lip of the blastopore, which, of course, is situated in the vegetative hemisphere. The neural folds therefore arise from tissue which would normally not have formed them, and neural folds are not formed from the presumptive neural fold material which has been rotated away from the meridian of the dorsal lip of the blastopore.[2] Something of the nature of what Herbst (1901) called a "formative stimulus" appears thus to be associated with the dorsal lip of the blastopore.

As to the time when the dorsal lip region exerts its organising action, there are two possibilities. The first is to imagine a transmission of stimuli through the tissues from the region of the organiser before gastrulation; the second possibility is to attribute its action to the transmission of stimuli from underneath the surface

---

[1] Spemann, 1916.   [2] Spemann, 1906 B, 1918.

layer after gastrulation, at which time the organiser has been in-vaginated, and forms the primitive gut-roof, i.e. notochord and axial mesoderm (future myotomes). In both cases, the hinder part of the presumptive neural fold region will be affected before the front part.

It appears that the organiser acts in both these ways. That it can exert its inducing action from below, after gastrulation, is demon-strated by the fact that when a graft is made from the dorsal lip of the blastopore of one gastrula into the flank of another blastula or gastrula, it brings about the formation of the essential structures (so-called axial structures) of an embryo. This embryo is called the secondary embryo in order to distinguish it from the primary embryo formed from the tissues of the host in the ordinary way.[1]

The secondary embryo arises from tissue which had very differ-ent prospective fates. The grafted organiser invaginates beneath the surface of the tissues of the host and itself gives rise to part or all of the notochord and axial mesoderm of the secondary embryo. The other structures of the secondary embryo are usually formed of host tissue only, but may contain an admixture of graft tissue also. These can be easily distinguished by performing the experiment with material derived from two species of *Triton*, *T. cristatus* and *T. taeniatus*, which differ in the pigmentation of their tissues.

There is therefore no doubt that the organiser can bring about the determination of tissues by the transmission of stimuli from underneath after gastrulation. This is further proved by the fact that pieces of the primitive gut-roof (notochord and mesoderm, which of course are derived from invaginated organiser material) are capable of inducing the formation of axial structures.[2] As we shall see later, the main activity of the organiser in normal develop-ment is to induce the formation of the neural plate and tube. This

[1] In all cases, portions of central nervous system, notochord, and axial meso-derm (somites) are formed; in addition, brain and spinal cord, eyes, ears, kidneys, peripheral mesoderm (lateral plate), gut-roof and heart may be produced. Such embryos have not been kept beyond the tail-bud stage. Whether certain organs of the secondary embryo are formed or not depends on several factors: (1) the level of the host's main axis at which the graft is made; (2) the region of the organiser which is used as a graft; (3) the distance of the primary from the secondary embryo, resulting in a greater or lesser degree of mechanical inter-ference. (Spemann and Mangold, H., 1924.)

[2] Marx, 1925; Bautzmann, 1926.

it does by contact. All of the gastrular ectoderm underlain by noto-chord and axial mesoderm will become neural plate (see p. 155).

But these facts, however, do not preclude the possibility of the organiser exerting some effect in earlier stages also. And, as a

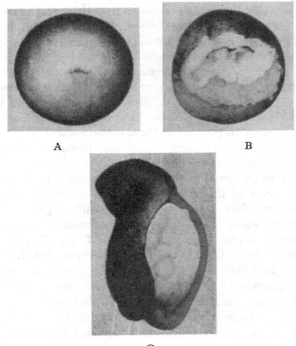

A   B

C

Fig. 62

Labile determination of neural folds in Urodeles. A, Early gastrula of *Pleurodeles*, from which, B, the entire dorsal lip region is extirpated. C, Resulting embryo showing *spina bifida* and neural folds prevented from reaching mid-dorsal line. (From Goerttler, *Zeitschr. f. Anat. u. Entwick.* LXXX, 1926.)

matter of fact, other experiments have shown that during the period before the irrevocable determination of the presumptive neural fold material, it is nevertheless not wholly indifferent, and possesses a labile determination[1] to develop into neural folds. This can be tested *in situ* in an embryo by removing small portions of the

[1] "Bahnung", Vogt, 1928A; "competence", Waddington, 1932.

organiser before gastrulation,[1] by preventing the organiser from invaginating, which can be effected either by removing it entirely
by killing part of it and so preventing invagination on one side;[2]
or by reducing its activities by means of exposure of the organiser
region to cold or deprivation of oxygen.[3] In spite of the absence of
an organiser or of any invagination, distinctive but somewhat

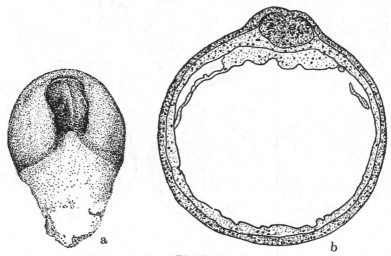

Fig. 63

a, Embryo of *Pleurodeles* in which gastrulation has been prevented by reducing
oxygen-access to the region of the dorsal lip; neural folds are nevertheless
formed. *b*, Transverse section through the same embryo, showing neural tube,
but absence of notochord; the lining of mesoderm and endoderm has been derived from the floor of the blastocoel, which, here, is the large central cavity.
(From Vogt, *Verh. deutsch. Zool. Ges.* xxxii, 1928.)

imperfect neural folds and tubes are developed. It is of interest
to note that in the absence of an underlying organism, the brain
achieves a more perfect differentiation than the spinal cord.[1]

In experiments of a different nature, in which developing Urodele eggs are subjected to a lateral temperature-gradient (see p. 342),
it is found that on the warmed side structures appear in the ectoderm resembling neural material in cell structure, but may differ
considerably from neural folds in form.[4] These structures arise in

<hr>

[1] Lehmann, 1926, 1928A.   [2] Goerttler, 1925, 1926.
[3] Vogt, 1928A.   [4] Gilchrist, 1929.

positions where they are not underlain by mesoderm. If, on the other hand, they are situated in regions where mesoderm does underlie them, they become typical neural folds (see fig. 64).

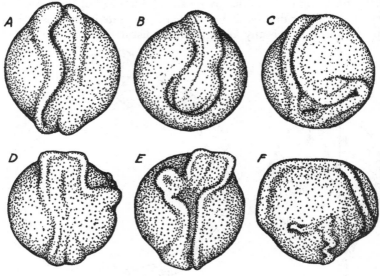

Fig. 64

Effects of a lateral temperature differential on development in Urodeles. A–F, *Triturus torosus*, treatment by temperature-gradient in blastula stage, with 5° C. temperature difference between the two sides of each egg. A, B, Warmed on left; A, dorsal view; B, anterior view; the warmed neural fold is much larger and more differentiated. C–E, Dorsal views. F, Lateral view, showing secondary neural structures on the previously warmed side, either connected with the main neural folds, or F, isolated from them. In C, D, E, the secondary formations appear to be underlain by mesoderm, and have differentiated into structures of neural fold type. In F they are not underlain, and do not show typical morphogenesis. (From Gilchrist, *Quart. Rev. Biol.* IV, 1929.)

Another method of testing this labile determination is by interplantation, i.e. the grafting of portions of blastulae (i.e. portions of tissue taken from an embryo before the invagination of the organiser) into the eye-sockets or coelomic cavities of other larvae.[1] The differentiation of various structures can be obtained in this way (fig. 148, p. 316). The fact that interests us here is that neural tube may be differentiated in these circumstances from tissue which

---

[1] Dürken, 1926.

has never been acted upon by an invaginated organiser. Or we may adopt the method known as explantation, in which the pieces of blastulae, after being enclosed in epidermal jackets, are grown *in vitro* in suitable media. Differentiation of neural tube and of noto-chord can be obtained in this way also, from tissue which has never been acted upon by an invaginated organiser[1] (see fig. 18, p. 49).

There is therefore some determinative agency at work in addition to the invaginated organiser. The labile determinations thus induced are presumably due to the transmission of stimuli from the organiser before gastrulation, in relation to the main axes of the egg, in a manner which will be considered below in connexion with gradient-fields[2] (see p. 310).

In any case, it is clear that the labile determination of the blastula stands in some relation to the bilateral symmetry imposed upon the egg at the moment of fertilisation.

The action of the organiser, then, must be considered as taking place in two phases. First, working as part of the gradient-field, the organiser may be figuratively said to sketch out the presumptive regions in pencil, and then, after invagination, the organiser goes over the same lines with indelible ink. At the same time, the organiser is capable of roughing out the sketch straightway in ink, without any previous pencil work, as in those experiments in which the organiser is grafted into the flank of another embryo. Neural folds can arise from the pencilling alone, and from the inking alone, and this duplicity of methods whereby neural folds can be formed is another example of the principle of "double assurance".

But there is another point to notice here. When an organiser is grafted into the flank of another embryo, the host-tissues are

[1] Bautzmann, 1929 B, C; Holtfreter, 1929 A, B.
[2] These examples have been mentioned in order to show that determination and differentiation can take place in the absence of an invaginated organiser. But several of these experiments introduce a new complication, since the tissue which is differentiated in interplantation and in explantation frequently is of a nature quite different from the presumptive fate of the region from which the piece was taken. Presumptive neural tube material, for instance, has been found to differ-entiate into notochord, muscle, mesenchyme, and glandular epithelium (Kusche, 1929; Holtfreter, 1931 A; Erdmann, 1931); presumptive epidermis can give rise to neural plate, especially, for some unknown reason, when interplanted into the coelomic cavity. Pieces of tissue from any part of the blastula have been seen to differentiate into notochord and muscle (Bautzmann, 1929 B) (see p. 317).

induced to differentiate in particular ways under its influence, and the labile determinations of these host tissues, whatever they may have been, are obliterated[1] and overridden. A cell-region which possesses a labile determination to become epidermis may be made to become neural folds. The organiser can, as previously mentioned (p. 46), even override the presumptive distinction between the germ-layers. For instance, a piece of presumptive ectoderm (epidermis) implanted just below the dorsal lip will be carried into the interior of the embryo, and there may give rise to a portion of any of the following organs: vertebral centrum, myotome, lateral plate, pronephros (mesodermal), notochord, or gut-wall (endodermal).[2] Presumptive neural folds can also form myotomes and pronephros. Similarly, pieces of presumptive mesoderm grafted into the region of presumptive ectoderm will (provided of course that they are taken at the stage prior to chemo-differentiation) form epidermis. The determination of epidermis, however, appears to be less rigorous, and already differentiated epidermis can be made to form conjunctiva (p. 178).

§ 2

It must be remembered that in the production of an end-result, such as a differentiated structure, two sets of factors are involved: first, the causal agent, in this case the organiser; second the material acted upon, the tissues. Examples of this resultant effect will be given in the following paragraphs.

The action of the amphibian organiser is not species-specific, i.e. it can induce the formation of axial structures when grafted into

[1] Another example of the overriding of a previous labile determination is provided by the Gephyrean worm *Bonellia*. This form shows extreme sexual dimorphism, the female being about the size of a plum with a proboscis a yard long, while the male is only a few millimetres in length, and lives parasitically in the uterus of the female. The larvae which hatch from the eggs all pass through an indifferent stage. If such larvae do not come into contact with an adult female, they themselves undergo development into females, by means of processes for which the larva must presumably possess some sort of determination. But this determination can be overridden if the larva comes into contact with an adult female and settles on her proboscis. The proboscis secretes a substance which induces in the larva the development of the male characters, involving reduction of the anterior end of the body, and differentiation of the male reproductive organs (Baltzer, 1931).

[2] Mangold, 1924.

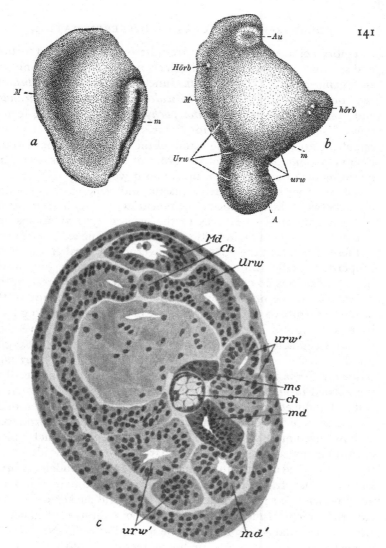

Fig. 65

Anuran organisers in Urodele hosts. A piece from the dorsal lip of the blastopore
of a gastrula of *Bombinator pachypus* grafted into a young gastrula of *Triton
taeniatus* induces the formation of a secondary embryo. *a, b,* Two stages of
development of an embryo thus obtained. *c,* Transverse section through *b.* Capital
letters refer to structures of primary embryo, small letters to secondary embryo.
Au, optic-vesicle; *Ch,* notochord; *Hörb,* ear-vesicle; *M,* neural folds; *Md,* neural
tube; *Urw,* mesodermal somites; of primary embryo. *ch,* grafted notochord;
hörb, ear-vesicle; *m,* neural folds; *md,* neural tube formed from graft tissue;
md′, neural tube induced from host-tissue; *ms,* undifferentiated mesoderm of graft
tissue; *urw′,* mesodermal somites induced from host-tissue. (From Geinitz, *Arch.
Entwmech.* CVI, 1925.)

an embryo of a species different from its own. We can go further, and say that its action seems singularly non-specific. Not only can an organiser from *Triton cristatus* function in *Triton taeniatus*, but also organisers from *Pleurodeles waltli*, *Amblystoma mexicanum*, and even the Anuran *Bombinator pachypus*, can induce the formation of secondary embryos in *Triton taeniatus*.[1] It is therefore established that the inducing action of the organiser is not impeded by a taxonomic difference of the order of value of a sub-class between its own tissue and that on which it works (fig. 65).

These experiments of heteroplastic and xenoplastic organiser grafts between different species demonstrate the fact that the action of the organiser is specific as to the general type of organs and structures produced by induction, but non-specific as to the details of these structures; these latter are governed by local and intrinsic properties and determinations of the tissues themselves, over which the organiser has no control. For instance, a piece of presumptive neural fold tissue of *Triton taeniatus* grafted on to the side of the head of an embryo of *Triton cristatus* will differentiate into gills in its new position. But, gills though they are, they preserve their *taeniatus* character in being larger than the normal *cristatus* gills on the other side of the embryo.[2] Conversely (fig. 15), *cristatus* tissue on *Triton taeniatus* gives rise to gills which are smaller than the normal *taeniatus* gills.[3] The retention of specific characters in spite of induced determination to develop into structures other than those which a piece of tissue would normally have produced, is shown even more strikingly in those experiments in which a piece of Anuran presumptive epidermis (from the ventral side of the trunk) is grafted over the future mouth-region of a Urodele embryo. In its new and strange position, the Anuran tissue differentiates into mouth-parts, and it also gives rise to a ventral sucker of Anuran pattern[4] which is functional and secretes an adhesive substance. It also appears that horny teeth can be formed as well. No Urodele normally possesses a sucker or horny teeth (fig. 66).

As a further illustration, we may take the results of experiments in which a Urodele organiser (from *Triton alpestris*) is grafted into an Anuran embryo (*Bufo vulgaris*). The induced secondary embryo

---

[1] Geinitz, 1925 B; Schotté, 1930.   [2] Spemann, 1921.   [3] Rotmann, 1931.
[4] Spemann, 1932, 1933; Spemann and Schotté, 1932.

possesses a ventral sucker, although the organiser which induced it comes from a species which does not possess one.[1] As the matter has been figuratively put: the organiser disposes of the fates of the

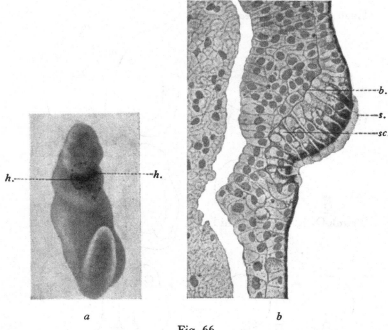

a   *b*

Fig. 66

The preservation of specific characters by a tissue, in spite of its having been induced to undergo differentiation into structures other than those representing its presumptive fate. *a*, A piece of ventral epidermis from a gastrula of the frog *Rana esculenta* is grafted into the mouth-region of an embryo of the newt *Triton taeniatus*, where it differentiates into mouth-parts in accordance with its position, but, in addition, gives rise to ventral suckers (*h.*, *h.*). *b*, Section through such an embryo, showing: *b*. basal membrane of grafted epidermis; *sc*. typical secreting cells of ventral sucker; *s*. functional secretion. A sucker is never formed by a newt. (From Spemann and Schotté, *Naturwiss.* xx, 1932.)

tissues in a general way, but as regards the details of their differentiation, the tissues already possess their instructions.[2]

The age at which an organiser first acquires its power of induction is not known, but constriction experiments on the egg of

[1] Spemann, 1932, 1933; Spemann and Schotté, 1932.   [2] Spemann, 1921.

144

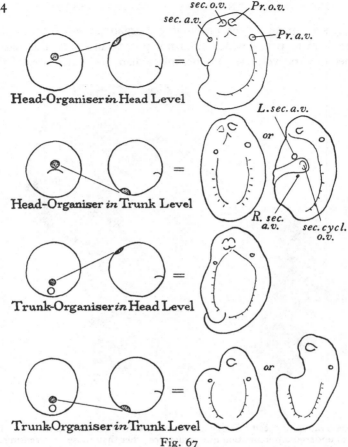

Head-Organiser *in* Head Level

Head-Organiser *in* Trunk Level

Trunk-Organiser *in* Head Level

Trunk-Organiser *in* Trunk Level

Fig. 67

Diagram showing the results of experiments testing the inductive capacities of head-organiser (invaginated early: the anterior region of the primitive gut-roof) and of trunk-organiser (invaginated late: the posterior region of the gut-roof), and the reacting capacities of the host-tissues at head level and trunk level (see also fig. 68). Head-organiser at head level forms only the head of a secondary embryo with eyes and ear-vesicles; head-organiser at trunk level may form a complete secondary embryo, and the cephalic structures may arise at a level considerably behind those of the primary embryo; trunk-organiser at head level may form a complete secondary embryo with cephalic structures at levels more or less corresponding to those of the primary embryo; trunk-organiser at trunk level produces the trunk of a secondary embryo; ear-vesicles are formed if the secondary embryo reaches to the level of those of the primary embryo. Head-organiser can thus form a head in both head and trunk levels, but trunk-organiser can only form a head in head level; the reaction to trunk-organiser of the host-tissues at head level is to form a head, and at trunk level to form a trunk. *Pr.o.v.* eye; *Pr.a.v.* ear, or primary embryo; *sec.o.v.* eye; *sec.a.v.* ear (left, *L.* or right, *R.*); *sec.cycl.o.v.* cyclopic eye, of secondary embryo. (Original, based on Spemann.)

the newt show that its site is already determined and localised ten minutes after fertilisation.[1] Rather later, portions of the blastula in the region of the grey crescent have been found to possess the inductive property.[2] As to the time at which this property is lost, it has been shown that the notochord, which of course is formed from the invaginated organiser, retains for a considerable period the power of inducing the formation of neural folds.[3]

It has also been shown that in the neurula stage, myotome material, which of course was originally derived from the organiser region, still retains the capacity of inducing neural tube formation from presumptive epidermis when grafted into an early gastrula. However, slightly more lateral mesoderm material, which had differentiated into pronephros, in similar experiments only induced other pronephric tubules.[4]

This is known as "homoiogenetic induction", to contrast it with the heterogenetic power of the organiser, which induces the formation of structures different from itself. It is found that the neural plate, once underlain by the organiser, possesses and retains for a very long time—certainly up to the free-swimming larva—this power of inducing the formation of structures of its own type. This is proved by grafting portions of neural tube into blastulae, where secondary neural folds are induced.[5] It is of interest that the hindmost portion of the neural fold region of the neurula induces the formation of mesoderm, which agrees with the fact that this region gives rise to the muscles of the tail in normal development (Chap. II, p. 28).[6] Accordingly, this induction also is homoiogenetic. Lens rudiments implanted into blastulae have no power of induction, either hetero- or homoio-genetic.[7]

Spatially, the region of the blastula and early gastrula which has organising capacities appears to coincide with the region which will become invaginated at gastrulation, i.e. the presumptive notochord and axial mesoderm regions.[8]

This is a large area, and it might be expected that there would be

[1] Fankhauser, 1930.   [2] Bautzmann, 1926.
[3] Bautzmann, 1928, 1929 A.   [4] Holtfreter, 1933 B.
[5] Mangold and Spemann, 1927; Mangold, 1929 B.
[6] Bytinski-Salz, 1931.   [7] Krüger, 1930.
[8] Bautzmann, 1926.

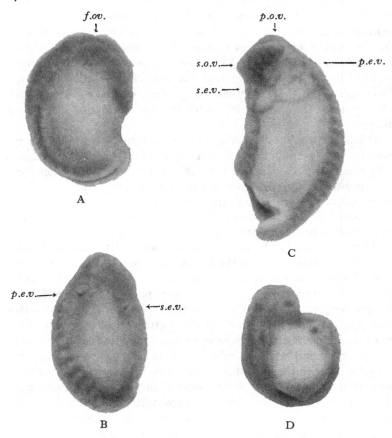

Fig. 68

The regional inductive properties of the organiser and the regional reactive properties of different levels of host-tissue in Urodele embryos. Photographs of the embryos on which the diagram fig. 67 is based. A, Head-organiser grafted at head level; the secondary embryo (on the right) consisting only of a head with ear-vesicles, and eyes fused with those of the primary embryo (*f.ov.*). B, Head-organiser grafted at trunk level; the secondary embryo (on the right) is nearly complete but its anterior end is imperfect, it lacks eyes, and its ear-vesicles (*s.e.v.*) are at a lower level than those of the primary embryo (*p.e.v.*). C, Trunk-organiser at head level; the secondary embryo (on the left) is complete, its cephalic structures (*s.o.v.* eyes, *s.e.v.* ear-vesicles) on a level with those of the primary embryo (*p.o.v.*, *p.e.v.*). D, Trunk-organiser at trunk level; the secondary embryo (on the right) consists only of a trunk, ending anteriorly with ear-vesicles on a level with those of the primary embryo. (From Spemann, *Arch. Entwmech.* CXXIII, 1931.)

regional differences in different portions of it. It will be realised that that portion of the organiser area which is the first to become invaginated at the rim of the dorsal lip of the blastopore will reach furthest forward and come to underlie the head, while that portion which becomes invaginated later will come to underlie the trunk. It has in point of fact been found that these two portions of the organiser show a regional difference as regards their power of induction. For instance, "head-organiser" (invaginated early), grafted at head level in the host, will form the cephalic axial structures (brain, eyes, ears) as might be expected, and the secondary embryo so formed may lack the trunk region. On the other hand, "trunk-organiser" (invaginated late), grafted at trunk level in the host, will form the axial structures characteristic of the trunk, and such secondary embryos will lack brain and eyes, and in many cases ears as well (figs. 67, 68; see also Appendix).[1]

Similarly, with regard to homoiogenetic induction by the neural tube, it is found that anterior portions induce the formation of anterior cephalic structures (e.g. eye), middle portions induce posterior cephalic structures (e.g. ear), while posterior portions induce structures characteristic of the trunk and tail.[2]

These facts make it clear that there exists a regional differentiation within the organiser area itself. The result of induction, however, is also dependent on the level along the main axis of the host of the tissues upon which the organiser exerts its action. This is shown by the following experiments. Head-organiser grafted at trunk level in the host will induce the somewhat imperfect formation of cephalic axial structures, including brain, eyes, and ears. On the other hand, trunk-organiser grafted at head level in the host can also produce these cephalic structures, but eyes will only be formed if the anterior end of the neural tube of the secondary embryo reaches forward as far as the level of the eyes of the primary embryo.[1]

Thus, as noted above (p. 140), the host-tissues are not without influence on the formation of the secondary embryo. As a general rule, it is found that the secondary embryo is arranged with its long axis roughly parallel with that of the primary embryo, or, in other

[1] Spemann, 1927, 1931; Bautzmann, 1929 A.
[2] Mangold, 1929 B; 1932.

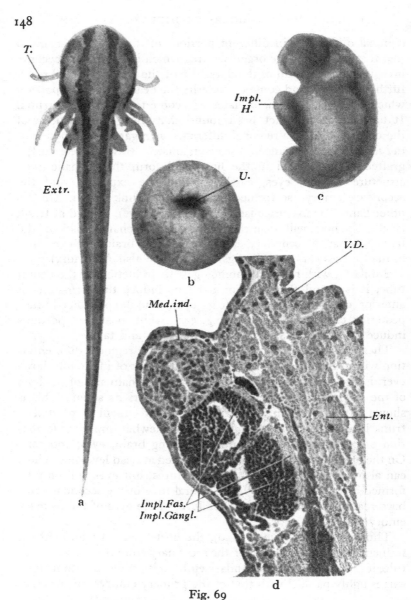

Fig. 69

Homoiogenetic induction of neural folds by brain tissue. *a*, A free-swimming larva of *Triton taeniatus*, with limbs (*Extr.*) and balancer (*T.*), from which a portion of brain tissue was grafted into *b*, a young gastrula of the same species; *U.* blastopore. *c*, The same embryo, 68 hours after the operation, from the left side, showing the graft (*Impl.H.*). *d*, Section through the graft and the induced neural tube (*Med.ind.*); *Impl.Fas.* nerve fibres, and *Impl.Gangl.* grey matter, of the highly differentiated graft; *Ent.* endoderm, *V.D.* foregut of host embryo. (From Mangold, *Ergebnisse der Biol.* III, 1928.)

words, meridional with reference to the host, and with its head facing in the same direction as that of the host.[1]

The axis of the secondary embryo is determined by the direction taken by the mass of material which is invaginated beneath the surface in relation to the grafted organiser fragment. It appears that the direction in which this invagination occurs is determined in part by the orientation of the grafted organiser,[2] but in part also by the activities of the host-tissues: the invaginated mass tends to bend round towards the animal pole of the host. This has been discovered by grafting portions of organiser with their original polarity rotated 90° or 180° relative to that of the host, so as to lie either transversely or reversed. In almost all cases, some influence of the host is to be observed, but the precise degree varies a great deal in individual instances. In some cases, the axis of an embryo derived from a reversed organiser may be completely deflected so as to coincide with the main host axis, but in other cases it may be almost precisely opposed to that of the host.[3] The orientating influence of the host is greatest in the region surrounding the blastopore, and least at the opposite pole. On the other hand, what we may call the invaginating power of organisers varies, and is greater in organisers from old than in those from young gastrulae. Consequently, reversed orientation of the secondary embryo is most often to be observed when

Fig. 70

The orientation of the secondary embryo is dependent partly on the polarity of the host-tissues, and partly on that of the grafted organiser and the direction in which it is implanted. The secondary embryo shown in this figure was induced by an organiser grafted with reversed orientation into the host; its hind end is towards the head of the primary embryo; its anterior end is curled round to the right and lies transversely to the host. (From Spemann, *Arch. Entwmech.* cxxiii, 1931.)

---

[1] Geinitz, 1925 A.

[2] The determination of the organiser to become invaginated is an instance of what has been called "dynamic determination" (Vogt, 1923), leading to form-changes which in turn result in the processes of gastrulation and neurulation (see p. 26). The possible relation between dynamic and chemo-differentiation is discussed below (pp. 163, 250, 301).   [3] Spemann, 1931.

an old organiser is grafted, reversed, into the antero-ventral region of the host.[1]

Those cases in which the secondary embryo fails to adapt itself to the polarity of the primary embryo are of interest because certain of the paired structures of the secondary embryo, such as ear-vesicles, lie at different levels in the host. In these cases it is

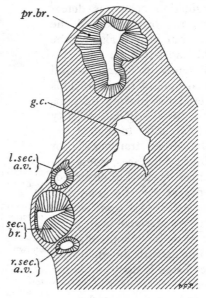

pr.br.

g.c.

l.sec. }
a.v.

sec. }
br.

r.sec. }
a.v.

Fig. 71

Section through an organiser-graft in *Triton*, in which the anterior end of the secondary embryo lay at right angles to the long axis of the primary embryo. The left ear-vesicle of the secondary embryo, *l.sec.a.v.*, which lies nearer the anterior end of the host embryo, is larger than the right, *r.sec.a.v. pr.br.* brain of primary embryo; *g.c.* gut cavity; *sec.br.* brain of secondary embryo. (After Spemann, *Arch. Entwmech.* CXXIII, 1931, simplified.)

found that the vesicle nearer to the anterior end (animal pole) of the host is larger than the other, and this shows that there is in the tissues of the host a stratification of capacities to react to the organiser (figs. 70, 71; and see pp. 147, 319).

In addition to the regional difference between head-organiser and trunk-organiser, it seems, however, that (contrary to previous

[1] Lehmann, 1932.

indications[1]) the organiser region is not divisible into right and left portions possessing predetermined laterality; for a lateral piece of primitive gut-roof, taken well to the left of the middle

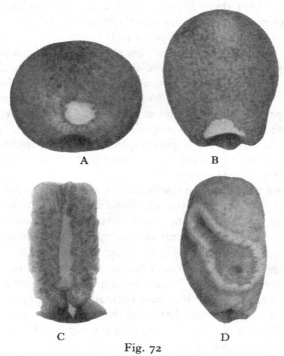

Fig. 72

The "infective" properties of the organiser region in Urodeles. A, A piece of presumptive ectoderm from the roof of a blastula of *Triton cristatus* is grafted into the dorsal lip of the blastopore of a gastrula of *T. alpestris*, where it is plainly visible on account of its light colour. B, The graft participates in the normal gastrulation process of the host and becomes invaginated. C, When gastrulation is completed, the embryo is cut open and the graft is found forming part of the gut-roof in the mid-dorsal line, in the position of the notochord. D, The graft is cut out and grafted a second time into a gastrula of *T. taeniatus*, where it induces the formation of neural folds. (From Spemann and Geinitz, *Arch. Entwmech.* CIX, 1927.)

line, can induce the formation of a bilaterally symmetrical secondary embryo when grafted[2]. One organiser region can thus induce several embryos (see also p. 310).

[1] Goerttler, 1927.   [2] Spemann, 1931.

The facts also permit of the interpretation that the quantitative potency of inductive capacity falls off in a graded way from the dorsal lip region, although this gradient appears to be steep.[1] In birds (p. 161) there appears to be a definite gradation of inductive potency along the organiser (primitive streak), this being highest anteriorly and lowest posteriorly.

The properties of the organiser are not intimately associated with any particular type of cell. If ordinary presumptive epidermis is grafted into the region of the organiser before gastrulation has started, it becomes "infected" with the power to organise. This has been proved by heteroplastic grafting of a piece of epidermis from *Triton cristatus* into the organiser region of *T. taeniatus*. Such a piece of tissue, originally presumptive epidermis, treated in this way, is found when grafted into another embryo to possess all the qualities of a normal organiser.[2]

Thus, the properties of the organiser seem to be attached to a certain region of the embryo, regardless of the identity of the cells which occupy it. This region, which owes its localisation to the egg-axis and the plane of bilateral symmetry, must be determined in the outermost or cortical layer of cytoplasm of the egg. For even when an egg is forcibly inverted and its contents stream about inside, the dorsal lip of the blastopore appears in the region of the grey crescent, i.e. where it would normally have appeared on the surface of the egg.[3] Since, however, the cells of this region divide more rapidly (see p. 39), it seems that some physiological activity is set up in this region of the cortex which later affects the dividing cells of this region, to a considerable depth below the surface.

In passing, it is of interest to note that in certain experiments, e.g. those in which myotome and pronephros material from a neurula were grafted into an early gastrula (p. 145), and those referred to on p. 191, show that the morphogenesis of artificially induced structures may differ considerably from that shown by the same structures in normal development. Thus the epidermis may be induced to form pronephric tubules without passing through a nephrotome-like stage (see p. 32): portions of brain-like structures may be induced to form from the epidermis by thickening and subsequent delamination without the formation of neural folds.[4] A

---

[1] Bautzmann, 1926, 1933.   [2] Spemann and Geinitz, 1927.
[3] Weigmann, 1927.   [4] Holtfreter, 1933 B.

corresponding set of facts is known from the study of normal events in Ascidians, where the same organ may be formed by quite different morphogenetic processes and sometimes even from different germ-layers, in development from the egg and development by budding. Similar cases are also known in regeneration.

§ 3

Concerning the physico-chemical aspect of the method of action of the organiser, little can be said, although the results so far obtained are of the greatest interest. In the first place, it is clear that the inducing tissue does not require to be alive in order to exert its effects. After an organiser has been subjected to a narcotic (trichlorbutyl alcohol) for a certain length of time, the tissues of the organiser may be so heavily damaged that they disintegrate after being grafted, but a secondary embryo is nevertheless induced.[1]

Even more drastic treatment, such as desiccation, or killing with high temperatures, or immersion for $3\frac{1}{2}$ minutes in 96 per cent. alcohol, does not destroy the inductive capacity of the amphibian organiser region.[2] (See also p. 497.)

It would seem therefore that the inductive effects of the organiser are due to some chemical substance which is elaborated by it, and support for this view is provided by the fact that pieces of agar jelly, or of gelatine, after being in contact with inductive tissue (neural folds) are themselves capable of inducing.[3]

The question next arises as to whether the initiation of the inducing effect, and therefore the production of the necessary chemical substance, is in any way dependent on the intimate structure of

---

[1] Marx, 1930.

[2] Here a new complication is introduced by the fact that certain tissues which possess no inductive capacities when alive, such as epidermis and endoderm, are able to act a organisers when killed. While the detailed significance of this fact is still obscure, it is of interest in suggesting that the normal living organiser differs only in some physical degree, and not in kind, from the tissues of the remainder of the embryo. (Spemann, 1929; Bautzmann, Holtfreter, Spemann and Mangold, 1932; Holtfreter, 1933 c.)

It may here be noted that living regenerating amphibian tissue (adult newts 12-day limb regeneration-buds) is capable of inducing neural folds in blastulae of the same species when introduced into the blastocoel (Umanski, 1932 B). Similar results have been obtained with insertions of mammalian and avian malignant tumour tissues (Woerdeman, 1933 C). No control experiments have yet been made with non-malignant tissues of the same species.

[3] Bautzmann, Holtfreter, Spemann and Mangold, 1932.

the organiser. If three extra organisers are grafted into the close
vicinity of an organiser in an intact embryo so that their polarities
all converge to a point in the centre of the host-organiser, there is
no inductive effect of any kind.[1] This annihilation of the inductive
effect is difficult to understand. It can scarcely be that an intact
structure, or an unimpeded gastrulation-process, are essential pre-
liminaries to the production of the chemical substance responsible
for the organising effect; for even if a piece of the organiser is
made to wait for some time before it is grafted, when it rolls up into
a ball, and the arrangement of its cells is markedly altered, its
organising power is not affected or reduced.[2]

The possibility that the organiser effect in birds is in some way
dependent on the normal tissue-movements which take place in
gastrulation, i.e. on so-called "dynamic determination" (Vogt), will
be discussed below (pp. 163, 250).

Recently, the decisive discovery has been made that cell-free
fractions of a liquid extract of whole neurulae can exert an organising
action, as evidenced by neural tube induction. The liquid is coagulated by heat and portions of the resultant solid material implanted
into the blastocoele. The active substance is certainly ether-
soluble, and probably lipoidal.[3]

Meanwhile, some interesting results have emerged from investigations into the glycogen-content of the cells of the amphibian
embryo. This is high in the cells of the animal hemisphere; low in
those of the vegetative hemisphere, and intermediate around the
equator. But as soon as the cells of the organiser have become invaginated, they immediately lose what glycogen they contained. It
is not improbable that this sudden disappearance of glycogen con-
notes an expenditure of energy connected with the physiological
activities characteristic of the organiser.[4]

§ 4

The fact that the organiser, in the form of the primitive gut-roof,
is capable of organising the epidermis overlying it so as to induce
it to give rise to neural folds, explains a number of phenomena

[1] Goerttler, 1931.   [2] Holtfreter, 1933 B.
[3] Waddington, Needham and Needham, 1933.
[4] Woerdeman, 1933 A; Raven, 1933 B.

which would otherwise be obscure. As regards the ordinary data of comparative embryology, this property of the organiser makes it possible to understand why there is a correlation between the width

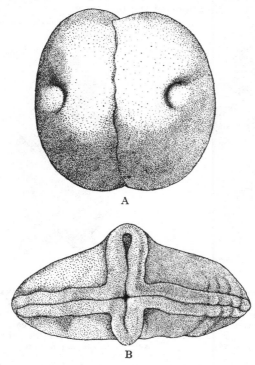

A

B

Fig. 73

A, Two dorsal gastrula-halves of *Triton* grafted together so that the directions of invagination of their blastopores are directly opposed. B, The resulting embryo, showing crossed doubling, or *duplicitas cruciata*; each half-gastrula has produced a posterior trunk region with spinal cord, but two heads and brains are formed, at right angles to the axis of the trunks, each formed partly from both half-gastrulae. (Redrawn from Morgan, *Experimental Embryology*, New York, 1927, after Spemann.)

of the neural plate and the width of the primitive gut-roof in different groups of Vertebrates:[1] the former is dependent on the latter.

Turning to experimental results, the production by operative

[1] Marx, 1925.

treatment of monstrosities which conform to the teratological types known as anterior, posterior, and crossed doubling (*duplicitas anterior*, *duplicitas posterior*, and *duplicitas cruciata*), is explicable only in terms of these functions of the organiser.

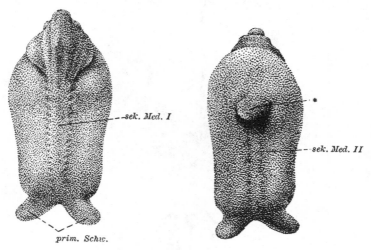

Fig. 74

Duplicitas cruciata, obtained by grafting together two gastrula-halves (see fig. 73); nearly the whole extent of each embryonic rudiment (*sek.Med.I, II*) is composite and derived partly from each of the half-gastrulae; only the tips of the tails (*prim. Schw.*) are uncrossed, i.e. each formed from one of the half-gastrulae. One of the trunks (*sek. Med. II*) is less well developed than the other, and ends anteriorly in a knob(*). (From Wessel, *Arch. Entwmech.* cvii, 1926.)

If the egg of a newt is partially constricted in the plane of bilateral symmetry during the period of gastrulation, the resulting embryo will show anterior doubling, i.e. it will have two more or less perfectly formed anterior ends joining on to a single posterior end.[1] The explanation is that when the primitive gut-roof becomes invaginated, it finds an obstacle in the constriction and has to fork, one portion going forward on each side of the constriction. The organiser or primitive gut-roof is therefore Y-shaped, and its anterior prongs underlie tissue which would normally not have given rise to neural folds. But the action of the organiser induces the

[1] Spemann, 1903; Hey, 1911.

formation of neural folds in these strange positions, with the result that two perfect heads and anterior trunk regions are formed. Further, it may be noted that it is impossible by the method of partially constricting gastrulae to obtain *duplicitas posterior*, or

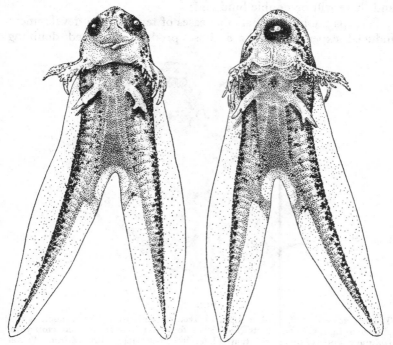

Fig. 75

Duplicitas cruciata, obtained by grafting together two gastrula-halves (see fig. 73); the heads and anterior regions of the trunk have a plane of symmetry which is at right angles to that of the posterior regions of the trunk; the former are seen in ventral, the latter in side view. One of the heads has a cyclopic eye. (From Wessel, *Arch. Entwmech.* CVII, 1926.)

doubling of the hind end. This is obviously because the constriction forces the anterior but not the posterior part of the primitive gut-roof to fork (fig. 169).

On the other hand, both anterior and posterior doubling can be obtained by grafting together halves of gastrulae in such a way that their original planes of symmetry (and therefore, directions of organiser-invagination) either diverge or converge anteriorly.[1] In the former case, the compound embryo will have a Y-shaped primitive gut-roof with the divergence anterior, and will develop double heads; in the latter case the divergence will be posterior, and there will be double hind ends.

Perhaps the most remarkable cases of teratological development induced experimentally are those producing crossed doubling

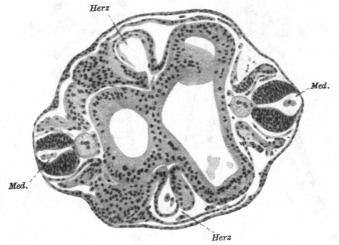

Fig. 76

Transverse section through a *duplicitas cruciata* embryo of *Triton*, such as that shown in fig. 75. The hearts (*Herz*) are formed partly from each embryonic rudiment, and are therefore situated laterally. *Med.*, neural tube. (From Wessel, *Arch. Entwmech.* CVII, 1926.)

(*duplicitas cruciata*). These result from the grafting together of two gastrula halves each containing the dorsal lip, in such a way that the directions of organiser-invagination are directly opposed to one another. Invagination takes place in each half, and the primitive gut-roofs meet one another, head on. Being unable to make any further progress forwards, they move to each side. The result is that the primitive gut-roofs together form a cross, two (opposite) branches of which are formed each from one of the two invaginations, and the other two branches are composite, half of each being

[1] Spemann, 1916, 1918; Koether, 1927.

formed from each of the two invaginations.[1] The former two branches represent the posterior portions of the primitive gut-roof: the latter two branches represent the anterior portions (figs. 73–76). Overlying the cross-shaped gut-roof, neural folds arise, and a monstrous double embryo is formed, the hinder portions of which have each been induced by a single organiser, while the anterior portions have been induced by tissue derived from two organisers. Furthermore, these anterior portions are formed from induced tissues which had very different normal presumptive fates. The relative lengths of the arms of the cross, or of the composite anterior and of the simple posterior portions of the double embryo, can be controlled by varying the distance which separates the two blastopore lips at the start of gastrulation. If they are far apart, the primitive gut-roofs will travel a long way forwards before they meet and cross, and the anterior composite portions will be short: if they are close together, the gut-roofs will meet and cross very soon, and continue their invagination as parts of the composite anterior ends. Crossed doubling can also be obtained by grafting an organiser into a normal embryo in such a way that the anterior ends of the primary and secondary embryos meet and obstruct one another.[2]

§ 5

Experiments on the blastoderm of the chick and duck have produced results of the greatest interest. They have shown that the primitive streak has organising powers similar to those of the amphibian dorsal lip of the blastopore (with which it is morphologically homologous), and they have confirmed and extended the results obtained from experiments with amphibian material.[3]

In these experiments, the method of tissue culture has been used. The embryonic rudiment of the bird at a very early stage consists of an upper layer (ecto-mesoderm), and a lower layer (endoderm). These layers can be separated from one another, and cultured separately *in vitro*. The upper layer will differentiate into neural folds, notochord, and mesodermal somites, but the lower layer will not differentiate at all. This is due to the fact that the lower layer

---

[1] Wessel, 1926.   [2] Bautzmann, 1926.
[3] R. Wetzel, 1929; Hunt, 1929; Waddington, 1930, 1932, 1933 A, B, C.

lacks the primitive streak which the upper layer possesses. The
lower layer is therefore in the same case as a ventral half of an
embryo of an amphibian. The organising action of the primitive
streak on the *lower layer* is shown by the fact that the upper layer is
capable of inducing the lower layer to give rise to the fore-gut in
the correct position with regard to the notochord, from tissue
which would normally not have given rise to fore-gut at all. This
is shown by experiments in which an upper and a lower layer are
cultivated together in such a way that the primitive streak overlies
a region of the lower layer other than that which represents the
presumptive fore-gut.

Fig. 77

Induction by organiser in birds. Two blastoderms of the chick grafted together.
u.n.g. normal neural plate of upper blastoderm; *i.n.g.* secondary induced neural
plate in upper blastoderm, formed in relation to *l.n.g.* normal neural plate in
lower blastoderm. (From Waddington, *Phil. Trans. Roy. Soc.* B, CCXXI, 1932.)

It is clear, therefore, that the primitive streak is an organiser. It
has further been found that it possesses regional differences of
potency, both as regards self-differentiating capacities and inductive power. Anterior pieces of the primitive streak differentiate
into neural tube, notochord, and mesodermal somites; middle
pieces produce mesoderm with or without neural tube; posterior
pieces never produce neural tube. In other words, there is a
gradient in developmental potencies along the primitive streak.[1]

It should, however, be noted that when portions of primitive
streak are cultivated in isolation, they give rise to considerably
more than their presumptive fate[2] (fig. 78).

[1] See also Hunt, 1932.
[2] Waddington and Schmidt, 1933.

Anterior and middle pieces of primitive streak (corresponding to the dorsal and lateral lips of the amphibian blastopore) grafted beneath an *upper layer* induce the formation of neural folds from host-tissue, but posterior pieces seem to be unable to do this. Thus, in the primitive streak, there appears to be a graded distribution of organising power. Since the induced neural tube is usually situated immediately above the mesodermal tissue of the graft (corresponding to the primitive gut-roof of the amphibian organiser), the latter is probably responsible for the inductive effect. The notochord in the bird is apparently unable to induce.[1]

When a graft which in the normal course of development would have formed trunk mesoderm is implanted into the head region, it produces only head mesoderm there, whether or not it succeeds in inducing the formation of a secondary embryo. This shows that there must be some influence of the host-tissues on the fate of the grafted organiser.[2]

The homoiogenetic power of the neural tube has been demonstrated in birds, for a grafted portion of neural tube will induce the formation of neural tube[3] (fig. 77).

The organising action of the avian primitive streak is not species specific, for the primitive streak of the duck is functional when grafted into the blastoderm of the chick, and *vice versa*.[2]

The orientation of the avian embryo is found to be dependent on the polarities of both the primitive streak and of the lower layer. The influences of the upper and of the lower layer are tested by rotating the one relatively to the other through 90° or 180°, and culturing them together. As in the comparable experiments in Amphibia (p. 149), in which rotated or reversed organisers are grafted, the results vary considerably in different individual cases. In some, the orientation of the primitive streak, and therefore of the upper layer, determines that of the embryo. In other cases, however, the embryo is developed in relation to the polarity of the lower layer or endoderm. The polarity of the upper layer is then either deflected or obliterated. This is very remarkable, for, as already stated, the lower layer lacks the primitive streak from which all the axial structures of the embryo are formed.[4]

[1] Waddington, 1933 B. See also Umanski 1932 A.   [3] Waddington, 1933 B.
[2] Waddington and Schmidt, 1933.   [4] Waddington, 1933 C.

162

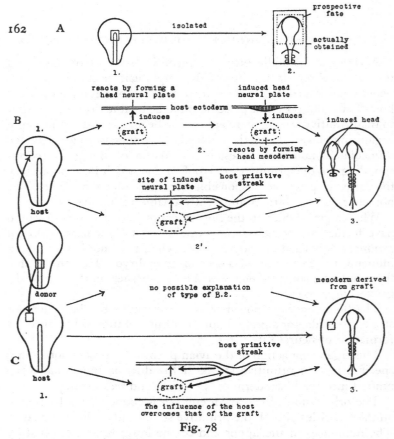

Fig. 78

Diagrams illustrating some of the properties of the organising centre in birds. A, The developmental potencies of a portion of the organiser region are greater than its prospective fate. B, Analysis of the problem presented by the fact that when a piece of the organiser region, the prospective fate of which is trunk mesoderm, is grafted into the head region of another blastoderm, it itself gives rise to head mesoderm, while at the same time inducing the formation of neural folds (B 1, B 3). The conversion of the graft into head mesoderm may be explained by assuming either: B 2, that after the graft has induced the formation of a head neural plate the latter in turn acts upon the graft and determines it to give rise to head mesoderm; or B 2¹, that the conversion of the graft into head mesoderm is due to a process of interaction between the graft and the host's own organising centre, to which latter the property must be ascribed of exerting an influence over an area of given extent, termed an "individuation-field", in which the whole complex of tissues are controlled in such a way as to lead to the formation of a complete individual. It is, further, an effect of the host's individuation-field that the neural plate which trunk mesoderm induces out of the host-tissues in the head region is head neural plate. That alternative B 2¹ is the correct interpretation follows from the cases, C 1–C 3, in which the grafted trunk mesoderm in the head region of the host becomes converted into head mesoderm without inducing the formation of a neural plate at all: here, the graft can only be under the influence of the host organising centre. (From Waddington and Schmidt, *Arch. Entwmech.* cxxviii, 1933.)

It seems that the endoderm can determine the polarity of the embryo by determining the localisation and polarity of the primitive streak itself, in the blastoderm overlying it. We are here confronted with a phenomenon which seems to be nothing less than the determination of the organiser itself. The primitive streak is dependent in some way ultimately on the endoderm, and it would seem that we have to look for the morphogenetic expression of this determination in certain streaming movements which take place in the blastoderm. The direction of these movements is backwards along the periphery on each side, and forwards along the central line, immediately along which line the primitive streak is formed.[1] In some as yet undetermined way, the endoderm seems to control these movements.

If this should turn out to be correct, we have here an example of the effects of "dynamic determination" referred to on p. 154. From the theoretical point of view, the interest and importance of these facts lies in the question whether dynamic determination can be regarded as the causal antecedent of "material" (chemical and histological) determination. The answer to this question appears to differ in different groups of vertebrates. In the amphibian embryo the early stages are characterised by well-marked movements (dynamic effects) of tissues; and attempts made to test the power of chemo-differentiation of other tissues which have been prevented from undergoing such movements[2] have yielded results which can only be regarded as negative (see Chap. VII, p. 250). For the moment, therefore, the general significance of dynamic determination in birds must remain an open question.

With regard to the physico-chemical nature of the action of the organiser in the bird, it is interesting to note that it retains its organising capacity although coagulated as a result of having been dipped in a thin glass tube into boiling water for 30 seconds.[3]

From all these results, it is abundantly clear that the dorsal lip of the amphibian blastopore, and its homologue the avian primitive streak, possess the function of an organiser, and it is probable that these structures will be found to have similar properties in other groups of Vertebrates.

[1] R. Wetzel, 1925, 1929; Gräper, 1929.
[2] Goerttler, 1927; Holtfreter, 1933 A.   [3] Waddington, 1933 A.

§6

Attention may now be turned to Invertebrates, and the question naturally arises whether regions with similar or comparable organising capacities exist among them. This is found to be the case, although the details, not unnaturally, vary considerably.

In *Hydra*, the hypostome, or region surrounding the mouth, is an organiser of simple type. When grafted into the proximal end of another polyp it induces the formation of tentacles. It further

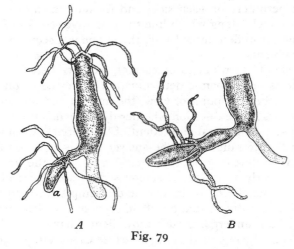

A   B

Fig. 79

Organiser grafts in *Hydra*. Induction of a bud by grafting an oral end of one individual (*a*) into the flank of another. The polarity of the bud is the reverse of that of the graft. (From Mutz, *Arch. Entwmech.* CXXI, 1930.)

causes an outgrowth of host-tissues in which the original polarity is overridden, and a new polarity established in relation to that of the graft.[1] Although grafts of organisers in *Hydra* between different species rarely succeed, an organiser from *Pelmatohydra* has been found to produce an inductive effect in *Hydra* (figs. 79, 80).

In another Coelenterate, *Corymorpha*, pieces of stem have the power of inducing the formation of new polyps when grafted into other stems.[2] This case is particularly interesting, for the facts

[1] Browne, 1909; Rand, Bovard and Minnich, 1926; Mutz, 1930.
[2] Child, 1929 B, 1932.

indicate that the organiser in *Corymorpha* is not a specific tissue or structure, but any level of the stem will act as an organiser, although pieces from distal levels are more potent. The bearing of these facts on the theory of gradient-fields and the interpretation of the mode of action of organisers will be discussed at greater length in Chap. VIII (fig. 138).

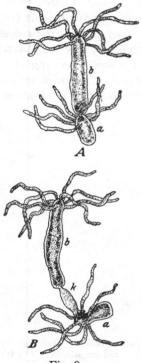

In *Planaria*, the head of one worm grafted into the posterior region of the body of another induces the formation of a pharynx and brings about the re-organisation of the host-tissues so as to make them conform to the new polarity set up by the graft. Here again, the effect is not species-specific, for a head of *Planaria dorotocephala* will act as an organiser in the tissues of *Planaria maculata*[1] (figs. 81, 82).

These last experiments merely extend previous work on regeneration in Planarians and various worms. In *Planaria*, for instance, Child had shown[2] that the reorganisation of the old tissues of a posterior fragment, of which the most obvious effect is the production of a new pharynx, only occurs if a head is regenerated. He further showed, in experiments where the size of the regenerated head was varied and controlled by the use of anaesthetics in varying concentrations, that the size of the new

Fig. 80

Organiser grafts in *Hydra*. Bud (*k*) induced from stock (*b*) by grafting an oral end of another individual (*a*) on to the aboral end of the stock. (From Mutz, *Arch. Entw-mech.* CXXI, 1930.)

pharynx and its distance from the anterior end of the piece were correlated with the size of the regenerated head (see Chap. VIII, pp. 287, 290).

As noted in Chap. VIII (p. 288) the head segments of the Polychaete worm *Sabella* act during regeneration as an organiser capable

[1] Santos, 1929.   [2] See Child, 1915 A, pp. 102, 138.

of transforming more posterior segments from the abdominal to the thoracic type.[1] Similar facts have been noted for the Oligochaete *Stylaria*.[2] These results will be further considered in connexion with gradient-fields (fig. 137). They are of great importance in showing that the processes at work in the organiser phenomena in the early stages of Vertebrate development are similar in essentials to those operating throughout life in regeneration and grafting experiments in lower forms.[3] There are, however, certain differences, in that the vertebrate organiser works mainly by contact, whereas

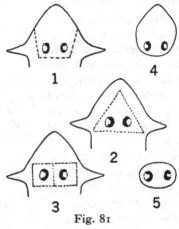

Fig. 81

Head-grafts in *Planaria*. 1, 2, 3, showing portions employed as grafts; 4, 5, isolated fragments of the type of 1 and 3 respectively, 12 days after operation. (From Santos, *Biol. Bull.* LVII, 1929.)

these invertebrate organisms can effect a reorganisation of tissues at a distance. For a further discussion of this point, see Chap. VIII, p. 310.

Organiser phenomena in normal ontogeny, though again of a less specialised type than in Amphibia, have been found in Echinoderms. A curious result (referred to in Chap. V, p. 102) of the isolation of animal halves of eggs and blastulae of the sea-urchin *Paracentrotus*, is that such halves do not develop a stomodaeum if they have been isolated from their vegetative counterpart at a stage earlier than 20 hours after fertilisation. A stomodaeum is,

[1] Berrill, 1931.   [2] Harper, 1904.   [3] See Child, 1928 C, 1929 A.

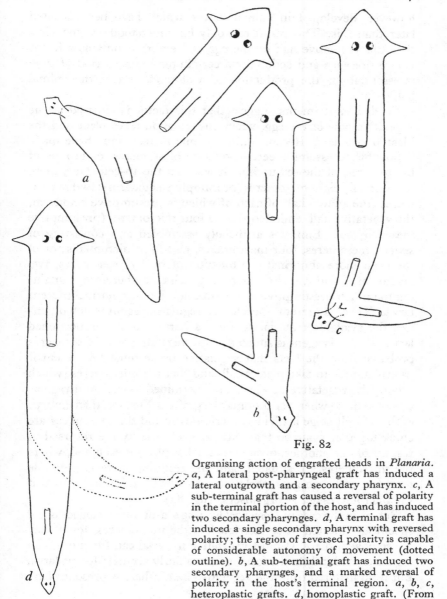

Fig. 82

Organising action of engrafted heads in *Planaria*.
a, A lateral post-pharyngeal graft has induced a
lateral outgrowth and a secondary pharynx. *c*, A
sub-terminal graft has caused a reversal of polarity
in the terminal portion of the host, and has induced
two secondary pharynges. *d*, A terminal graft has
induced a single secondary pharynx with reversed
polarity; the region of reversed polarity is capable
of considerable autonomy of movement (dotted
outline). *b*, A sub-terminal graft has induced two
secondary pharynges, and a marked reversal of
polarity in the host's terminal region. *a*, *b*, *c*,
heteroplastic grafts. *d*, homoplastic graft. (From
Santos, *Biol. Bull.* LVII, 1929.)

however, developed in animal halves which have been isolated later than this. This result can only be understood on the view that the vegetative half of the egg and embryo contains a factor whose presence and action for a certain minimum period of time is essential for the production of a stomodaeum in the animal half.[1]

Further experiments suggest that this factor is situated at the vegetative pole of the egg, where invagination takes place and the blastopore arises. Recent improvements in technique have made it possible to assemble certain definite blastomeres, or groups of blastomeres, of the sea-urchin, at will. At the 16-cell stage, there are normally eight mesomeres (presumptive ectoderm): four macromeres (the animal half of each of which is presumptive ectoderm, the vegetative half, endoderm): and four micromeres (presumptive mesenchyme). Embryos artificially assembled and consisting of sixteen mesomeres, four macromeres, and four micromeres; or of the even more abnormal combination of twelve mesomeres, two macromeres, and two micromeres (in each case, therefore, containing too much presumptive ectoderm), develop into normal pluteus larvae. There is present, therefore, a regulating agent which organises the available material to form a harmoniously proportioned larva. That this agent is situated at the vegetative pole of the egg is probable from the facts that vegetative tissue must be present if gastrulation is to take place at all, and that the micromeres (which occupy the vegetative pole) are predetermined to initiate invagination, and do so wherever they may be grafted. Further, if an embryo at the 16-cell stage is divided meridionally and the two halves are stuck together again so that their axes of polarity are reversed in respect of one another, invagination takes place at each end, where the micromeres are situated, and the resulting larva is a double monster, with two guts, skeletons, etc. This can be understood if the micromeres act as organisers[1] (fig. 83).

But, as in the case of *Corymorpha*, this sea-urchin organiser is not specifically located in or restricted to the micromeres, for if they are removed, the next most vegetative material can function as an organiser, and induce the formation of a fairly normal pluteus larva. But if no material from the vegetative hemisphere is present, there

[1] Hörstadius, 1928.

169

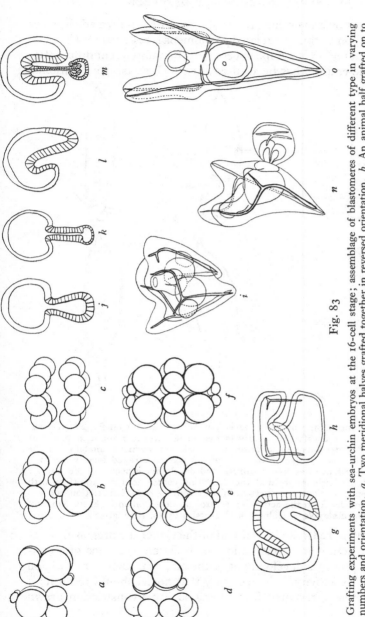

Fig. 83

Grafting experiments with sea-urchin embryos at the 16-cell stage; assemblage of blastomeres of different type in varying numbers and orientation. *a*, Two meridional halves grafted together in reversed orientation. *b*, An animal half grafted on to an inverted vegetative half. *c*, Two animal halves grafted together (i.e. eight mesomeres only). *d*, An animal half and a meridional half grafted together (i.e. twelve mesomeres, two macromeres, two micromeres). *e*, An animal half added to a normal 16-cell stage (i.e. sixteen mesomeres, four macromeres, four micromeres). *f*, A vegetative half grafted inverted on to a normal 16-cell stage. *g*, *h*, *i*, Results of combinations *a* and *f*; invagination and organisation has proceeded from each group of micromeres, acting as an organiser. *j*, *k*, *l*, *m*, Result of combination *b*; the animal-most material of the inverted vegetative half (presumptive epidermis) is extruded from the invagination, and in the fully-formed pluteus, *n*, gives rise to an ectodermal vesicle near the anus. *o*, Normal pluteus, resulting from combinations *d* and *e*. Combination *c*, being devoid of vegetative-pole material, produces a blastula with two apical organs. (From Hörstadius, *Acta Zool.* IX, 1928, slightly modified.)

is no regulation and nothing resembling a pluteus is formed. Further consideration of this and related phenomena is reserved for Chap. IX.

Lastly, a region of cytoplasm essential for subsequent differentiation of the embryo has been discovered in Insects. As mentioned

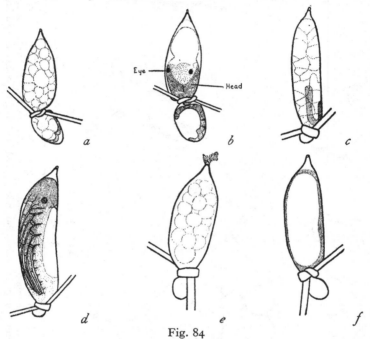

Fig. 84

The activating centre in early development of the dragon-fly *Platycnemis*. *a*, The egg was constricted tightly (some nuclei already present in posterior portion). No development anterior to constriction, owing to inability of the activating substance to reach it. *b*, The egg was constricted loosely, and the activating substance was able to diffuse, and development occurs in the anterior portion. *c*, Constriction behind the activating centre does not interfere with development. *d*, Normal embryo resulting from *c*. *e*, *f*, Constriction and exclusion of nuclei from the activating centre: no development. (From Morgan, *Experimental Embryology*, Columbia University Press, 1927, after Seidel.)

in Chap. V (p. 128), the hinder end of the egg of the dragon-fly contains a region, whose destruction, or isolation by means of a constriction, prevents development of the embryo. Before this region can exert its activity, it is necessary that nuclei should reach it; if the nuclei are prevented from doing so (by a constriction drawn only just tight enough to prevent their passage), there is no development. After the nuclei have reached this region, a substance appears to be given off from the activating region, and to diffuse through the egg in an anterior direction. As time goes on, increasingly large portions of the hinder part of the egg can be destroyed without interfering with development, which shows that as soon as a region has received the diffused substance, it is no longer dependent on the activating region[1] (fig. 84).

This activating region differs, however, from an organiser in that it is not concerned with the differentiation of this or that structure in any particular position: it is merely a starter or activator, conferring on the remaining regions of the egg the power to undergo development.

Farther forward in the insect egg, in the region which will normally give rise to the thorax, the existence of a differentiating centre has been established. For the differentiation of the regions anterior to this, it is necessary not only that the activating substance from the activating centre should have reached the differentiating centre, but there must be cellular continuity between the differentiating centre and the regions of the blastoderm anterior to it. The activating substance, on the other hand, diffuses freely through the egg, whether the cells are in continuity or not. It follows, either that the differentiating centre absorbs this substance in its cells and distributes it from cell to cell, or that it initiates a new chain of reactions. At all events, the differentiating centre is responsible for the localisation and determination of the various regions of the embryo, and its presence is necessary if a properly and harmoniously proportioned embryo is to result from an egg in which an anterior portion has been isolated by constriction, or by a discontinuity between the cells of the blastoderm.[2]

Further results must be awaited before the question can be answered as to whether the mode of action of the Insect differentiating centre is comparable with that of the organiser in other groups.

It has been mentioned (Chap. v, p. 113) that in *Chaetopterus* and in *Tubifex*, when the polar lobe or pole-plasms are equally distributed to the first two blastomeres, instead of being restricted to

[1] Seidel, 1929, 1931.   [2] Seidel, 1931.

the blastomere *CD*, double monsters, each member of which possesses a complete set of trunk organs, are produced. It might therefore seem as though the somatoblasts (for the formation of which the polar lobe or pole-plasms are essential) of Annelids and Molluscs deserve the title of organiser.[1] It is, however, unlikely that these exert an effect similar to that of the amphibian dorsal lip, or of an engrafted *Planaria* head: it is more probable that the growth processes initiated by a single somatoblast automatically lead to the production of a more or less complete set of trunk organs.

A hitherto unique type of determination is found in the wings of moths. As is well known, in *Lymantria*, intersexual types can be produced by appropriate crosses of local races (see Chap. XII, p. 409). The normal wing pigmentation is white in females, dark in males. In male intersexes (i.e. animals which begin adult differentiation as males but continue it as females) the wing shows a mosaic of white (female) and dark (male) pigmentation. The quantity (total area) of female-type pigmentation is directly proportional to the degree of intersexuality as measured by other secondary sexual characters, but the pattern is irregular and varies from specimen to specimen. Careful observation shows that the limits of the male and female areas are defined in reference to the course of the veins. The appearance is as if there had been a flow of a certain quantity of dark pigment through the veins.[2] However, from other work we know that pigment deposition occurs in relation to the determination of the scales. The scales, if determined as female, develop quickly; if as male, develop slowly. The visible determination of sexual type can be seen to occur long before the wings become pigmented. Meanwhile the processes leading to the deposition of white pigment occur some time before the end of pupal development, and those leading to the deposition of melanin occur later. Pigment can only be deposited during a certain stage in the development of a scale. Thus in the female the white pigment-precursors find the female-determined scales at the right stage, while later, when the processes leading to melanin-production occur, no scales are available in which it can be deposited. The reverse is true in males. The result is brought about by interaction of two independent sets of processes.

[1] E. B. Wilson, 1929.   [2] Goldschmidt, 1923, 1927.

The intersexual males demonstrate that the determination of the scales must in them occur as the result of the streaming out of some chemical agent, responsible for initiating male-type determination, from the body over the wings along the course of the veins. There exists what Goldschmidt calls a "stream of determination". Slight variation in the resistance of the various veins will lead to large individual variation in the precise course of the flow. Normally after a time the flow reaches every part of the wing. But in the intersexes, if the switch-over from male to female metabolism occurs during the time occupied by the outstreaming of this substance, all the parts which it has not yet reached will develop as female, and the male-determined areas, later becoming coloured with melanin, remain as a record of the early course of the flow.

We may presume that there is some passage of a determining substance from the organiser to other regions in the pre-gastrulation stage of the amphibian egg (see p. 139); but this is the only case known where one must postulate a flow of such a substance along anatomically differentiated channels. Much remains to be cleared up as regards this phenomenon. For instance, it manifests itself in certain rare cases among female intersexes, but the wings of these are usually whole-coloured and of male (dark) type.

§ 7

Organiser phenomena are clearly special cases of what Roux termed *dependent differentiation*. As noted in Chap. III, p. 54, we will use the term in its restricted sense for cases in which the differentiation of one part depends, in one way or another, upon the presence and previous differentiation of another part. The factors concerning dependent differentiation fall into several rather different categories.

It will be convenient first to give brief consideration to those effects which commonly begin to operate after the functional period of development has started. These are of various distinct types. First, there are the morphological effects of hormones, such as the influence of the gonad hormones upon secondary sexual characters in vertebrates. Another example, this time from invertebrates, concerns the differentiating capacities of rudiments of insect organs, which have been tested by means of explantation experiments. If the leg imaginal discs of mature blow-fly larvae are cultured *in vitro* in media of inorganic salts or of larval body-fluid, they will remain healthy for several days, but will not develop. If now the larval body-fluid is replaced by pupal body-fluid, or if the cultures had been put up in this medium straightaway, the leg imaginal discs become evaginated and grow into segmented limbs.[1] They do not, however, develop beyond the stage corresponding to the fifth day of pupal life, and this is possibly due to the absence from the culture medium of some substances necessary for further development. At all events, it is clear that the differentiation of the leg-rudiments is dependent on changes which occur in the body-fluid at the onset of pupal life. Other experiments have shown that the process of moulting, so characteristic of insect development, is a reaction of the epidermis to a substance in the body-fluid, amounting to a hormone.[2]

Next, there are the effects of other substances carried in the blood stream. The classical example of this concerns the pigmentary pattern of the late embryo of the fish *Fundulus*. The pattern is due to the pigment-cells arranging themselves along the blood-vessels of the yolk-sac, i.e. in situations where the maximum amount of oxygen is available.[3]

Then, there are the trophic effects of nerves, such as the dependence of the differentiation of taste-buds in the fish *Amiurus* upon contact with the nerve endings of the facial nerve[4] (see p. 430). Finally, there are the moulding effects of pressure and tension upon the form, size, and intimate structure of such organs as sinews and blood-vessels (see p. 432).

With regard to differentiation of the type seen in the prefunctional period of development, there are, at the opposite extreme from the organiser phenomena, effects primarily mechanical in nature. An example of these is seen in the dependence of the arms of the pluteus larvae of sea-urchins upon the growth of the larval skeleton. In the absence of the skeleton, no arms are produced: if an abnormal number of skeletal spicules are formed, a corresponding number of arms are produced: if the spicules are abnormal in

[1] Frew, 1928.   [2] von Buddenbrock, 1930; Bodenstein, 1933.
[3] J. Loeb, 1912.
[4] Olmsted, 1920; G. H. Parker, 1932 A, B.

position, so are the arms.[1] It appears that the formative stimulus consists in the continuous pressure exerted on the epidermis by the growing tips of the skeletal spicules. But, as we shall shortly see, the position of the skeletal spicules is itself under the control of the epidermis, and therefore arms and spicules are, in a measure, mutually dependent.[2]

Other examples of dependent differentiation are seen in the adjustment of the skeleton of Vertebrates to the underlying organs. For example, if the rudiment of the optic-cup is extirpated in early amphibian embryos, when the cartilaginous cranium comes to be formed, the skeleton of the orbital region is markedly smaller on the operated side, and, in certain respects, irregular.[3] When foreign structures, e.g. mesonephros, are grafted in place of the mid-brain, the cartilaginous cranium is distorted by the increased intracranial pressure due to the graft.[4] When the rudiment of the nasal sac is extirpated, the cartilages of the nasal region arise by self-differentiation, but the nasal capsule is completely collapsed: the normal form of the nasal capsule is attained through the cartilage adjusting its growth to the form of the nasal sac.[5]

Of a rather different nature, however, is the relation of the cartilaginous auditory capsule to the primary ear-vesicle. In this case, the cartilaginous capsule wholly fails to develop if the vesicle has been extirpated at an earlier stage. Conversely, a grafted ear-vesicle may induce the formation of a cartilaginous capsule around it. The dependence has been shown to obtain both in amphibian[6] and in avian[7] embryos. Here it would appear that a chemical stimulus from the ear-vesicle is necessary to initiate cartilage production by the neighbouring mesenchyme, though, doubtless, mechanical factors play a part in the later growth of the capsule. The effect is not species-specific, for an ear-vesicle of *Rana* can induce the formation of a cartilaginous capsule from tissues of *Amblystoma*, when grafted into an embryo of that animal.[8]

In the fish *Acipenser*, the relations between the ear-vesicle and the cartilaginous capsule are slightly different, but resemble those

[1] Herbst, 1912.   [2] Runnström, 1929.
[3] Steinitz, 1906.   [4] Nicholas, 1930.
[5] Burr, 1916.
[6] Filatow, 1916; Luther, 1925; Guareschi, 1928.
[7] Reagan, 1917.   [8] Lewis, 1906.

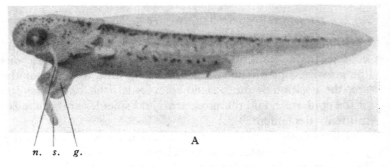

A

n. s. g.

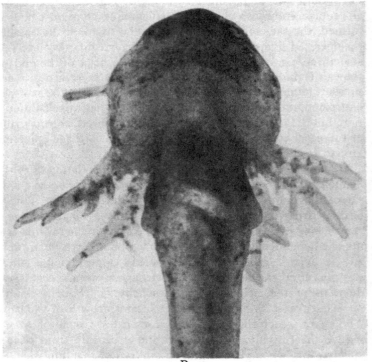

B

Fig. 85

The formation or non-formation of a given structure depends not only on the presence of an inductive or formative stimulus (organiser), but also on local specific factors, intrinsic to the fields. *Triton taeniatus* normally develops a balancer, the axolotl does not. Nevertheless, a piece of gut-roof of axolotl grafted into an embryo of *Triton* can induce the balancer field of the latter to develop supernumerary balancers, A. Conversely, B, a piece of trunk epidermis of *Triton* grafted on to the head of an axolotl embryo, gives rise to a balancer (left side of photograph), while no balancer is formed from the normal axolotl epidermis on the other side. The axolotl therefore possesses the necessary formative stimulus for balancer formation, but its epidermis fails to react to it. *g.* grafted axolotl tissue; *n.* normal balancer; *s.* supernumerary balancer. (From Mangold, *Naturwiss.* XIX, 1931.)

between the nasal sac and olfactory capsule described above. After removal of the ear-vesicle in *Acipenser*, no auditory capsule is formed, but a shapeless chondrification appears in its place. Here, then, the actual formation of cartilage is independent of the ear-vesicle, but the differentiation of the cartilage into an auditory capsule is dependent.[1]

Of a different nature again is the response of the uterine mucosa to the presence of foreign bodies in the uterus.[2] Any foreign body— glass, platinum wire, paraffin, etc.—causes a proliferation of the mucosa essentially similar to that which it shows as a result of the implantation of the embryo. In both cases, the proliferation will only occur provided that certain of the ovarian hormones are present in the blood-stream. Here, the response is not the *direct* result of mechanical forces, as with the arms of the pluteus. A somewhat similar case from the early stages of development is the effect of grafting foreign objects under the flank ectoderm of Urodele embryos: these in some cases induce the formation of supernumerary limbs. These experiments were first performed with ear-vesicles,[3] but it has since been found that inorganic objects, such as celloidin beads, have the same effect[4] (see also Chap. x, p. 362, for the effect of nerve-endings on limb-induction). The type of structure induced thus appears to be determined by local regional factors, regardless of the specific nature of the graft, which acts as a releasing mechanism (see Chap. VII, p. 231).

Another case which appears to be comparable is the induction of supernumerary balancers in *Triton* within the balancer field (see p. 236) by means of grafts of neural crest cells of *Rana*,[5] or of anterior neural plate cells, or even of fore-gut wall-cells of *Amblystoma tigrinum*, which donors possess no balancer.[6] These cases serve as a further illustration of the fact noted above (p. 140), that local regional properties of the tissues acted upon, as well as the properties of the releasing mechanism, do play a part in determining the quality of the induced structure. The relative importance of these two sets of factors varies in different cases: the amphibian organiser is capable of overriding nearly all the local

[1] Filatow, 1930.   [2] L. Loeb, 1908.
[3] Balinsky, 1925–6; Filatow, 1927. [4] Balinsky, 1927.
[5] Raven, 1931 A.   [6] Mangold, 1931 C.

properties of the tissues acted upon: the various grafts mentioned in the previous paragraphs do not override the local regional potencies, but merely evoke them (fig. 85).

We may now return to cases in which the dependent differentiation appears definitely to be due to chemical effects arising from proximity with some other organ. A classical example is the dependence of the conjunctiva[1] upon the presence of the eye.

In the absence of contact with an optic vesicle, the epidermis of the presumptive conjunctiva region remains pigmented and opaque. If, however, contact is established, it loses its pigment and becomes transparent.[2] This effect is exerted not only by the whole optic vesicle, but also by portions of the retina, by the lens alone, and even by disorganised fragments of the optic vesicle grafted under the skin. It even appears that an engrafted limb occupying the place of an eye is capable of inducing the differentiation of the conjunctiva.[3] Pieces of already differentiated epidermis from other regions grafted over the eye, or when eye or lens is grafted under them, can be induced to undergo modification into conjunctiva.[4]

A case which is in many ways comparable with that of the conjunctiva is provided by the Anuran tympanic membrane. This structure is differentiated out of the epidermis at metamorphosis by means of processes involving histolysis and reconstructions of certain layers. Here, the annular tympanic cartilage is the structure on which the differentiation is dependent. Epidermis from other regions of the body will differentiate into tympanic membrane if grafted over the tympanic cartilage, and if the cartilage is extirpated, no membrane forms. Tympanic cartilage grafted under the skin of the back induces the formation of a tympanic membrane in that place.[5] A similar case is the differentiation of an articular cup on the palato-quadrate to fit the base of the balancer. This is dependent on the presence of the balancer, and its formation can

[1] The conjunctiva is of course the epidermal, and the cornea the mesodermal layer of tissue overlying the pupil and lens. These terms have often been used very carelessly, the conjunctiva being called the cornea, and *vice versa*. In most cases, the experiments have not been carried on long enough for the cornea to become properly differentiated.

[2] Spemann, 1901 A; Lewis, 1905.

[3] Dürken, 1916.

[4] Fischel, 1917; W. H. Cole, 1922; Groll, 1924.

[5] Helff, 1928.

be induced by balancer grafts even in species which normally have no balancer.[1]

In Anura, the nervous portion of the pituitary (infundibulum and pars nervosa) is dependent for its full differentiation and growth upon contact with the epithelial or hypophysial invagination, which originates from the epidermis of the front of the head, and later gives rise to the pars anterior, intermedia, and tuberalis.[2] If the hypophysial rudiment be extirpated or destroyed, the infundibulum and pars nervosa fail to develop normally, both as regards size and qualitative differentiation.

In chick embryos, it appears probable that contact of the heart rudiment with the endodermal gut-floor is necessary for the latter to undergo differentiation into a liver.[3] (In Amphibia, however, the liver appears to possess marked powers of self-differentiation: see Chap. VII, p. 203.)

In tissue cultures, it has been conclusively shown that the differentiation of kidney-epithelium into characteristic tubules is dependent on the presence of connective tissue. When cultivated alone, the kidney-tissue merely forms an undifferentiated sheet.[4] Similarly, tissue-cultures of mammary gland carcinoma may be induced to redifferentiate into structures resembling the acini of mammary gland by addition of connective tissue. Again,[5] epithelial tissues grown in culture tend to dedifferentiate unless connective tissue is present also,[6] and cultures of chick-epithelium can be induced to differentiate into structures resembling salivary glands by the addition of fibroblasts.[7]

The perforation of the mouth in Urodele embryos is preceded by a reduction of the ectoderm from a two-layered to a one-layered condition, and by the sinking in of the stomodaeal depression. It has been found that these processes, and the consequent perforation of the mouth aperture, is dependent on the establishment of contact between the ectoderm and the underlying endoderm of the fore-gut. The latter is capable of inducing these changes even when ectoderm from other regions is grafted in place of the normal

[1] Harrison, 1925 B.   [2] Smith, 1920.
[3] Willier and Rawles, 1931 A.   [4] A. H. Drew, 1923; see also Rienhoff, 1922.
[5] A. H. Drew, 1923.   [6] Champy, 1914.
[7] Ebeling and Fischer, 1922.

stomodaeal ectoderm.[1] [See also p. 498.] The perforation of the choanae, on the other hand, is dependent on the establishment of contact between the nasal rudiment and the endodermal roof of the mouth. Even a rudimentary nasal pit is capable, provided it establishes contact with the endoderm, of inducing the latter to give rise to a typically normal choana.[2]

A curious case is that of the perforation of the operculum in Anuran larvae at metamorphosis. This occurs on the right-hand side, allowing the right fore-limb to emerge (the left fore-limb emerges through the spiracle, an aperture which has been present since the first formation of the operculum). It was at first supposed that this was due to mechanical pressure exerted by the growing limb. Then it was discovered that perforation took place even if the rudiment of the fore-limb had previously beeen extirpated.[3] Finally, it has been established that the perforation is caused by a substance produced by the gills as they degenerate during metamorphosis.[4] The degenerating gills will cause perforation of the skin in any region if grafted beneath the surface (see p. 429 and figs. 208, 209).

In Echinoderms, it has been shown that the formation of the amnion and of large portions of the rudiment of the adult sea-urchin are dependent on the presence of the hydrocoel. This follows from the cases in which the abnormal presence of a right hydrocoel is accompanied by the formation of a right amnion and echinoid rudiment, with dental sacs, perihaemal rudiments, oesophagus, and mouth.[5] The size of this echinoid rudiment is correlated with that of the hydrocoel;[6] and in those cases in which by experimental treatment the position of the left (normal) hydrocoel is altered, it is found that the amnion and adult echinoid rudiment arise immediately over the hydrocoel wherever it happens to be, and not from their presumptive tissues.[7] At the same time, it seems that the presence of the amnion is necessary for the complete differentiation of the hydrocoel, so that we are here confronted with a case of mutual dependence.[8]

Another example of this, also from Echinoderms, concerns the

[1] Adams, 1924, 1931.   [2] Ekman, 1923.
[3] Braus, 1906.   [4] Helff, 1924, 1926.
[5] MacBride, 1911, 1918.   [6] von Ubisch, 1913.
[7] Runnström, 1918.   [8] Runnström, 1929.

location of the skeleton in the larva. This skeleton arises from groups of primary mesenchyme cells, which are normally to be found on either side of the blastopore at the close of invagination. If these cells are scattered through the blastocoel by shaking, they return to their original position.[1] It would appear that the ectoderm near the line of its junction with the endoderm exercises a specific attraction on these mesenchyme cells, and this view is further supported by the following experiments. When sea-urchin larvae are made to develop in water to which lithium salts have been added, the proportion between the relative amounts of tissue devoted to the formation of endoderm and ectoderm is altered, to the advantage of the endoderm and at the expense of the ectoderm, with increasing concentrations[2] (see p. 334). The ectoderm may be reduced to a tiny region occupying the animal pole, and in such larvae, the skeleton-forming cells are to be found there, and not in their normal position near the vegetative pole.

The ectoderm is thus responsible for the localisation of the skeleton-forming cells, and, in addition, it appears to control certain details of the growth of the skeleton itself. The mesenchyme cells secrete a triradiate spicule, apparently as an act of self-differentiation. The type of the spicule is also a result of self-differentiation, as is clearly seen in those experiments in which micromeres (presumptive skeletogenous mesenchyme cells) of *Echinocyamus* (which normally possesses complex spicules) are grafted into the animal half of a blastula (presumptive ectoderm) of *Echinus* (which normally possesses simple spicules). The spicules ultimately developed in such larvae are of the complex type.[3] But the growth of the various spicules and struts characteristic of the pluteus skeleton is dependent on the ectoderm. This has been shown by experiments similar to those described above, in which the relative proportions of ectoderm and endoderm are varied. If the ectoderm is very deficient, skeleton production goes no further than the triradiate stage, in spite of the fact that the mesenchyme cells are present in ample quantity. With increasing development of ectoderm, and particularly of the ciliated band, there is progressive development of the skeletal arms.[4] As we have already seen (p. 174), the pressure

[1] Driesch, 1896.   [2] Herbst, 1895.
[3] von Ubisch, 1931.   [4] Runnström, 1929.

of the tips of the skeletal spicules against the ectoderm in the region of the ciliated band is necessary for the formation of the arms. The ectoderm and skeleton are therefore mutually dependent in the formation of the arms.

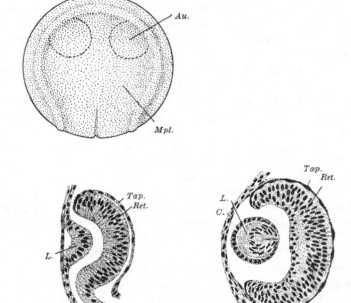

Fig. 86

Eye development in Amphibia. Above, early neurula showing neural plate (*Mpl.*) and limits of presumptive eye-rudiments (*Au.*). Below, left, section of early optic cup, with tapetal (*Tap.*) and retinal (*Ret.*) layers, and epidermis proliferating to form the lens rudiment (*L.*). Below, right, eye at onset of functional stage. *C.* cornea. The central portion of the lens has differentiated into lens fibres. (From Mangold, *Naturwiss.* XVI, 1928, figs. *b, e, f.*)

It should be noted that under the influence of abnormal environmental agencies, the course of local differentiation may be markedly modified. One example is the formation by frog embryos, markedly retarded by being kept in solution of KCl, of an almost solid neural tube, recalling that normally found in the development of *Petromyzon* and Teleosts. Another, of extreme interest, is the development, in tadpoles arising from eggs kept in urea solutions, of patches of tissue, within the nerve-cord or the gut, whose histological structure is identical with that of the notochord. These patches of ectopic notochordal tissue are always adjacent to the true notochord. It would appear that there has been some spread of the factors responsible for this particular histo-differentiation, possibly by the diffusion of specific substances from the notochord-rudiment. As Lehmann (*Naturwiss.* xxi, 737) has recently shown, lithium treatment results in differential reduction of the trunk-notochord in *Triton*.

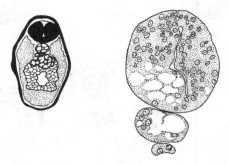

Fig. 87

Spread of notochordal type of histo-differentiation to neighbouring organs in frog tadpoles reared in 1·5 per cent. urea. Left, notochordal differentiation in the gut-roof. Right, notochordal differentiation in the nerve cord. Below the noto-chord in each case is the sub-notochordal rod. (Redrawn after Jenkinson, *Arch. Entwmech.* xxi, 1906.)

§ 8

We have left to the last what is the most celebrated example of dependent differentiation—the formation of the lens of the vertebrate eye from the epidermis under the influence of the eye-cup (fig. 86). The matter, however, is not simple, and is worth going into at some length.

In *Rana temporaria* (*fusca*) the lens is dependent for its development on contact with the eye-cup. If the latter is removed (at the tail-bud stage), the lens is not formed.[1] Further, the eye-cup in this species is capable of inducing the formation of a lens out of

[1] Spemann, 1901 A, 1905.

epidermal tissue which would normally not have given rise to a
lens at all. This can be effected either by grafting the eye-cup under

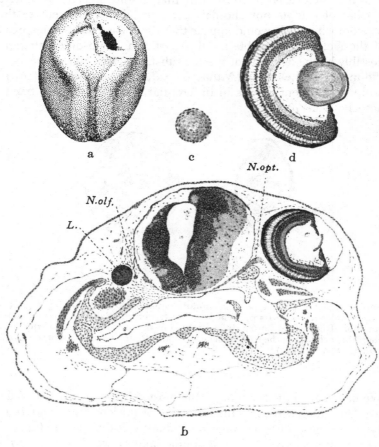

Fig. 88

Self-differentiation of the lens in *Rana esculenta*. *a*, Extirpation of the presumptive eye-rudiment at the early neurula stage. *b*, Transverse section through
resulting larva 14 days after operation; in spite of the absence of an eye-cup, a
lens (*L.*) has developed by self-differentiation. *c*, This lens, at the same scale as
d, normal eye and lens, for comparison. (From Mangold, *Ergebn. der Biol.* VII,
1931, after Spemann.)

the skin in an abnormal position, or by grafting a piece of foreign epidermis over the eye-cup *in situ*. In both cases, a lens is formed.[1]

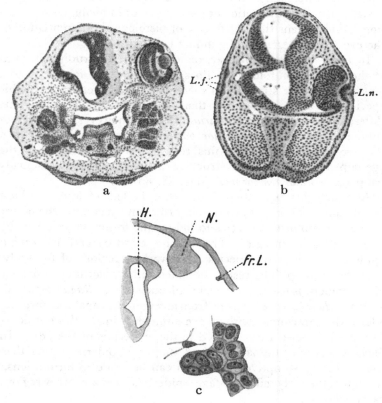

Fig. 89

Development of the lens in *Bombinator pachypus*. *a*, Transverse section through larva from which the presumptive eye-rudiment was removed at the early neurula stage: result, no lens. *b*, Transverse section through larva from which the optic vesicle was removed at the early tail-bud stage: result, small lentoid thickening of epidermis (*L.f.*). *L.n.* normal lens of unoperated side. (From Mangold, *Ergebn. der Biol.* VII, 1931, after Spemann.) *c*, Enlarged view of lentoid (also seen at *Fr.L.*) developed after eye-cup removal and rearing at 23° C. *N.* nose; *H.* brain. (From von Ubisch, *Zeitschr. Wiss. Zool.* CXXIII, 1924.)

[1] Filatow, 1924, 1926.

Bufo, Triton, and the chick[1] agree with *Rana temporaria* in the conditions of formation of the lens (see fig. 21, p. 55). In the chick, the interesting observation has been made that the optic vesicle, as well as the optic cup, is capable of inducing lens-formation.[2] This means that the degree of histological differentiation of the eye is immaterial for the inductive effect.

In *Rana esculenta*, however, removal of the eye-rudiment, even at the early neural fold stage, does not prevent the formation of a lens, which latter structure is therefore self-differentiating in this species at a stage even earlier than that at which it is dependent-differentiating in *Rana temporaria*.[3] The lens, however, is sometimes subnormal in size. *Bombinator pachypus* is intermediate between the two species of *Rana* in this respect, for after removal of the eye-cup a small lens-like structure develops. This occasionally happens in *Rana temporaria*[4] (figs. 88, 89).

Although this experiment shows that the lens of *Rana esculenta* is self-differentiating, it gives no information concerning the power of the eye-cup of this species to induce the formation of a lens by dependent differentiation. This can be tested by grafting foreign epidermis of the same species, from various regions of the body, over the eye-cup. The results obtained differ according to the age of the epidermis used. At the late tail-bud stage in *Rana temporaria* and in *Hyla arborea*, epidermis from any region is capable of forming a lens when in contact with an eye-cup, while in *Bombinator*, lens-forming potencies are restricted to the epidermis of the head. In *Rana esculenta* at the late tail-bud stage, no epidermis other than that of the presumptive lens region can be made to form a lens,[5] though at the early tail-bud stage, epidermis from any other regions can do so.[6]

The inducing power of the eye-cup of *Rana esculenta* may be further tested by grafting over it some epidermis from another species, in which the lens is normally dependent in its differentiation, such as *Bufo vulgaris*, and such experiments invariably result

---

[1] Danchakoff, 1924.   [2] Hoadley, 1926 B.
[3] It appears that the lens in *Rana esculenta* is not invariably self-differentiating, especially at low temperatures. Further experiments on the modifiability of lens induction and lens differentiation in different species are much to be desired. See von Ubisch, 1924.
[4] von Ubisch, 1927.   [5] Spemann, 1912 B.   [6] von Ubisch, 1927.

in the induction of a lens.[1] Thus at this stage, the lens of *Rana esculenta* is self-differentiating, but the eye-cup also possesses the inductive power of forming a lens, so that there is here another example of the principle of "double assurance."

This state of affairs can be interpreted as follows. In *Rana esculenta*, the lens is already determined irrevocably at a stage (early tail-bud) when in *Rana temporaria* it is usually still plastic. We may conjecture, therefore, that the determination of the lens occurs precociously in *Rana esculenta*. In this form, the lens is presumably determined by the presumptive eye-rudiment while this is still an invisibly determined region of the neural plate.

We may also assume, however, that, just as with the presumptive neural plate before gastrulation, there has been a preliminary labile determination of the lens, so that the lens-forming potencies will be more easily called forth at a certain spot, viz. the presumptive lens region.[2] When definitive determination occurs, we must assume that some influence, presumably of a chemical nature, diffuses from the eye-area, and affects the region of optimum lens-forming potency. In a similar way (as will be seen in Chap. VII, p. 223) we may note the limb is actually formed at a region of maximum limb-forming potency, in a much more extensive potential limb-area.

However, when the eye-rudiment of *Rana esculenta* has become converted into an optic cup, it still retains its lens-inducing power. Indeed, it would seem that in some forms this power is retained throughout life, for in many Urodeles it has been shown that the adult eye can regenerate a new lens from its own margin if the lens has been removed[3] (see p. 237). It is interesting in this connexion to note that the eye can resort to this method of lens-formation in embryonic development and form a lens from its own margin if it is deprived of contact with epidermis[4] (fig. 90).

The apparent "double assurance" found in *Rana esculenta* thus apparently means (a) that there exists a region of optimum lens-forming potency in the epidermis of the neurula, and (b) that the power of the eye-cup to induce a lens persists after the lens has differentiated, and after the remaining epidermis has been determined to form epidermis and has ceased to be capable of responding

[1] Filatow, 1925.   [2] Spemann, 1912 B.   [3] Colucci, 1891; G. Wolff, 1895.
[4] Spemann, 1905; Beckwith, 1927; Adelmann, 1928.

to induction. There is thus an overlap in time between these two phases of differentiation ordinarily spoken of as dependent differentiation and self-differentiation. There seems to be little reason to doubt, however, that both the methods concerned in "double assurance" are ultimately referable to one and the same causative agent: in this case presumably situated in the rudiment of the eye-cup and in the fully formed cup which later arises from it.

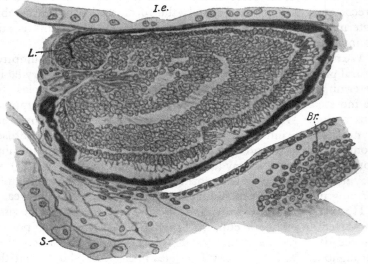

Fig. 90

Lens-formation from the margin of the optic cup in ontogenetic development. The presumptive eye-rudiment of an embryo of *Triton* was grafted into the side of the body of another embryo, and developed by self-differentiation, deep beneath the epidermis. Under these circumstances it has given rise to a lens from the margin of its own cup, in the manner characteristic of regeneration experiments. *Br.* portion of grafted brain tissue; *I.e.* wall of intestine; *L.* lens; *S.* epidermis of ventral side. (From Adelmann, *Arch. Entwmech.* CXIII, 1928.)

The divergent results obtained with different species are apparently to be accounted for by differences in the rates at which the two processes, of capacity of the eye-cup to induce and of the epidermis to differentiate, occur.

The proliferation of cells from the epidermis is not, however, the only process involved in lens-formation: the cells require to become converted into the characteristic lens-fibres. While the proliferative effect may, as we have seen above, be more or less independent, the subsequent differentiation of lens-fibres appears to be always dependent, usually on the eye-cup. But the action of the latter in this case does not appear to be specific, for experiments in which lens-rudiments are allowed to develop in proximity with portions of brain or nose tissue show that the latter are also capable of inducing the formation of lens fibres.[1]

Recent work on the American bull-frog, *Rana catesbiana*,[2] has given additional results. This species shows an extreme of dependent differentiation for the lens, rivalling or exceeding *Rana temporaria* in this respect. Of greatest interest is the fact that here the continued presence of the optic vesicle or eye is necessary for the lens to achieve full differentiation and full size, even after it has been initially determined. The lens-rudiment, however, once determined, has a certain power of self-differentiation. After determination but before visible differentiation, the lens-rudiment, by itself, is only capable of producing lentoid structures without differentiation of fibres. After visible thickening has occurred, however, the rudiment left *in situ* after removal of the underlying eye-cup will produce a true lens, but this is small and slightly abnormal. There is thus a complementary action of inherent potencies and external induction (see p. 264 for what may be a similar effect with the avian gonad). Another interesting point is that if the visible lens-rudiment at the same stage is separated from the eye-cup and grafted heterotopically, it undergoes a certain amount of regression and never reaches the same degree of differentiation as if left *in situ*, though in both cases it is removed from the inductive influence of the eye. Thus in this species, although epidermis from any region can be made to form a lens, potencies favourable to lens-differentiation are highest in the area of the normal lens-field.

The crystalline fibres of the lens in Amphibia are oriented in a definite manner, normally converging to a sutural line which is dorso-ventral on the outer surface, and antero-posterior on the inner surface of the lens. It is to be noted that the plane of the external sutural line coincides with that plane of the eye-cup in which the choroid fissure is situated, for this structure occupies the most

---

[1] Balinsky, 1930.   [2] Pasquini, 1933.

ventral region of the cup. It has been found that the orientation of the fibres of the lens is also dependent on the eye-cup[1], and, in particular, on the position of the choroid fissure. This is proved by experiments on embryos of *Rana esculenta* at the early neurula stage, in which the presumptive lens epidermis is rotated through 90°: the lens-fibres are nevertheless normally oriented. On the other hand, if the eye-rudiment is rotated so that the choroid fissure comes to occupy an abnormal position, the lens-fibres are also abnormally oriented.[2] Therefore, while the lens is normally self-differentiating as regards its general formation in *Rana esculenta* at this stage, the determination of the orientation of its fibres is still dependent on the eye-cup. At later stages, rotation of the lens-area of epidermis shows that this orientation becomes self-differentiating also.

Lastly, it may be noticed from those cases in which a lens can be induced by an eye-cup of a different species, that the lens-inducing capacity of the eye-cup, like the organising capacity of the amphibian dorsal lip, the avian primitive streak, the hypostome of *Hydra*, the head of *Planaria*, and the capsule-inducing capacity of the amphibian ear-vesicle, is not species-specific.[3] In this respect, the action of organisers and inducing structures has much in common with that of hormones. Many if not all hormones are similar or identical in widely separated groups: thyroxin from a mammal will metamorphose amphibian larvae; testis hormones from a bull will cause the comb of capons to grow; adrenalin from a fish will excite vaso-constrictor effects in man. On the other hand, the precise effect produced depends on the reacting tissues, just as it does with the organiser effects during development. The tail and limbs of Anuran larvae react to thyroxin, while those of Urodela do not; the larval epidermis of most Amphibia reacts to thyroxin, while that of the adult never does. The relation of organisers to induced organs, as of hormones to reacting tissues, is thus much less specific than the interaction of hereditary outfits (genomes) from different species, where a difference of generic degree is usually more than sufficient to prevent co-operation.

[1] Dragomirow, 1930.   [2] Woerdeman, 1932.
[3] Woerdeman (1933 B) finds marked changes in glycogen content in the eye-rudiment before and during the period of lens-differentiation. The precise meaning of these, as of similar changes in the organiser region (p. 154), remains a subject for future investigation.

§ 9

In reviewing the various aspects of dependent differentiation it is clear that the organiser phenomena occupy a special place. The part which organisers play is of supreme importance. From the theoretical point of view, they present a biological property of the first order, and had Roux known of their existence he would undoubtedly have classified them among the "complex components" of development (see p. 9).

However, the precise mode of action of organisers cannot be understood except in relation to the properties of gradient-fields: this problem will be considered in some detail in Chaps. VIII and IX. Meanwhile, attention may be turned to the general result of the presence in an embryo, such as an amphibian, of an organiser and other structures, exerting effects of hetero- and homoio-genetic induction, and some of them showing local regional differentiation. The main result is that almost everywhere in the body formative stimuli are found capable of inducing plastic tissues to undergo this or that type of differentiation, according to their position. Normally, of course, the tissues cease to be plastic as soon as they have undergone the inductive action of their organiser. But the existence, distribution, and local regional characters of the various inductive influences in the amphibian embryo can be studied by grafting portions of plastic early gastrula tissue into older hosts (at the neurula stage), thanks to the fact that the inductive effects persist for a longer time than is necessary for the normal determination of the embryo's own tissues.

It has been found, using pieces of presumptive epidermis or neural fold tissue as grafts, implanted into the dorso-lateral region of neurulae, that the quality of the differentiations which the grafts then undergo is dependent on their position in the host embryo. In the head, grafts may differentiate into portions of brain with epiphysis, nasal sacs, and eyes: in the gill region, into portions of hind-brain: in the trunk region, into portions of spinal cord. The grafts may be induced to form ear-vesicles, sense-organs, visceral cartilages, and ganglion cells in the head; gills in the gill region; fore-limbs in the fore-limb region; pronephric tubules in the pronephric region, etc.[1] (fig. 91).

[1] Holtfreter, 1933 B.

A fact of great importance is that the various determinations are regional, but the regions are not very closely circumscribed. There is as it were a certain amount of latitude as to exactly where a particular structure will arise, although it is bound to be within

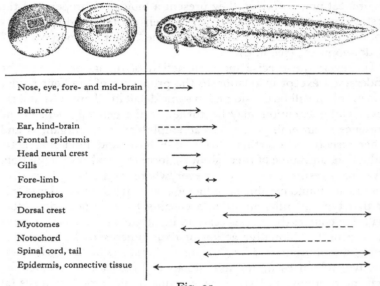

Nose, eye, fore- and mid-brain	--- →
Balancer	--- →
Ear, hind-brain	---- →
Frontal epidermis	---- →
Head neural crest	→
Gills	↔
Fore-limb	↔
Pronephros	← →
Dorsal crest	← →
Myotomes	← ----
Notochord	← ---
Spinal cord, tail	← →
Epidermis, connective tissue	← →

Fig. 91

Diagram showing the results of grafting portions of presumptive epidermis or presumptive neural plate from a gastrula into various regions of the flank of a neurula (*Triton*). The grafts are induced to undergo differentiation into the structures enumerated in the left-hand column; the effect of the position of the graft in the host upon the type of structure resulting is indicated by the extent of the lines in the right-hand column, imagined as projected on to the larva shown above. (From Holtfreter, *Arch. Entwmech.* cxxvii, 1933.)

a certain region. The following table will show the frequency with which particular structures are induced from ectodermal grafts in different parts of the body:

	Ear region	Gill region	Fore-limb region	Phrone-phros region
Balancer	10	6	—	—
Ear-vesicle	72	15	4	—
Pronephros	—	2	4	28

Each of the various inductive effects accordingly covers a wide area, or *field*, and the intensity of the induction decreases with increasing distance from a sub-central point in each field (see Chap. VII, p. 223).

If, now, we stop to inquire which structures are responsible for the inductive effects, the answer appears to be in most cases that each field is dependent, not on one but on several other structures. The organiser for the neural tube induction in the trunk region appears to be the segmented mesoderm; this, which of course is derived from the invaginated organiser, is known to induce neural tube when grafted beneath strange epidermis. Here, the converse experiment has been performed, and strange epidermis has been grafted over the derivative of the organiser. The inductive action which produces portions of brain, etc., in the head, appears to proceed from the neural crest, which is also capable of inducing cartilage, ganglion-cells, sense-organs, and ear-vesicles.[1] At the same time, the induction of ear-vesicles can be performed homoiogenetically, by ear-vesicles, just as fore-limb and pronephros can induce fore-limb and pronephros respectively.

The formation of a tail is the combined and coordinated result of a number of inductive influences. The elongation and stretching of the notochord and musculature, and the metameric arrangement of the latter, are dependent on the presence of neural crest mesenchyme; dorsal and ventral fins are formed when neural tube is present; in the absence of neural crest mesenchyme, the initial elongation of the tail-bud stops, and regression sets in.

We see, in general, that as a result of the inductive capacities of the organiser and of certain other structures (themselves the result of induction by the organiser), the amphibian embryo at the neurula stage is already what may figuratively be called a physiological mosaic of formative stimuli, leading to the demarcation of fields, each of which represents the sphere of action of a particular type of inductive effect. We shall see in the next chapter that these fields constitute one of the most important features of the next or mosaic stage of development.

[1] In these experiments the grafts do not appear to have come into close contact with the brain itself of the host-embryo. The homoiogenetic inductive capacity of the brain has, however, been established by other work (see above, p. 147).

Chapter VII

THE MOSAIC STAGE OF DIFFERENTIATION

§ 1

It has been seen in previous chapters that, after a certain stage, the various regions of the early embryo are irrevocably determined to

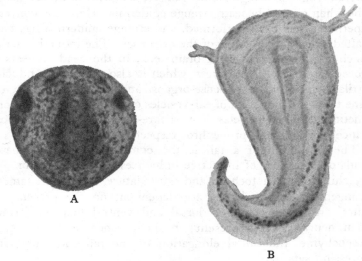

Fig. 92

Mosaic development in explanted tissues. A, Isolated head, containing brain, eyes, nasal pits, tip of notochord, cartilage, and functional jaw-muscles, differentiated in inorganic culture medium from the dorso-anterior portion of an early neurula of *Amblystoma*: 25 days after explantation. B, Isolated trunk, containing spinal cord, notochord, muscles, pronephric tubules, gut, fore-limbs, and tail with dorsal fin, differentiated in inorganic culture medium from the postero-ventral portion of an early neurula of *Triton*: 25 days after explantation. (From Holtfreter, *Arch. Entwmech.* CXXIV, 1931.)

undergo some particular type of development, although at the time that they are thus determined there is no visible differentiation of any kind. This determination is presumably to be ascribed to the local elaboration of specific chemical substances, and may be referred to as chemo-differentiation (see p. 46).[1] It now becomes necessary to consider this phase of development in greater detail.

Experiments and operations on early stages (early tail-bud) of embryos of Urodeles (*Amblystoma*) have now shown that, beyond mechanical wound healing, no regeneration or regulation occurs.[2] If, for instance, the embryo is cut into two by a transverse section, the two portions continue their prospective development, the front portion forming a head and neck region, the posterior portion a trunk and tail. The number of external gills on the one or the other portion depends upon the precise position of the cut (fig. 92).

Similarly, it has been found that the anterior third of a 24-hour blastoderm of a chick embryo grafted on to the chorio-allantois of another egg gives rise to just those organs which it would have produced in normal development[3] (fig. 93). Two half-embryos of frogs grafted together will develop into a single frog, even if the halves belong to different species. Each half retains its specific characteristics (see p. 406, and fig. 196).

In Amphibians, such fragments of course cannot develop far beyond hatching. If, however, in the early tail-bud stage, the tip of the tail is cut off, the organism develops into a healthy larva, but with a permanently shortened tail.[2] Removal of rudiments of eyes, gills, limbs, heads, or snouts at this stage results in permanent absence of these structures in the later embryo and larva. Similarly, experiments on the fish *Fundulus* have shown that removal of portions of the embryonic shield results in permanent absence of the structures whose rudiments have thus been affected[4] (fig. 94).

More recent and detailed work has shown that in *Triton* at the stage when the tail-bud is hemispherical, complete or almost complete removal of the mesodermal contents of the bud results in completely tailless larvae, whereas in only slightly later stages complete regeneration can and does occur.[5] Further, by appropriate operations, more localised defects can be obtained, e.g. absence of ventral fin membrane, of mesodermal somites, of notochord, or of nerve tube. When regeneration experiments are carried out on such partially defective tails in the larval stage, it is

[1] Huxley, 1924; Goldschmidt, 1927; Bertalanffy, 1928.
[2] Schaxel, 1922 B.   [3] Murray and Huxley, 1925.
[4] Eycleshymer, 1914; Nicholas, 1927; Hoadley, 1928.   [5] Vogt, 1931.

196

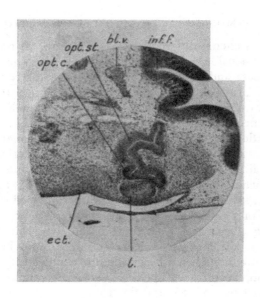

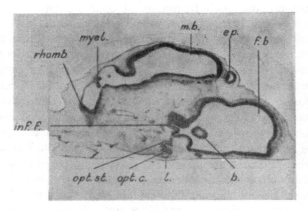

Fig. 93

Self-differentiation in fragments of the vertebrate embryo. The head-region of a 24-hour chick embryo was grafted on to the chorio-allantois of another egg and allowed to develop for 4 days. Above, eye-region enlarged. Below, longitudinal section. It differentiated the main regions of the brain (*rhomb*. rhombencephalon; *myel*. myelencephalon; *m.b.* mid-brain; *f.b.* fore-brain; *ep*. epiphysis; *inf.f.* infundibulum), together with an eye showing optic stalk (*opt.st.*), optic cup (*opt.c.*), and lens, *l*; *bl.v*. blood-vessel. The histogenesis was normal, but the form-differentiation, notably of the eye, and also of the brain (note *b*. bar across cavity of fore-brain), abnormal in many respects. (From Murray and Huxley, *Brit. Journ. Exp. Biol.* III, 1925.)

found that, although they have the capacity for regeneration, the regenerate still shows the defective organisation (e.g. with regard to the ventral fin membrane). It appears in these cases that the organism cannot regenerate a structure which has never been formed in its own ontogeny: a fact of great interest in itself and

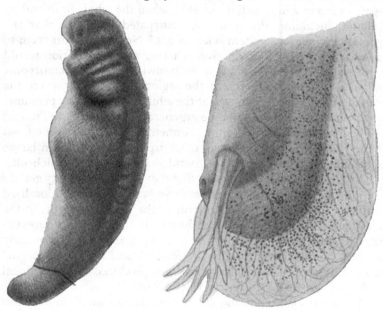

Fig. 94

Effect of extirpation of tail-rudiment at early stage, in *Triton*; left, the operation; right, the resulting larva, with total absence of tail. (From Schaxel, *Arch. Entwmech.* L, 1922.)

with an important bearing on the problem of gradient-fields, to be discussed in Chap. x. It is still, however, uncertain whether this limitation is universal.[1]

---

[1] Cases are known in which an abnormal limb (which owes its abnormality to the fact that its rudiment was grafted at an early embryonic stage and failed to develop normally) can after amputation regenerate a normal limb (Swett, 1924). Clearly, the conditions here are different from those in which a structure is abnormal, imperfect, or absent as a result of *removal* of its rudiment. The abnormality of the grafted limb is a consequence of some local conditions due to the experiment, and does not reflect any intrinsic restrictions of potency in the limb-rudiment. Consequently, when a new set of conditions supervenes as a result of amputation, these potencies are present and able to control the regeneration of a normal limb.

A similar total and permanent absence of a whole organ has been obtained by extirpation of the presumptive limb-area, both in Urodela[1] and in the chick.[2] In these cases, the presumptive limb-area is a discoid region of mesoderm and ectoderm, with no visible differentiations. As will be seen later (p. 420), in adult Urodela, regeneration of a limb will occur even when the whole limb and its skeleton, including the girdle, is extirpated, provided that the sympathetic nervous system is left intact.[3] No experiments seem to have been carried out to discover whether any regeneration would occur in an animal lacking a limb owing to early embryonic extirpation of the limb-area, if the region on the flank where the limb ought to be were removed at the adult stage; we may presume, however, that there would be no regeneration. (See fig. 22, p. 56.)

The hypophysis arises from a rudiment of ectoderm on the front of the head. This rudiment can be extirpated from Anuran larvae at the tail-bud stage, and it is found that the larvae[4] which ultimately develop are normal except that they lack the pituitary gland.[5]

Even the blood in the Anuran embryo has a definite and localised rudiment, situated in the mesoderm of the splanchnopleur, in the mid-ventral line, anterior to the heart. If this rudiment is extirpated completely from embryos of *Rana temporaria* at the early tail-bud stage, no erythrocytes are formed, and in cases of partial extirpation the quantity of erythrocytes produced is proportional to the amount of the rudiment which is left.[6]

The Ascidians provide another case of animals which in the adult state are capable of extensive and far-reaching regeneration and reorganisation, but which in the early stages of embryonic development are unable to make good any loss which the various determined regions may sustain.[7]

---

[1] Harrison, 1915.   [2] Spurling, 1923.
[3] Bischler, 1926.
[4] Incidentally, it may be mentioned that such larvae are of great interest also from another point of view, for they are incapable of producing the pituitary hormones, and are therefore permanently light in colour, and incapable of normal metamorphosis.
[5] Smith, 1920.   [6] Frederici, 1926.
[7] Conklin, 1905, 1906; Huxley, 1926.

§ 2

This mosaic predetermination of various regions in the chemo-differentiated stage in development is also demonstrated by numerous experiments in which a region continues its presumptive development even after grafting into an abnormal position. As previously mentioned, the presumptive eye-region of amphibian embryos has the power of self-differentiation at the early neurula stage (p. 46). Limb-discs of *Amblystoma* grafted on to the flank or into other abnormal situations will still continue to form limbs.[1]

The presumptive ear-region will differentiate semi-circular canals, etc., when grafted into abnormal situations.[2] The mosaic nature of this power of self-differentiation is further shown by the fact that if a neurula of *Rana esculenta* is divided transversely by a cut passing through the presumptive ear-region, it is found that both halves develop auditory vesicles, but they are incomplete, the details varying with the precise position of the cut. On each side of the body, there is only one ductus endolymphaticus developed on each side, and this may be either in the anterior or the posterior half.[3]

Other experiments of grafting and extirpation have shown that the gill-region, the balancer, nerve placodes, portions of the neural crest, and various other amphibian organ-rudiments possess this capacity for self-differentiation (for references, see later sections). The outgrowth of the glomerulus from the aorta has been shown to be due to self-differentiation, independent of the presence or absence of the pronephros with which it normally comes into functional relation.[4]

In the chick, grafting of embryonic rudiments on to the chorio-allantois of another embryo has been the main method employed, e.g. with the ear-region, eye-region, complete limb-rudiments, fractions of limb-rudiments, presumptive thyroid-region,[5] meso-nephros (see also below), metanephros, adrenal,[6] spleen, portions of brain and spinal cord,[7] lung[8], etc. (figs. 95, 96).

[1] Harrison, 1918; Detwiler, 1918.
[2] Streeter, 1906, 1907; Sternberg, 1924.
[3] Spemann, 1910.   [4] Howland, 1916.
[5] Rudnick, 1932.   [6] Willier, 1930.
[7] Rienhoff, 1922; Danchakoff, 1924; Hoadley, 1924, 1925, 1926 A, 1929; Murray and Huxley, 1925.   [8] Rudnick, 1933.

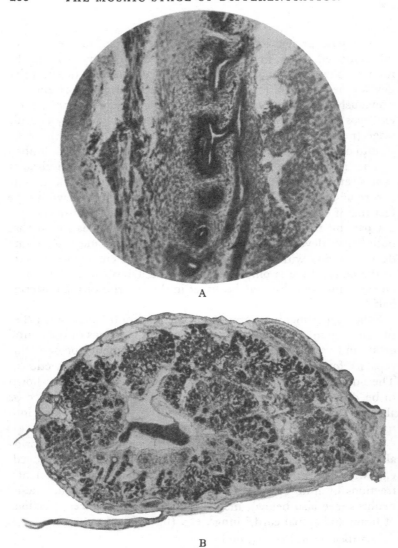

A

B

Fig. 95

Self-differentiation of grafted chick metanephros. A, Metanephric rudiment at the time of grafting (5-day chick). B, Differentiation after 5 days on the chorio-allantois of another egg. (From Danchakoff, *Zeitschr. f. Anat. u. Entwgesch.* LXXIV, 1924; B, after Atterbury.)

In fish, optic cups and other organ-rudiments grafted into the yolk-sac of other embryos show self-differentiation.[1]

The capacity for self-differentiation in mammalian embryos has been tested in rabbits by grafting portions of the embryonic area on to the omentum of other rabbits, where they show a degree of differentiation comparable to that of normal embryos of the same age.[2]

The self-differentiating capacity of mammalian tissues has also been tested by grafting thirds of 11-day rat embryos on to the

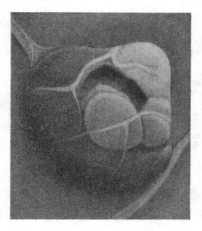

Fig. 96

Mosaic development and self-differentiation of the eye of the chick, grafted on to the chorio-allantoic membrane. The rudiment was removed from an embryo incubated for 48 hours, and grafted for 7 days. (From Hoadley, *Biol. Bull.* XLVI, 1924.)

chorio-allantoic membrane of the chick, where, in spite of the wide taxonomic difference between donor and host (involving as it does, among others, the difference between the temperatures of normal uterine and incubatory development), they are able to differentiate. In these conditions, different structures vary greatly as regards their capacity for self-differentiation; endoderm and nervous tissue show hardly any differentiation, but epidermis with its included hair-follicles, cartilage, and bone, possess it to a high degree, and reach a stage comparable to that of the corresponding structures

---

[1] Mangold, 1931 B.   [2] Waterman, 1932.

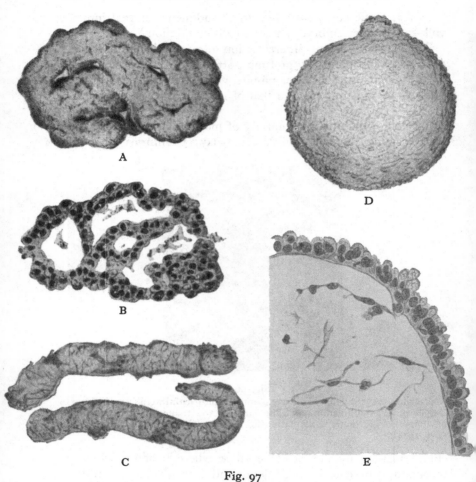

Fig. 97

Mosaic development and self-differentiation of isolated regions of the neurula (Urodele) explanted in inorganic culture medium. A, Epidermal vesicle, 9 days after explantation. B, Section through A; note ridges and folds (cf. fig. 13). C, Notochords, 11 days after explantation. D, Endodermal vesicle, representing an everted gut with the endothelial cells directed outwards. E, Section through D showing the endodermal cells secreting outwards into the medium, 13 days after explantation. (From Holtfreter, *Arch. Entwmech.* cxxiv, 1931.)

in a normal rat of similar age.[1] Nasal sacs and mesonephros achieve a less but considerable degree of differentiation.

Explantation methods have also been applied to the problem. Presumptive rudiments of organs, as yet without any visible differentiation, are removed from the body and allowed to develop in culture media. In some cases they are enclosed within jackets of epidermis, but this is not an essential condition. In addition to

| a | b | c | d |

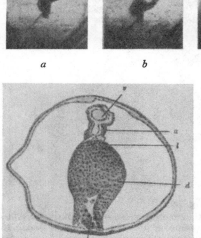

Fig. 98

Self-differentiation of median heart-rudiments *in vitro*. The heart-rudiments together with some of the neighbouring ento-mesoderm were removed from early tail-bud stages of *Bombinator*, and cultivated as explants in epidermal jackets. Above, part of a micro-cinema film of an explanted heart, 11 days after operation; *a.* systole; *d.* diastole. Below, longitudinal section through a similar explant showing differentiation into *a* auricle; *v*, ventricle. In addition, *l*, liver; *d*, yolk-sac; *D*, gut, also present in explant. (From Stöhr, *Arch. Mikr. Anat. u. Entwmech.* CII, 1924.)

notochord, neural tube, muscle-segments, epidermis and kidney tubes,[2] auditory vesicles, gut, liver, and pancreas-rudiments of amphibian embryos treated in this manner develop for considerable periods of time, and produce their appropriate structures, including functional ciliated epithelium and secretory tissue with actual secretion: portions of gut thus differentiated may even show peristaltic action.[3]

Paired heart-rudiments of Urodele embryos at the neurula stage,

[1] Hiraiwa, 1927; Nicholas and Rudnick, 1931.
[2] Erdmann, 1931.   [3] Holtfreter, 1931 A, B.

before they have united in the middle line, will, when explanted
singly, form vesicles of heart-tissue, and some (those from the left
side, see p. 77) may show pulsations.[1] Heart-rudiments taken at
later stages, when they have united in the mid-ventral line, give
still more elaborate self-differentiation, showing sinus, auricle,
ventricle, and bulbus[2] (fig. 98).

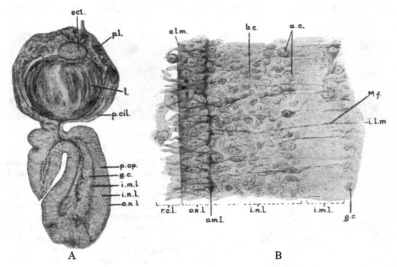

A   B

Fig. 99

Mosaic development and self-differentiation of the chick eye-rudiment cultivated
in vitro. The eye-cup and lens were removed from an embryo incubated for
66 hours and cultured for 8 days in plasma with embryo extract. Histological
differentiation has proceeded at almost the normal rate, in spite of the fact that
the morphological differentiation of the structures is highly abnormal and that
they are subnormal in size. Histological differentiation is therefore independent
of morphological differentiation, and of the normal rate of cell multiplication.
A, Section of the whole explant. B, Section through the retina of a 17-day explant,
in which all the layers are normally developed. *a.c.* amacrine cells; *b.c.* bipolar
cells; *ect.* ectoderm; *e.l.m.* external limiting membrane; *g.c.* ganglion cell;
i.l.m. internal limiting membrane; *i.m.l.* inner molecular layer; *i.n.l.* inner
nuclear layer; *l.* lens; *M.f.* Müller's fibres; *o.m.l.* outer molecular layer; *o.n.l.*
outer nuclear layer; *p.cil.* pars ciliaris retinae; *p.l.* pigment layer; *p.op.* pars optica
retinae. (From Strangeways and Fell, *Proc. Roy. Soc.* B, c, 1926.)

In vitro cultivation of rudiments of presumptive regions has also
been practised with chick material. The optic cup (fig. 99), por-

[1] Goerttler, 1928.   [2] Stöhr, 1924.

tions of limb,[1] ear,[2] metanephros[3] and other rudiments[4] thus treated have shown successful histological self-differentiation. Interesting examples of chemical self-differentiation are found in isolated portions of the skeleton. The cartilages of the palato-quadrate and of the femur normally undergo ossification, whereas the distal portion of Meckel's cartilage does not. The future histological structure is already determined by the sixth day of incubation, although there is then no visible distinction. The difference between these two types of cartilage is revealed by cultivation *in vitro*, where rudiments of the palato-quadrate and of the femur show a marked synthesis of phosphatase, while that of the distal portion of Meckel's cartilage does not: phosphatase activity is correlated with ossification.[5] If cultured long enough, ossification of a normal type supervenes in the rudiments. It is worth mentioning that even in the abnormal conditions provided by tissue-culture, in which the organs are without blood supply, the volume of a chick femur will increase up to about thirty times.

Even extra-embryonic regions, such as the presumptive blood-islands, develop histologically differentiated blood when cultivated *in vitro*.[6]

In some cases, at least, the determination imposed upon regions in the mosaic stage of development concerns even the *duration* of progressive differentiation and growth. The mesonephros of the chick embryo normally undergoes regression at about the tenth day of incubation, and if its rudiment is grafted on to the chorio-allantoic membrane of another egg, it will first differentiate the typical mesonephric tissue, and then proceed to regress at about the same time as regression would normally have occurred if it had been left in place in the embryo.[7] The time of regression in these cases is, of course, in no way determined by the age of the host-egg on to the chorio-allantois of which it is grafted (fig. 100).

The specific growth-capacities of the rudiments may also be determined. In the intact bird, the right ovary is rudimentary and the left is well developed. Four-day rudiments of the ovaries

[1] Strangeways and Fell, 1926.   [2] Fell, 1928.
[3] Rienhoff, 1922.   [4] Hoadley, 1924.
[5] Fell and Robison, 1929, 1930.   [6] Murray, 1932.
[7] Danchakoff, 1924.

grafted on to the chorio-allantoic membrane show the same specific differences between the growth-capacities of the right and left sides.[1] In the mammal, also, the tissue culture of embryonic material provides evidence of self-differentiation.[2] Portions of rabbit embryos 9 to 12 days old, cultured *in vitro*, reveal the mosaic character of development: the various rudiments differentiate independently,[3] just as similar fragments do when grafted. The growth-partition coefficients of Urodele limbs are inherently determined.[4]

A striking case of independent differentiation is provided by the silkworm. The wing rudiments in Lepidoptera are protruded from the surface during pupation, and the pupal case has pockets into which the wings fit snugly. In the silkworm, a mutation has been found which results in the animal being wingless. Nevertheless, the pupal cases of such mutants possess the characteristic pockets, although no wings project into them.[5] Similar occurrences have been observed in *Papilio dardanus* where in the female the wings may have no tails, but pockets for them are provided in the pupal wing-cases.[6]

This case is of considerable theoretical interest, for, in general, when two structures are closely associated topographically, it is found that the differentiation of the one is often dependent on the other. Numerous examples have been given in Chap. VI: we may recall the eye-cup and the lens of *Rana fusca*; the eye and the conjunctiva; the tympanic cartilage and the tympanic membrane; the skeleton and the arms of the pluteus; the hydrocoel and the amniotic cavity of the echinoid rudiment. In *Rana esculenta*, however, as we have seen (p. 186), the eye-cup and the lens are independent from a very early stage, and in this they resemble the wing and wing-case of the silkworm.

Further evidence of the self-differentiating capacity of the wing-rudiment in Lepidoptera is provided by the experiments of grafting the wing-rudiments of caterpillars from one sex into the other. The fully developed wing is markedly different in the two sexes, and it is found that regardless of the sex of the host into which it has

---

[1] Willier, 1927.
[2] Waddington and Waterman, 1933.
[3] Maximow, 1925.
[4] See Huxley, 1932, Chap. VI.
[5] Goldschmidt, 1927, p. 203.
[6] Lamborn, 1914.

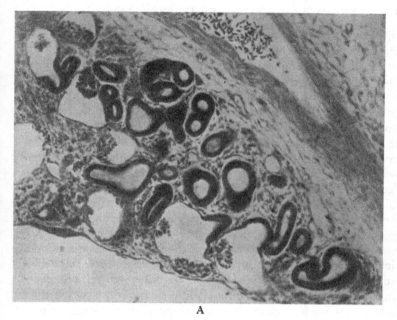

A

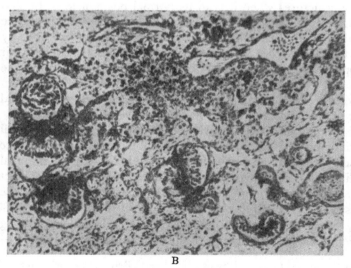

B

Fig. 100

Self-determination of degenerative development in the chick mesonephros.
A, After 5 days as a chorio-allantoic graft, the mesonephros-rudiment shows
marked progressive differentiation. The figure is of a grafted short section of the
trunk; similar differentiation is obtained with isolated mesonephros-rudiments.
B, After 7 or more days, the graft shows regression. All secreting tubules have
disappeared. The malpighian capsules persist, as in normal development. (From
Danchakoff, *Zeitschr. f. Anat. u. Entwgesch.* LXXIV, 1924.)

been grafted, the wing-rudiment differentiates according to the sex of its donor.[1] Recent work on various Insects indicates that after a certain stage the embryo is a mosaic of chemo-differentiated regions, although the details of the determination-process differ considerably from those found in Amphibia.[2]

In Cephalopods it has been shown that fragments of embryos cultivated by explantation methods continue their differentiation as if they formed part of the whole organism.[3] These experiments were undertaken after visible differentiation had appeared; others, however, indicate that the embryo passes into the mosaic chemo-differentiated stage just before visible differentiation occurs.[4] This would, in general, be similar to the state of affairs in Amphibia.

Another remarkable case of self-differentiation during the mosaic stage of development concerns the self-orientating properties of the auditory vesicle in Amphibia. If at the stage when it is a simple vesicle, the auditory sac is turned upside down and left *in situ*, it often rights itself by rotation, so that its dorso-ventral axis conforms to that of the whole animal.[5] The suggestion that the ear-vesicle rights itself because it only fits properly into the neighbouring structures when it is in its normal position must be discarded, because a right ear-vesicle, grafted upside down in the space vacated by an extirpated left vesicle, rotates and becomes right way up and right way out, but as the vesicle retains its laterality, it develops with its normally anterior side pointing backwards in the animal. It thus rights itself in respect of its dorso-ventral axis in spite of the evident misfit which results. Further, an inverted vesicle of *Rana* will right itself in *Amblystoma*, and *vice versa*.[6]

The rotation of the ear-vesicle may be impeded by special local conditions of the experiment, but when it occurs it takes place gradually, and, to all appearances, in relation to gravitational stimuli. The ear is, of course, an organ whose function it is to detect the direction of maximum gravitational attraction, and, should the supposition be verified that the righting effect is directed by gravitation, the ear-vesicle in Amphibia may be regarded as determining its orientation independently of the rest of the organism. Un-

[1] Kopeć, 1911, 1913.   [2] Seidel, 1929, 1931; Reith, 1932; Pauli, 1927.
[3] Ranzi, 1931.   [4] Ranzi, 1928.
[5] Streeter, 1906, 1914; Spemann, 1910.   [6] Ogawa, 1921.

fortunately, it has not yet been found possible to test the directional effects of gravity on the developing ear-vesicle by forcing the embryo to adopt abnormal positions, for the embryo invariably rights itself also, and explantation methods have not been applied to this interesting problem.[1]

§ 3

The principle of self-differentiation is further illustrated by experiments of tissue-culture, from which it emerges clearly that the cells of any particular tissue are permanently determined (except in so far as metaplasia may occur: see below). Mesenchyme, smooth muscle, heart-muscle, striped muscle, epithelium, endothelium, kidney-epithelium, and blood-corpuscles of adult birds and Mammals have been shown to preserve their specific character in a wide range of media, and experiments have now been conducted long enough to show that they can preserve them indefinitely. Fibroblasts of the fowl have been cultured *in vitro* for over 20 years (a much longer period than the maximum length of life of the fowl) and show unchanged characters and an unchanged rate of growth.

In many cases, particular characteristics assumed by a cell are a function of the environment or medium in which it finds itself. Epidermis which, like that of the chorio-allantoic membrane of the avian embryo, does not normally show keratinisation, may do so as a reaction to grafts of tissue placed upon it.[2] Under certain conditions of the medium, an apparent loss of specific characters, or dedifferentiation, may occur, and the tissue reverts to an undifferentiated type. Such dedifferentiation is, however, a reversible phenomenon. Cartilage-cells[3] or kidney-epithelium[4] may undergo dedifferentiation and grow as sheets of embryonic cells, but on restoration of the original conditions, the cells readopt the differentiated character typical of the tissue to which they belong. This may take place *in vitro*, or after interplantation subcutaneously under the wing of a young chick. Cartilage-cells, epithelial cells,

---

[1] A further interesting fact is that in those cases in which the ear-vesicle has been inverted and has failed to rectify its position completely, the resulting tadpoles have an altered sense of balance, which they show by swimming in abnormal attitudes and upside down (Spemann, 1906 A).

[2] Huxley and Murray, 1924.

[3] Strangeways, 1924.   [4] A. H. Drew, 1923.

and intestinal endothelial cells which had completely dediffer-
entiated *in vitro* were found to possess equally complete powers of
redifferentiation.

The various strains of cells differ not only in their structural
characters, but determined physiological differences may also be
observed between cells which are morphologically indistinguish-
able. Thus, strains of fibroblasts have been found differing from
one another in their nutritional requirements, and differing also
from epithelial cells and macrophages.[1] The differences show themselves in the rate of proliferation of the cells in any given medium,
and by specific reactions, such as cytolysis,[2] to certain induced
pathological conditions.

Tissue-culture methods have also thrown certain new light upon
the problem of differentiation. It has been found in the first place
that fibroblasts, isolated from different organs of the same embryo,
exhibit different growth-rates and other physiological characteristics
such as resistance to acidity and capacity to digest fibrin: these
differences appear to be persistent. For instance, fibroblasts
isolated from the skeletal muscle of a 17-day chick embryo have a
growth-rate nearly three times as high as that of fibroblasts from
the thyroid of the same embryo, and nearly ten times as high as those
from the heart.[3] A further and more surprising result is that comparable physiological differences exist between fibroblasts isolated
from the same organ of embryos of different age. For example,
fibroblasts from the skeletal muscle of the leg of the 17-day chick
embryo have a growth-rate about 60 per cent. higher than those
from the same tissue of 8-day embryos.[4] These differences continue to be shown even when the strains have been subjected to
marked environmental changes, and are returned to standard
conditions.

There seems no escape from the conclusion that the primitive
mesenchymatous tissue, from which the fibroblasts of the body are
derived, receives some impress affecting its physiological characteristics from the regions in which it happens to find itself, and this
impress changes with age. As regards the regions, the process is
doubtless an aspect of the self-differentiation which we have been

[1] Carrel, 1931.   [2] Horning, 1932.
[3] R. C. Parker, 1932 A.   [4] R. C. Parker, 1932 B.

considering: but the reactions of the fibroblasts to this process are purely passive; and if removed from the local influence, they simply retain the characteristics impressed up to the time of isolation. This we may regard as a new type of dependent differentiation: mesenchyme is predetermined to differentiate into fibroblast tissue, but the detailed characteristics of the fibroblasts are impressed from without.

It is of interest to note that many of the characteristics of tissues are dependent on specific physiological characters of the cells themselves. In tissue-cultures, fibroblasts form an irregular matted tissue; epithelial cells associate with one another in an orderly manner; amoebocytes remain separate and never form a compact tissue. These determined types of cell behaviour persist indefinitely *in vitro*.

Cells which normally form part of a more highly differentiated tissue possess and retain the type of behaviour which leads to the formation of such tissue. Thus, kidney cells can redifferentiate into kidney tubules,[1] and capillary cells can redifferentiate into capillaries *in vitro*.[2]

Under certain circumstances, however, it appears to be possible for cells to undergo a permanent and irreversible change in type and characters, comparable in its way to the changes observed in somatic mutations *in vivo*. This phenomenon, known as *metaplasia*, has from time to time been claimed to occur in many cases of regeneration, when it has been asserted that certain structures have been formed from cells of a different tissue. It is, however, often difficult if not impossible to be sure that undifferentiated and embryonic cells were not present, and that the differentiation of the structure in question did not proceed from them. This possibility seems to be excluded in the regeneration experiments performed on Nemertines. In these animals, there is a certain region at the anterior end of the body in front of the mouth, which contains no endodermal tissue at all. If such a piece be isolated, it will reconstitute itself into a complete worm, with an alimentary canal which quite certainly, therefore, is derived from cells of an entirely different tissue.[3]

---

[1] A. H. Drew, 1923.   [2] Lewis, 1931.
[3] Nussbaum and Oxner, 1910.

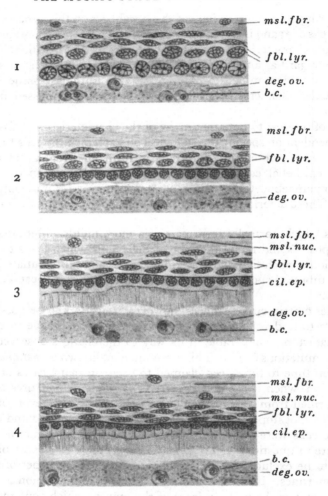

Fig. 101

Metaplasia of fibroblasts of *Pecten* into ciliated epithelium. A piece of ovarian tissue was grafted into the adductor muscle where a cyst was formed round it, lined by an epithelium formed of fibroblasts. 1–4, stages in the transformation of the fibroblasts into ciliated epithelium; 1, after 23 days; 2, 26 days; 3, 30 days; 4, 98 days. *b.c.* blood corpuscles; *cil.ep.* ciliated epithelium; *deg.vo.* degenerating ovarian grafted tissue; *fbl.lyr.* fibroblast layer; *msl.fbr.* muscle fibres; *msl.nuc.* nuclei of muscle cells. (From Gray, *Experimental Cytology*, Cambridge, 1931, after G. H. Drew.)

Metaplasia has been observed to occur as a result of certain graft-ing experiments. If a small piece of the ripe ovary of the Mollusc *Pecten* is grafted into the adductor muscle of another individual, the implant rapidly becomes surrounded by a layer of fibroblasts. The grafted tissue degenerates and is destroyed by phagocytosis, but the fibroblasts remain, forming the lining of a cyst containing the debris. After three weeks, the fibroblasts begin to take on the appearance of columnar epithelium, which eventually becomes ciliated.[1] It is almost impossible to believe that undifferentiated ciliated cells were originally present in the muscle, and we are ac-cordingly forced to regard this case as one of true metaplasia (fig. 101).

Tissue-culture experiments likewise provide evidence for meta-plasia. Monocytes which have been treated with filtered extracts of a particular type of tumour (the Rous sarcoma) become transformed into fibroblasts.[2] The crowding of the cells in the culture often produces the same effect, whereas various modifications of the medium fail to do so. The change into fibroblasts is of an adaptive nature, occurring when conditions are becoming impossible for the continued existence of monocytes.[3] This transformation may be permanent.[4] On the other hand, fibroblasts treated with plasma containing liver extract may become transformed into macrophages with all their physiological characteristics, which they now keep indefinitely.[2]

Cultivation of fibroblasts in a plasma medium which only permits of their slow growth may also induce metaplasia into macrophages. Here again, the change to the outwandering macrophage type is probably adaptive. Even Carrel's 20-year old strain of fibroblasts has been made to produce daughter-strains of macrophages in this way (fig. 102).

The rate of growth of the macrophages is markedly superior to that of their parent fibroblasts; they appear to retain their charac-teristics indefinitely[5] (fig. 103).

Lastly, it may be mentioned that the obscure changes which tissues undergo when tumours and cancers arise are of the nature of metaplasia. The morphological characters of the cells are lost to

---

[1] G. H. Drew, 1911.   [2] Carrel, 1931.
[3] Carrel and Ebeling, 1926.   [4] Fischer, 1925.
[5] R. C. Parker, 1932 c.

a greater or lesser extent, and these transformations are accompanied by irreversible physiological changes, as a result of which the tumour cell becomes capable of glycolysis (or fermentative in-

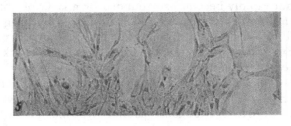

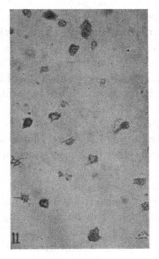

Fig. 102

Microphotographs of living cultures of chick fibroblast tissue. Above, typical fibroblasts (after 103 days' cultivation and 13 passages). Below, macrophages derived by metaplasia from a pure culture of fibroblasts (12 passages as pure fibroblast culture in optimum medium, then 29 days' growth in an unfavourable medium containing no embryonic tissue juice). (From R. C. Parker, *Journ. Exp. Med.* LVIII, 1932.)

tramolecular respiration) and is less dependent on normal aerobic respiration.[1]

These examples will be sufficient to demonstrate the real exist-

[1] See Warburg, 1926.

ence of what has here been called the mosaic stage of differentiation and development. This is of great theoretical interest, since it shows that the capacity for regulation, which has been regarded by some authors[1] as a universal property of life, does not hold at all for an important stage of development, universally passed through by all higher animals.

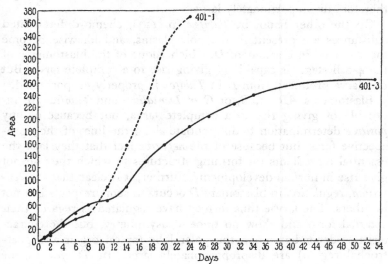

Fig. 103

Physiological changes accompanying metaplasia. Solid line throughout (401–3), growth-curve of a flask culture of fibroblasts from embryo chick muscle; although grown in an unfavourable medium (lacking embryo extract), the culture showed no metaplasia. 104–1, growth-curve of sister-culture under similar conditions, in which metaplasia fibroblasts to macrophages occurred on the eighth day. The subsequent growth (dotted line) was much more rapid. (From R. C. Parker, *Journ. Exp. Med.* LVIII, 1932.)

§ 4

Turning now to the question of the time of onset of the mosaic stage of development, we must refer to the classical experiments on so-called mosaic-eggs, referred to in Chap. v. They serve as a further illustration of the principle which is here under discussion; in their case, the onset of chemo-differentiation has merely been transferred to an earlier stage of development.

[1] Driesch, 1921; J. S. Haldane, 1929.

It will be remembered that in *Dentalium* (p. 110), chemo-differentiated substances are present in the polar lobe and become incorporated in blastomere *D*, with the result that of the cells of the 4-cell stage, only blastomere *D* is able to produce a complete larva, but that the structures to which the polar lobe gives rise (apical organ and post-trochal region with mesoderm) are full-sized, and therefore disproportionately large.

On the other hand, in *Tubifex* (p. 113), chemo-differentiated substances are present in the pole-plasms, and likewise become incorporated in blastomere *D*, which, alone of the blastomeres of the 4-cell stage, is capable of giving rise to a complete larva. But this larva produced from *D* in *Tubifex* is properly proportioned.

Blastomeres *AB*, *A*, *B*, or *C* of *Dentalium* and *Tubifex* are incapable of giving rise to a complete larva, not because of any *positive* determination to differentiate along the lines of their prospective fates, but because of the *negative* fact that they lack the essential ingredients for forming structures to which they do not give rise in normal development. Further, it is clear that in *Dentalium*, regulation in blastomere *D* occurs in some respects but not in others. The larvae thus formed have regulated as regards their external form and show no trace of asymmetry, but the characteristics dependent upon the polar lobe (apical organ and post-trochal region) are disproportionately large. In *Tubifex*, on the other hand, regulation in blastomere *D* seems to be complete. If we are to make a conjecture as to the meaning of this distinction, it would be that chemo-differentiation is more precocious in *Dentalium* and results in a complete determination, quantitative as well as qualitative, of the organ-forming substances contained in the polar lobe. The state of affairs in *Tubifex*, on the other hand, is more like that of the early half-gastrula of *Triton*, in which quantitative regulation of the neural folds is still possible (see p. 239).[1]

The most striking demonstration of the presence of organ-forming substances is that of the Ascidians, already referred to (pp. 119, 123). The fertilised egg possesses a yellow crescent and

---

[1] In the absence of experiments involving the removal of the pole-plasms of *Tubifex*, comparable to those in which the polar lobe of *Dentalium* is cut off, it is impossible to rule out the suggestion made by Morgan (1927, p. 379) that the pole-plasms of *Tubifex* may be indices of some underlying peculiarity of organisation, rather than organ-forming substances.

a clear crescent, and a large amount of yolk. The first cleavage takes place in a plane passing through the centre of the yellow and clear crescents, and in each blastomere the clear cytoplasm displaces the yolk from the animal hemisphere so that the latter now occupies the vegetative hemisphere. Immediately opposite the yellow crescent, and therefore marking the antero-dorsal side, a third region termed the grey crescent makes its appearance, containing slaty-blue coloured yolk. Eventually, the yellow crescent shows a darker and a lighter coloured region, and there are then at least six different organ-forming substances, which become sorted out between the various blastomeres during cleavage. The determinations which these substances represent are the following:[1]

Clear cytoplasm ...	Ectoderm
Dark yellow cytoplasm	Muscle
Light yellow cytoplasm	Mesenchyme
Yolk region	Endoderm
Slaty-blue	Notochord and neural plate

This distribution can be seen in normal cleavage; the causal connexion between the substances and the organs to whose rudiments they are distributed has been proved by the experiments involving killing and disarranging of blastomeres, referred to in Chap. v (pp. 97, 123).

As mentioned in Chap. v (p. 124), the visible inclusions in the various regions of cytoplasm, such as mitochondria, yolk, etc., are not themselves organ-forming substances, but merely cytological indices of the organisation of the egg.[2] In many other cases particular regions of the cytoplasm may be distinguished by their pigmentation, but it can in most cases be shown that the visible or coloured elements do not represent any qualitative determination. The egg of the sea-urchin *Arbacia* contains fat, yolk, pink granules, and clear cytoplasm, and as these materials differ in their specific gravities, they can be disarranged by the centrifuge. When the eggs of *Arbacia* are centrifugalised for 5 minutes at 10,000 revolutions per minute, the contents are stratified into four zones, quite regardless of the original egg-axis, which, it is found, may come to occupy any position in the centrifuge tube. Centripetally (with reference

[1] Conklin, 1905, 1906, 1924, 1931.   [2] Conklin, 1931.

to the centrifuge), the fat forms a layer, and beneath this, in succession, there are layers of clear cytoplasm, yolk, and pink granules, the latter occupying the centrifugal pole[1] (fig. 104).

But in spite of this complete restratification of the visible egg-contents along a new axis, the original axis of polarity has not been affected. The position of the original axis is indicated by the funnel in the jelly which marks the position of the original animal pole of the egg: the micromeres are formed and invagination begins at the opposite (original vegetative) pole, regardless of the visible contents

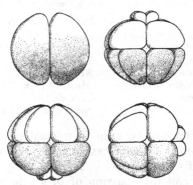

Fig. 104

Persistence of the primary axis in sea-urchin (*Arbacia*) eggs in spite of the rearrangement of visible substances in the cytoplasm. After centrifuging, the egg becomes stratified with fat at the centripetal pole, then clear cytoplasm, then yolk with increasing amounts of pigment. The first cleavage (left top) is always at right angles to the stratification, but the micromeres are always formed at the vegetative end of the original axis, whether this coincides with the centripetal pole of the centrifuged egg (top right), its centrifugal pole (bottom left) or its side (bottom right). (Redrawn after Morgan, *Experimental Embryology*, Columbia University Press, 1927.)

which happen to be situated there.[2] Development continues along the lines of the original axis and is normal, from which it follows that the various substances which have been disarranged are not organ-forming. (See also p. 69 as regards determination of bilaterality by centrifuging.)

Similar results have been obtained from centrifuge experiments on eggs of other animals. The egg of the Lamellibranch Mollusc *Cumingia* after centrifuging shows a stratification into four zones:

---

[1] Lyon, 1906; Morgan and Lyon, 1907.   [2] Morgan and Spooner, 1909.

an oil cap, a clear zone, a yolk field, and a zone of pigment; this stratification may bear any relation to the original axis of polarity. Nevertheless, normal larvae develop, regardless of the distribution of the visible contents.[1] The same is true of the egg of *Chaetopterus*, and the polar lobe may contain any of the visible materials without influencing normal development.[2]

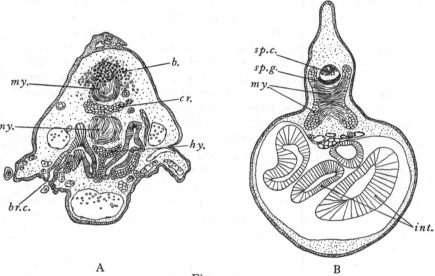

A   B

Fig. 105

Section through a frog tadpole (external gill stage) developed from an egg centrifuged for 5 minutes at about 1500 revolutions a minute. A, Through the head region. The brain (*b.*) is represented by a degenerate mass of pigment cells. The cranium (*cr.*) is rudimentary; *hy.* hyoid; *br.c.* branchial cleft. B, Through the trunk and the spinal cord (*sp.c.*); the distribution of cells is abnormal, and the spinal ganglia (*sp.g.*) are fused below it; *int.*, intestine. In both, the myotomes (*my.*) are fused in the middle line. (After Jenkinson, *Quart. Journ. Micr. Sci.* LX, 1915.)

By way of contrast with specific materials of the organ-forming type as seen in *Dentalium* and *Styela*, the preformed substances such as yolk and fat to be found in many eggs thus appear to play the part of raw materials only. Their importance as regards normal development is perhaps best illustrated by the experiments of centrifuging the eggs of the frog. When a frog's egg is thus treated,

---

[1] Morgan, 1910.   [2] Lillie, 1906.

its ability to rotate within its membranes and the greater weight of its yolk cause it to orientate itself in the centrifuge tube in such a way that the animal pole is centripetal and the vegetative centrifugal. The result of centrifugalisation is therefore an intensification of the stratification normally found along the primary egg-axis. The yolk is concentrated more densely than ever at the vegetative pole; above it is a layer of clear cytoplasm, and the animal pole is occupied by a layer of fat. If the centrifugalisation is heavy, development proceeds a certain way and then stops, largely owing to mechanical difficulties arising from the inertia of the abnormally dense mass of yolk. But if the centrifugalisation is light, development is normal except for the fact that the structures of the head contain an abnormally large amount of fat. The cells of the brain may contain many times the normal quantity of fat, but nevertheless the differentiation of the brain and the development of its form are normal. Similarly, some of the regions of the trunk can develop normally although their cells contain less than the normal quantity of yolk. It is obvious, therefore, that yolk and fat are only raw materials.[1] When, however, the amount of fat at the animal pole exceeds a certain proportion, normal development is impossible. Vacuolisation is the first sign, but in more extreme cases the brain and other head-structures are reduced to a small degenerated mass of cells[2] (fig. 105).

We may illustrate the part played by the yolk and fat in the frog's egg with the help of an analogy. The construction of a conservatory is of course conditioned by the availability of the necessary raw materials—wood and glass. There is an optimum proportion in which these materials should be present in order to give the best results, but this proportion may be altered in either direction up to

---

[1] Jenkinson, 1915.
[2] With slightly heavier centrifugalisation, curious malformations appear in the trunk region. The myotomes are frequently fused together beneath the nerve tube, with consequent absence of the notochord. The spinal ganglia may also be fused ventrally beneath the nerve tube. The latter has an abnormally thick floor and thin roof, with the white matter concentrated ventrally. From other experiments (see Chap. XI, p. 375), it is known that these effects are associated with notochord absence, and it is probable therefore that here absence of the notochord is the cause of the other observed changes, but the cause of this primary change remains for the present obscure. Possibly the centrifugalisation has resulted in an alteration in the composition of the organiser region: further research is needed on the question.

a certain point without preventing the construction. It may have too much wood and not enough glass, or too much glass and not enough wood, but, provided that the disproportion does not exceed a certain degree, it will still be a conservatory. But if the amount of glass be too great for the wood, the construction is mechanically impossible. The yolk and fat in the frog's egg may fancifully be compared with the wood and glass in the conservatory.

Other evidence of a similar nature is provided by centrifuge experiments on the eggs of echinoderms, in which centrifugation has been continued until the egg has separated into two or even four (unequal) portions along the direction of centrifugal force (which of course may bear any relation to the original polarity of the egg). The fragments differ considerably in colour and the type of their contained granules. We may call these halves A and B, and the quarters A_1, A_2, B_1, B_2, in order from centripetal to centrifugal region. In *Sphaerechinus granulatus*, a centrifugal half (B) or either of the two centrifugal quarters (B_1 or B_2) is capable of producing plutei. A centripetal half (A) on the other hand never goes further than the blastula, and the same is true for the most centripetal quarter (A_1). The other quarter (A_2), however, may in some cases produce a pluteus. We may provisionally assume that the fragments incapable of pluteus-formation contain an excess or defect of certain raw materials, as in the frog experiments described above. Presumably the excess substance responsible for failure to develop in the A halves was all contained in the A_1 quarters, thus permitting the A_2 quarters to develop. Results similar in principle have been obtained for several other genera: in one case (*Tripneustes esculentes*) the conditions are reversed, the A (centripetal) pieces being capable of fuller development.[1]

§ 5

The determination and localisation of organ-rudiments is revealed sooner or later by the presence of chemo-differentiated material or morphogenetic substances in certain places which constitute what may be called *fields*, or areas of differentiation of organs. Within the fields the presumptive rudiments become determined by *progressive* chemo-differentiation. As an illustration of this important

[1] Harvey, 1933.

principle, we may turn to the phenomena presented by some of the presumptive organ-regions in Amphibia, beginning with the limbs.

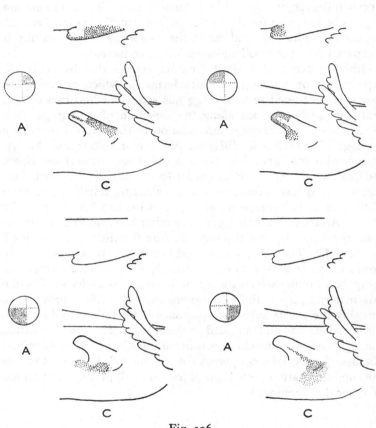

Fig. 106

Diagram of the fate of the four quadrants of the fore-limb field in *Amblystoma*. The limb-disc is shown at *A* in each case, with one quadrant stippled. The dotted lines intersect at the point of maximum limb-forming potency. At *C* in each case is shown the young limb, viewed dorsally (above) and laterally (below), showing the portions derived from the stippled quadrant. (From Swett, *Journ. Exp. Zool.* xxxvii, 1923.)

In *Amblystoma*, the presumptive fore-limb area or field occupies a discoid zone on the side of the body, extending from the anterior margin of the third trunk segment to the middle of the sixth. The limb potencies are restricted to the mesoderm of this region, the ectoderm not being predetermined in any way.[1] The first important point to notice is that within this limb-disc there is no definite spot or area which is necessarily destined to form a limb in normal development: all regions of the field have the power of forming a limb, and the extent of the field is greater than the region which actually does form the limb in normal development.[1] The limb-field is already determined at the middle gastrula stage.[2]

The limb-forming potencies are highest in a subcentral region of the field, situated near to its anterior and dorsal margins, and grade away from this.[3] A normal limb can be formed from half the

Fig. 107

Polarisation of the limb-field. Axolotl in which a limb-disc from the right side of the body has been grafted on to the same side, a little way behind the normal limb, the correct side out but with the antero-posterior and dorso-ventral axes reversed. It has developed into a limb (*TR*) with correct dorso-ventral relations, but with the preaxial border facing the tail of the larva; consequently it possesses left-handed asymmetry. (From Harrison, *Journ. Exp. Zool.* XXXII, 1921.)

limb-field, either from what is left *in situ* after removal of half, or from a half grafted elsewhere: a single field may therefore give rise to two perfect limbs. Conversely, a single perfect limb can be formed from two half-rudiments grafted together (provided only that their antero-posterior axes are coincident, see below, p. 224 and also pp. 357, 418). From a very early stage, therefore, the limb-field is irreversibly determined as a whole to give rise to limb-tissue, but there is as yet no regional determination within the field, of the constituent parts of the future limb. In addition to the fact that a limb will arise somewhere within the limits of the field, there is only one additional determination, and that is

[1] Harrison, 1918; Detwiler, 1918.   [2] Detwiler, 1929 A; 1933 A.
[3] Swett, 1923.

that the preaxial border of the limb (marked by the first digit and radius in the fully developed limb) will arise from the anterior portion of the limb-disc. Although the limb-field is regionally still undetermined, it is polarised along an antero-posterior axis from the first moment at which its existence can be detected: this is proved by grafting limb-discs in abnormal orientations [1] (figs. 107, 108, 173).

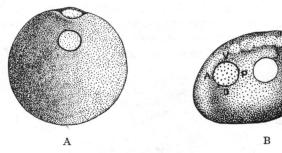

A   B

Fig. 108

Limb determination in *Amblystoma*. A, Middle gastrula stage showing presumptive limb area which was removed and grafted with reversed orientation into a neurula (B), where it developed into a limb with left asymmetry, though on the right side, like that shown in fig. 107. This proves not only that the limb is determined at the middle gastrula stage, but also that its antero-posterior axis is already determined. (From Detwiler, *Journ. Exp. Zool.* LXIV, 1933, figs. 2, 3.)

The hind-limb field in *Amblystoma* extends from the level of the sixteenth to the eighteenth trunk segments inclusive, and shows properties similar to those of the fore-limb field.[2] Its determination and differentiation takes place later than that of the fore-limb field.

---

[1] A common occurrence when portions of limb-discs are grafted is the fact that they give rise to reduplications, i.e. monstrous double or even treble limbs are formed connected with one another at some point along their length. This in itself is merely another example of the fact that the limb-area is as yet only a field and not a regionally determined rudiment. But these reduplications are of interest from another point of view, for the reduplicated member is as a rule a mirror image of the original member. They therefore supply an illustration of Bateson's rule, which may be formulated as follows: (1) the long axes of reduplicated structures lie in the same plane; (2) two reduplicated limbs are mirror images of one another about a plane which bisects the angle between the long axes of the members, and which is at right angles to the plane of these axes. The detailed explanation of reduplication and mirror-imaging has given rise to considerable controversy. See Harrison, 1921 A; Przibram, 1924; Mangold, 1929 A.   [2] Stultz, 1931.

By this stage, too, the growth-coefficient of the limb relative to the body has also been determined, as is shown by heteroplastic experiments in grafting limbs between slow-growing and fast-growing species of *Amblystoma*. Limbs of the fast-growing species on the body of the slow-growing one become disproportionately large, and *vice versa*.[1] (See fig. 203, p. 421.)

It is only at later stages that the unitary limb region, which forms one of the major pieces in the mosaic of the whole organism, itself becomes converted into a mosaic of invisibly determined sub-regions. The precise time of onset of this stage varies in different forms. In *Amblystoma punctatum* it appears to be reached when the visible limb-bud has attained a markedly conical form. The organism is then a larva with well-developed external gills and tail. In *Triton taeniatus*, on the other hand, it appears to set in relatively earlier, in the tail-bud stage.[2] In *Triton*, the limb also develops relatively earlier than in *Amblystoma punctatum*, but there is no correlation between time of determination and time of development, for in *Amblystoma tigrinum*[3] determination sets in earlier but development does not occur until later than in *Amblystoma punctatum*.

When the stage of regional determination of subregions within the field has been reached, division of the rudiment will no longer result in the formation of two limbs by regulation, but each portion will give rise to a partial structure. Progressive chemo-differentiation has taken place, and within the main limb-field a secondary mosaic has been formed, each region of which, however, is still indefinitely determined and therefore capable of regulation.

The analysis of these late stages has been undertaken in the limbs of the embryo chick. By grafting portions of the limb-bud of a 4-day chick on to the chorio-allantois of another egg, it is found that if the limb-bud is divided into pieces by cuts at right angles to its future long axis, the proximal piece differentiates into a perfect femur, the next piece into a perfect tibia and fibula, and the distal piece into a perfect foot. It is important to note that even the structure of the joints appears to be predetermined in almost all its details.[4] (See figs. 109, 111.)

---

[1] Harrison, 1924 A; Huxley, 1932.   [2] Brandt, 1924.
[3] Ruud, 1926.   [4] Murray, 1926.

It is impossible to imagine that the cuts which were made passed exactly in each case between the limits of the zones allocated to thigh, shank, and foot, and it is necessary to conclude that these

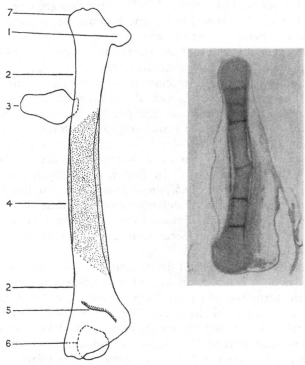

Fig. 109

Mosaic determination and partial regulation within the limb-rudiment of the chick. Differentiation of a small basal fragment of a very early (4-day) left hind-limb bud, grafted on the chorio-allantois of another egg after 5 days. Right, microphotograph of entire graft, in longitudinal section. The connexion with the chorio-allantois is seen on the right: the graft has differentiated into a femur, 7·5 mm. long, mesenchyme, and some muscle-fibres (right bottom). Left, re-construction of skeletal elements. The curve of the bone is in the same direction as in a normal left femur. 1, head; 2, shaft; 3, ectopic fragment of pelvis; 4, sheath of perichondral bone; 5, attachment of muscles (on far side); 6, patella; 7, tro-chanter. Being the basal region, the graft has formed only basal structures; there has however been some intra-regional regulation, leading to the formation of a femur complete at either end. (From Murray and Huxley, *Journ. Anat.* LIX, 1925.)

segments or constituent parts of the limb are roughly determined at varying levels down its length, but that they are determined *only* roughly, and their frontiers appear to overlap. There is no doubt that a cell which in one experiment forms part of the thigh, would, in another experiment with the cut in a slightly different

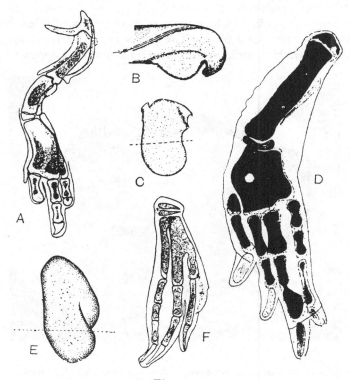

Fig. 110

Mosaic determination within the hind-limb rudiment of the chick. A, Complete hind-limb developed from a chorio-allantoic graft of a whole limb-bud in the stage shown at B. A graft of the distal half of the limb-bud shown at C resulted in a distal half-limb (D). A still smaller distal region (E) produced only a foot (F). In D and F, the sub-regions (shank, foot) are complete. (After Murray, from Wells, Huxley and Wells, *The Science of Life*, London, 1929.)

place, form part of the shank. The cuts must have roughly separated the sub-zones from one another, and each sub-zone, though irreversibly determined to give rise only to its own segment, is still capable of regulation to give a *whole* sub-zone. In a similar way,

15-2

some of the original presumptive limb-area does not give rise in normal development to limb, but merely to flank, skin, and muscle.

Similarly, if the 4-day leg-bud be divided longitudinally, so as to separate preaxial and postaxial halves, the fragment usually forms only those digits which it would have produced if left *in situ* and either a tibia or a fibula. The femur rudiment regulates to a miniature whole in each portion. The limb-bud thus appears to be

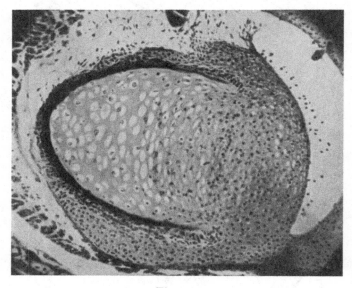

Fig. 111

Self-differentiation of the femur-head joint, without function. The femur shown in the section belonged to a limb from which all nerve supply was excluded. In spite of the fact that the limb never functioned at all, the cartilage cells and fibres of the femur-head show the normal configuration. (From Hamburger, *Arch. Entwmech.* CXIV, 1928.)

a thorough-going mosaic of predetermined but slightly overlapping regions (fig. 110).

The recognition of the existence of organ-fields, i.e. regions possessing a general determination for the production of certain structures, and undergoing progressive regional specification of detail, constitutes one of the major advances made in the analytical study of development. The organ-fields resemble the whole organ-

ism in the pre-mosaic stage, in combining a general determination
with an epigenetic mode of development.

The arm-field and the leg-field are each of them self-differentiating in a general way, and will produce only an arm or a leg, as the
case may be.[1] But certain details of the development of the field
are not independent of the im-
mediate environment, i.e. the remainder of the organism, and
in particular, its gradient-field.
We have seen that the arm-field
of *Amblystoma* is polarised from
its inception. The leg-field also
has some polarity from the start,
as is revealed when it is grafted
heterotopically in an abnormal
orientation: the original antero-
posterior axis of polarity of the
disc persists and becomes that of
the limb. After rotation of the
leg-disc at the original site (ortho-

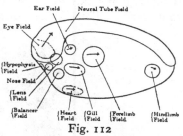

Fig. 112

Diagram of an amphibian neurula to
show the approximate localisation of
the main regional fields as yet discovered by experimental analysis. The
arrows indicate that the fields are
known to be polarised from their first
appearance. (Original.)

topic), however, it acquires a new polarity in relation to that of the
body as a whole, i.e. the leg-disc has a new antero-posterior axis
impressed upon it.[2] It may be hazarded that this result is due to the
size of the disc rotated not being large enough to cover the whole
limb-field. Later, however, the antero-posterior axis of the leg-disc
is entirely fixed, but the dorso-ventral axis is not. This is also the
condition of the arm-disc at the earliest stage studied (middle
gastrula). Rotation experiments with discs at this stage show that
the determination of the dorso-ventral axis is dependent on the body
of the host. Later still, the dorso-ventral axis of the discs is also
determined, and the limb-disc has by then proceeded far on the
way to becoming a mosaic of determined subregions.

The fact that a field, although qualitatively determined, is capable
of being quantitatively influenced in its development, is well shown
by those experiments in which an arm-disc of an *Amblystoma*
neurula is grafted in place of an extirpated leg-disc. The graft

[1] Harrison, 1918; Ruud, 1929, 1931; Stultz, 1931.
[2] Stultz, 1931.

develops into an unmistakable arm, but it may have five digits. This is the normal number of digits of the amphibian leg, while the normal arm only has four [1] (fig. 113). In this case, we may suppose that the larger nerve-trunk supplying the leg exerts a trophic effect on the growing rudiment, leading to a condition in which the distal region tends to be meristically divided into five instead of four digits. This is the converse of the results obtained after

Fig. 113

Modification of the arm-bud when grafted into the leg-region. The amphibian arm ends in four fingers (1, 2, 3, 4); the leg in five toes (1, 2, 3, 4, 5). An arm-bud of a white axolotl, grafted into the leg-region (after extirpation of host's leg) of a black axolotl, develops into a typical arm (*Tr*), except that it possesses five fingers. (From Ruud, *Arch. Entwmech.* CXVIII, 1929.)

inducing subnormal development in the centres of the mid-brain by extirpating the rudiments of fore-limb or eye: [2] in these cases the hind-limb was usually malformed and under-developed, and frequently possessed only four or even three toes. Curiously enough, totally denervated hind-limb rudiments, though their growth is reduced, are not malformed, and develop the normal complement of five toes. [3] It may be supposed that abnormal conditions in

[1] Ruud, 1929.   [2] Dürken, 1925, 1930.   [3] Hamburger, 1928.

the mid-brain exert a specific "negative" trophic effect in the hind-limbs, but the question cannot be regarded as settled. (See p. 430.)

Before leaving the limb-field, there is a further point which requires consideration. As early as the neurula stage, the mesoderm of the limb-field is found to be self-differentiating. At the same time, it is clear, if only from the fact that the limbs are symmetrically placed with regard to the plane of bilateral symmetry, that the localisation of limb-forming potencies is in some way dependent on something else, which we may at present call the general gradient-field of the organism (see Chap. ix). We are ignorant as to the causes which are normally operative in calling forth these limb-forming potencies in ordinary development, but we do know that these potencies may be experimentally called forth by a variety of agents. Grafted ear-vesicles,[1] celloidin beads, or the free termination of various nerves deflected so as to underlie the tissues of the field,[2] all result in the formation of limbs. The quality of the structure produced is therefore a specific property of the field, the activities of which may be "released" by a variety of non-specific agents. As we shall see, the same is true of other fields, and probably of all.

§ 6

A curious contrast to the regulative capacity of the limb-field is the mosaic nature of the rudiment of the shoulder-girdle. This rudiment consists of three centres of chondrification, representing the coracoid, precoracoid, and scapular elements, but they are not all contained within the $3\frac{1}{2}$-somite limb-disc, for when the latter is grafted it will give rise to a shoulder-girdle of about one-third normal size.[3] Conversely, after extirpation of a limb-disc, portions of the shoulder-girdle rudiment are left *in situ*.[4] Removal of, or grafting of, portions of the limb-girdle rudiment at the early tail-bud stage in *Amblystoma* results in the development of partial structures, and regulation to form a complete girdle does not take place, while regulation does take place to form a perfect limb.[3]

In some experiments on *Amblystoma* in which limb-discs

---

[1] Balinsky, 1925, 1926, 1927; Filatow, 1927.   [2] Detwiler, 1918.
[3] Locatelli, 1925; Guyénot and Schotté, 1926; Bovet, 1930.
[4] Harrison, 1918.

($3\frac{1}{2}$ somites in diameter) were grafted after rotation, it was found that the limb occasionally underwent a rotation at the shoulder-girdle, so as to become correctly oriented.[1] When the disc was rotated through an angle up to 235°, the limb might right itself by a rotation in the reverse direction: on the other hand, if the disc was rotated through 270°, the limb might complete the circle by rotating the remaining 90° in the same direction.[2]

It appears that the rotation of the limb is in some ways dependent on the shoulder-girdle. If the rotated disc is only $1\frac{1}{2}$ somites in diameter, it contains none of the girdle-rudiment, and no regulatory rotation takes place. In the case of rotated discs of the normal diameter of $3\frac{1}{2}$ somites, parts of the shoulder-girdle are formed, and regulatory rotation may take place. If the rotated disc is 5 somites in diameter, a complete girdle is formed and the limb conforms to it, without regulatory rotation.[3] Lastly, if in a graft 5 somites in diameter a $3\frac{1}{2}$-somite disc is separated from a peripheral ring, and then both central disc and peripheral ring are rotated independently, the limb undergoes postural regulation with reference to the ring.[4] Apparently, therefore, the portions of the girdle whose rudiments lie outside the $3\frac{1}{2}$-somite disc but within the 5-somite ring, act as determining factors on those portions of the girdle whose rudiments are included within the $3\frac{1}{2}$-somite disc. The girdle then brings about the rotation of the limb, but in a manner which is still obscure.

§ 7

Turning now to other examples of fields, we may take that of the amphibian ear. This occupies a region of ectoderm on each side of the head, behind the eye, and must in some way be dependent on the organiser, since dorsal lip grafts are capable of inducing the formation of ears. Here again, it is found that the field is more extensive than the normal presumptive rudiment, for if a portion of the presumptive ear-area be removed at the early neurula stage in *Rana nigromaculata*, a normal (though smaller) vesicle is formed, and this can be shown to arise from the neighbouring cells which have closed over the wound, though these would normally have

[1] Harrison, 1921 A.   [2] Nicholas, 1924 B.
[3] Nicholas, 1926.   [4] Nicholas, 1925.

given rise to epidermis.[1] This shows that more cells are capable of ear-formation than normally exert this capacity.

This is confirmed for Urodeles by experiments on *Amblystoma*. At later stages, when the rudiment has invaginated to form the ear-vesicle, the power of ear-formation is lost by the neighbouring epidermis,[2] for if the vesicle is extirpated it is not regenerated.

Just as in the case of the limb-field, the ear-field very soon shows a polarisation. If a piece of the ear-area of *Rana nigromaculata* (at the stage when the rudiment is just thickened) is rotated through 180°, the auditory vesicle which sub-sequently develops is reversed. Further, a piece of the ear-area of one side grafted on to the opposite side of the body develops with the asymmetry of its side of origin.[1] This shows that the rudiment was already determined as regards two at least of its axes.

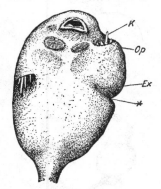

Fig. 114

Ventral view of larva of *Bombinator* in which the ectoderm of the gill-region on the left side had been rotated through 180°. * limit between rotated and normal ectoderm; *Ex*, position of fore-limb; *K*, gills; *Op*, operculum. (From Braus, *Zeitschr. f. Morph. u. Anthrop.* XVIII, 1914.)

The gill-region in Amphibia also constitutes a field in the ectoderm of the embryo. At the early neurula stage in embryos of *Amblystoma*,[3] *Rana fusca* and *esculenta*, and *Bombinator*,[4] rotation of a piece of the gill-area through 180° is followed by development of the gills (and operculum in Anura) in reversed orientation. This shows that the field is polarised along an antero-posterior axis. At the same time, the fact that at this early stage it is still only a field and not a spatially and regionally determined rudiment is shown by the capacity of two rudiments grafted together to regulate and give rise to a single normal set of gills (provided that the antero-posterior axes coincide)[3] (fig. 114).

Turning now to the heart, it is found in embryos of *Bombinator* at the neurula stage that if the presumptive heart-area (which occupies a region of the mesoderm) is extirpated, a heart is formed from

[1] Tokura, 1925.   [2] Kaan, 1926.
[3] Harrison, 1921 B.   [4] Ekman, 1913, 1922.

neighbouring regions:[1] it may therefore safely be concluded that the heart-field is more extensive than the actual presumptive heart

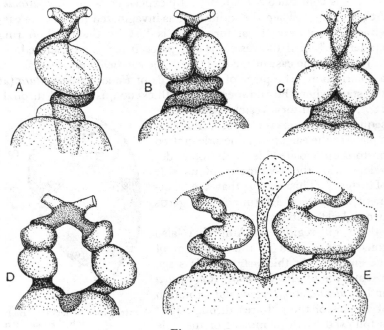

Fig. 115

Experiments on the development of the heart in *Bombinator pachypus*. A, Single normal heart formed from an enlarged rudiment, that of another embryo having been added in the mid-ventral line: the parts arising from the graft shown dark. B, C, Partially doubled hearts formed after grafting foreign tissue (pharyngeal wall) into the mid-ventral line; ventricle and bulbus show duplication; the graft has been used up in the formation of the heart in B, but in C a portion remains as undifferentiated material. D, Heart duplication almost complete, as a result of grafting a large piece of foreign tissue in the mid-ventral line; most of the graft has remained undifferentiated, and only a small portion (shown dark) has contributed to the heart in front and behind. E, Complete duplication as a result of grafting a piece of foreign tissue which has remained undifferentiated; the right-hand member shows *situs inversus*. (Redrawn after Ekman, *Arch. Entwmech.* CVI, 1925.)

region. Heart-forming potencies decrease with increasing distance from the normal presumptive heart region[2] and at the neurula

[1] Ekman, 1921.   [2] Ekman, 1925.

stage, a piece of the heart-field can be rotated through 180° and still give a normal heart, but this is no longer possible at later stages such as the tail-bud.[1] As in the case of the limb, ear, and gill-fields, therefore, the heart-field is polarised along an antero-posterior axis from an early stage.

A normal heart can be formed from a longitudinal half-rudiment; a single rudiment, split lengthwise, can be made to give rise to two or even three hearts; and two rudiments grafted together at

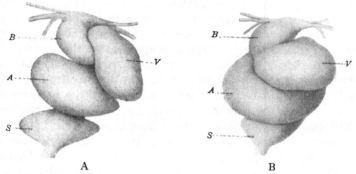

A   B

Fig. 116

Power of regulation of the heart-field, in *Bombinator*. At the neurula stage, the left half of the heart-field was extirpated; the right half, remaining *in situ*, has regulated to form a complete heart, with sinus venosus (*S*); atrium (*A*), ventricle (*V*), and bulbus (*B*). The histological differentiation of the parts is normal. Certain details of morphological differentiation are, however, abnormal; the ventricle projects to the left, and its long axis may be longitudinal (as in A) or transverse (as in B). Figures taken 14 days after operation. (From Stöhr, *Arch. Entwmech.* CXII, 1927.)

the neurula stage can regulate to form one normal heart, provided that both the antero-posterior axes are similarly oriented.[2]

All these results have been confirmed for Urodela by experiments on *Amblystoma*.[3]

The epidermis itself may be regarded as a large field, the determination of which is characterised not so much by any positive differentiation (for this is comparatively slight), but by the progressive incapacity to differentiate into other structures, e.g. lens. Nevertheless, the epidermis possesses a polarity, and this is ex-

[1] Stöhr, 1925.   [2] Ekman, 1924.
[3] Copenhaver, 1926.

pressed in amphibian embryos by the direction of beat (mostly antero-posterior) of the cilia which it bears. The cilia arise at the early neurula stage, and if a piece of epidermis of *Amblystoma* is rotated through 180° and replanted at this stage, the cilia beat in the normal direction. If, however, the epidermis is rotated at the late neurula stage, its polarity is then fixed, and the cilia beat in the reversed direction.[1]

The so-called balancer, present in some Urodela larvae (*Triton, Diemyctylus, Amblystoma punctatum,* but absent or extremely rudimentary in *Amblystoma tigrinum*[2]), is an organ of attachment in the form of a cylindrical projection of ectoderm with a mesenchymal core. Balancer-forming potencies occupy fields in the ventral ectoderm of the head, beneath the eyes; they are at a maximum at a central point, and decrease with increasing distance from it.[3] If a part of the rudiment is extirpated at an early stage, a balancer will be formed from the neighbouring regions.[4] A balancer rudiment grafted into other positions is self-differentiating, and induces the formation of the mesenchymal core from host-tissue. A rudiment from *Amblystoma punctatum* can be grafted on to *Amblystoma tigrinum*, or even on to the anuran *Rana sylvatica*, and develop into a balancer with induced core, although these hosts normally possess no balancer. As mentioned in Chap. VI (p. 177), a single balancer-field can give rise to as many as four balancers.[5] (See also p. 327.)

The fully formed balancer has radial but not bilateral symmetry, and it does not appear that the balancer-field is polarised. The facts already recorded in Chap. VI (p. 177), viz. that balancer-forming potencies can be evoked by neural crest cells, neural fold cells, and fore-gut-wall cells, even belonging to Urodeles or Anura which normally possess no balancer, serve as a further illustration of the principle enunciated above (p. 231), that the quality of the structure produced depends on intrinsic properties of the field, and not on specific stimuli of the releasing mechanism. At the same time, the fact that tissue from species which possess no balancer is capable of evoking balancer-forming potencies, shows that the absence of a balancer in these species is due to the absence of such a field in

---

[1] Twitty, 1928.   [2] Nicholas, 1924 A.
[3] Harrison, 1925 B.   [4] Bell, 1907.
[5] Raven, 1931 A.

their tissues. This has been verified by experiments in which epidermis from *Amblystoma tigrinum* is grafted into the appropriate position on embryos of *Amblystoma punctatum*, and is found to be incapable of forming a balancer.[1]

Evidence regarding the existence of a nose-field is provided by experiments on *Rana temporaria* in which the nose-rudiment is extirpated at a stage prior to the formation of a nasal pit; a nose is nevertheless formed from neighbouring tissue, which grows over to cover the wound. More distant epidermis will, however, not do this.[2] The nose-field is therefore more extensive than the presumptive nose-rudiment, and if a large area, representing the entire nose-field, is extirpated, no nose is formed. In some cases, the nose-field gives rise to a single median nasal organ in place of the normal paired two: this monorhiny is associated with and due to the same causes as cyclopia (see Chap. IX, p. 348).

§ 8

Another case in which the presumptive zones or fields of different organs appear to overlap is seen in the capacity which the adult newt possesses of regenerating a lens to its eye.

The material for the regenerated lens is derived from the dorsal margin of the iris of the eye-cup itself.[3] It will be remembered (see p. 187) that in *Rana esculenta* the eye-cup retains for a considerable time the power of inducing a lens, and that in many forms when an eye-cup is grafted into the body of another embryo in such a way that it is deprived of contact with epidermis, the eye-cup may form a lens from its own margin.[4] This appears to be what happens in the regeneration of the lens in the adult newt. The eye-cup is then of course separated from the epidermis by the cornea, and the epidermis itself is differentiated into the conjunctiva; the edge of the eye-cup is represented by the margin of the iris. This power and method of regeneration implies that the lens-inducing faculty and the lens-producing faculty have not been lost by the eye-cup even in the adult.

The fact that it is always the dorsal margin of the iris which provides the material for the regenerated lens requires considera-

---

[1] Mangold, 1931 C.   [2] Ekman, 1923.
[3] Colucci, 1891; Wolff, 1895.   [4] Spemann, 1905; Adelmann, 1928.

tion. The fact itself is attested by numerous experiments, in some of which the newt is made to lie on its back during the period of regeneration;[1] in others, the whole eye-cup is rotated *in situ* through 180°, so that the choroid fissure which is normally ventral comes to lie dorsally;[2] in others again, the lens-forming potencies of all parts of the iris margin are tested by grafting definite sectors representing one-sixth of the circumference of the iris into the cavity of the eye of another newt from which the lens has been extirpated.[3] In all cases there is found to be a gradient of lens-forming potencies extending dorso-ventrally through the eye-cup and resulting, in the intact eye-cup, in the regeneration of the lens invariably from its dorsal margin. Once lens-formation is initiated here, it inhibits the formation of lenses at other points. In this connexion, it may be mentioned that the presence of the normal lens inhibits such grafted fragments of iris from regenerating a new lens. The presence of the normal iris, however, does not act in this way, and it may sometimes regenerate a second lens from its own upper border.[4]

If, as already noted (p. 187), there is in the early neurula stage a labile preliminary determination of a lens-area prior to the definitive determination of a lens, we may suggest that some of this area overlaps the eye-area, and that a portion of it becomes incorporated in the eye-cup. Further, topographical considerations make it clear that if this were so, the region of maximum lens-forming potency in the eye-cup would be the dorsal part, since, on the analogy of the limb, lens-forming potencies must be assumed to decrease along a radial gradient from some central spot in the presumptive lens-area, and this spot lies nearer the dorsal than the ventral part of the future eye-cup. If this is so, then, in the absence of a lens, this dorsally situated tissue in the eye-cup may well be stimulated to exhibit its original lens-forming capacity.[5]

[1] G. Wolff, 1901.   [2] Wachs, 1920.   [3] Sato, 1930.
[4] Spemann, 1905; Wachs, 1914; Sato, 1930.
[5] Recent experiments have shown that the restriction of the site of lens-regeneration to the dorsal margin of the iris is *also* due to the fact that this is the region of the eye-cup which is farthest away from the choroid fissure, which is always formed ventrally, and appears to exert an inhibitory effect on lens-regeneration. If at the early tail-bud stage the optic vesicle is rotated *in situ* about its stalk through 180°, the choroid fissure will be formed ventrally, in tissue which was the presumptive dorsal part of the eye-cup. If, then, in later stages

§ 9

Turning now to the amphibian nervous system, a number of experiments show that the neural fold region is first determined as a field.

In *Triton*, a dorsal half of an embryo, even at comparatively late stages such as that of the early gastrula, will develop into a normally proportioned little embryo.[1] In such cases the neural folds of the miniature embryos are proportional to their reduced size, although the material which the half contains is that from which a full-sized neural plate would normally be formed.[2] If this experiment is repeated at the late gastrula stage, the neural folds produced from the half are full-sized, and therefore relatively too large. Irreversible determination of the neural fold region has therefore taken place towards the close of the period of gastrulation (fig. 117).

Experiments of another kind, likewise on *Triton*, lead to further results. Gastrulae can be produced which are deficient in median material lying in the plane of bilateral symmetry. This may be effected either by cutting out a median disc of tissue and sticking the two lateral portions together again, or by making paramedian cuts in two embryos, on the right of the mid-line in one and on the left in the other, and then exchanging halves and sticking them together, so that one composite embryo so formed will be deficient and the other overprovided as regards median material. In the case of the deficient embryos, the region involved includes part of the presumptive neural fold tissue. Normally, of course, this region is broad anteriorly in the region of the brain, and narrow posteriorly in the region of the spinal cord. The operation to which the deficient embryos have been subjected results there-

the lens is extirpated, a lens will be regenerated from the actual dorsal margin, which was the presumptive ventral part of the eye-cup. But the original dorso-ventral gradient of lens-forming potency persists, though masked by the inhibitory effect of the choroid fissure. For if sectors representing one-sixth of the circumference of the iris of such eye-cups developed from rotated optic vesicles be tested for lens-forming potencies, it is found (a) that the potencies of the actual dorsal (presumptive ventral) sector are low, and (b) that there is little difference between the potencies of this and other sectors. In other words, the original dorso-ventral gradient of lens-forming potency and the ventro-dorsal gradient of lens inhibition have here cancelled out. (Sato, 1933; see also Beckwith, 1927.)

[1] Spemann, 1903.　　　　　　[2] Ruud and Spemann, 1923.

fore in partial or complete removal of the presumptive neural fold tissue from the trunk region, and in removal of only a portion of it from the region of the head. Such embryos develop complete brains with properly proportioned eyes, but the spinal

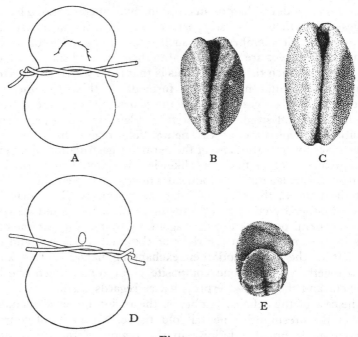

Fig. 117

Loss of power of regulation in the development of dorsal halves of newt embryos. A, The constriction and isolation of a dorsal half at the early gastrula stage leads to the development of (B) a diminutive embryo with neural folds of proportionately reduced size. C, Normal neurula for comparison with B. D, The constriction and isolation of a dorsal half at the late gastrula stage leads to the development of (E) a diminutive embryo with disproportionately large neural folds, unable to close. (From Ruud and Spemann, *Arch. Entwmech.* LII, 1923.)

cord may be absent. The important point to notice is that the neural tube was already determined at the time of the experiment, for the complete removal of its rudiment in the trunk region will result in absence of spinal cord altogether. But the neural tube was only determined as a whole and not in its detailed constituents,

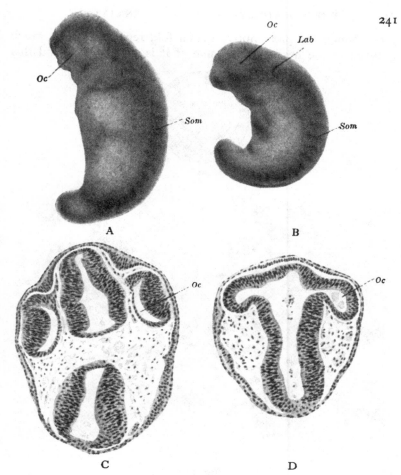

Fig. 118

Regulation in newt embryos deficient in or overprovided with median material. At the early gastrula stage, two embryos of *Triton taeniatus* are cut parasagittally into slightly unequal lateral halves, the cuts passing on the right of the middle line in one, on the left in the other. The larger right half is then stuck on to the larger left half; the smaller right half on to the smaller left half. In spite of the excess and deficiency of material lying in the plane of symmetry, both embryos develop with normal proportions, thus showing regulation. A, "Large" embryo. B, "Small" embryo: in this case, the neural fold material was partially and not completely removed in the trunk-region. C, Transverse section through A. D, Transverse section through B. *Oc*, eye; *Lab*, ear-vesicle; *Som*, somites. (From Spemann and Bautzmann, E., *Arch. Entwmech.* CX, 1927.)

for although the presumptive neural fold region has been much reduced in the head-region, some of it is still present, and this

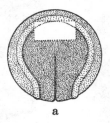

a

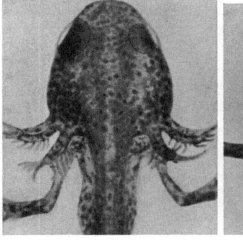 

b   c

Fig. 119

Power of regulation in differentiation and growth of the eyes in *Triton*. *a*, Operation performed at the early neurula stage: a piece of the presumptive eye-region is removed without interfering with the underlying tissues. *b*, Resulting larva 17 days after operation; left eye normal, right eye 1/3 normal size. *c*, The same animal just metamorphosed, 103 days after operation; both eyes approximately normal and equal-sized. (From Mangold, *Ergebn. der Biol.* VII, 1931.)

remnant has undergone regulation and has differentiated into a complete and normally formed brain with its attendant structures.[1] On the other hand, the overprovided embryos also regulate to

[1] Spemann and Bautzmann, 1927.

produce large but properly proportioned nervous systems; there may, however, in some cases, be a certain amount of duplication of the extreme anterior end (fig. 118).

In Urodeles, removal at the early neurula stage of a portion of the region of the presumptive eye-rudiment does not prevent the embryo from developing normally as regards its brain and paired eyes[1] (fig. 119).

All these experiments show that the neural fold region is deter-mined as a whole at these stages: within the neural fold region there also appears to be a determination along the antero-posterior axis of the levels of the various constituent subregions, much as in the limb. There is, however, no evidence that the neural fold region as a whole ever passes through a totipotent phase in which any part of sufficient size can regulate to produce a whole, as does the limb-field.

It is to be noticed in the experiments mentioned above, in which the presumptive neural fold region was removed altogether in the trunk region and only reduced in amount in the head, that regula-tion takes place within levels on the antero-posterior axis of the embryo, but not along that axis. The neural fold tissue of the anterior region regulates to form a brain, level for level, but it does not regulate longitudinally to form brain *and* spinal cord, i.e. structures characteristic of other levels.

As with the limb, ear, gill, and heart-fields, the nervous system is polarised. The existence of this polarisation or gradient is shown by the following experiment. If at the early neurula stage in an embryo of *Amblystoma* one presumptive eye-region is cut out, rotated through 90° and grafted back again, the resulting embryo possesses on the operated side an eye-cup which is subnormal and deficient in its development and differentiation. Had this region been extirpated completely, the remainder of the neural fold field would have regulated to give rise to perfect eyes (see above). It follows therefore that the region in question possesses a polarity which at this stage presents obstacles to regulation if it is interfered with.[2] The difference from such a field as that of the limb is that here the chemo-differentiation of subregions at different levels occurs at the first formation of the field, instead of later.

[1] Adelmann, 1929, 1930; Mangold, 1931 A.   [2] Woerdeman, 1932.

Progressive chemo-differentiation within the neural fold field may best be illustrated with reference to the eye-region. As we have seen (p. 243), experiments involving extirpation of the presumptive eye-region or of part of it at the early neurula stage in *Triton* and *Amblystoma* have shown that the eye-field is more extensive than the region which in normal development actually gives rise to the eyes.[1] Eye-forming potencies, as tested by grafting, are higher in the mid-line than more laterally.[2] The reason that a single median eye is not normally formed depends on the presence of the underlying gut-roof (see below). Further, regulation takes place most readily across the transverse axis of the neural fold region.[3] At the same time, experiments of grafting portions of the eye-region into other parts of the body show that the eye-rudiment is already invisibly chemo-differentiated.[4] If the entire eye-field is extirpated, no eyes are developed.[1]

At the same time, although the eye-region is determined as a subregion of the neural fold field at the neurula stage, it is still capable of regulation within itself, as is shown by those experiments on *Amblystoma* at the neurula stage in which a portion of presumptive eye-tissue is grafted into the belly wall and differentiates there into a more or less well-formed eye consisting of tapetum and retina. Curiously enough, these eyes lack a stalk, although the graft included the region which would normally have given rise to the optic stalk in the intact embryo. It follows, therefore, that the various regions of the optic complex—retina, tapetum, and stalk— are not rigidly determined within the eye-area at the stage in question in *Amblystoma*.[5]

In *Pelobates* (Anuran) at the tail-bud stage, grafted portions composed of tapetum only can regulate to form little optic cups with retina and tapetum in correct proportions.[6] The optic stalk, however, at this stage is already predetermined. In other Amphibia it has been found that two eye-rudiments, grafted together, regulate to form one.[7]

A complication in the analysis of the progressive determination

[1] Woerdeman, 1929; Manchot, 1929.   [2] Adelmann, 1930.
[3] Mangold, 1928, 1931 A.
[4] Spemann, 1919; Spirito, 1928; Adelmann, 1929, 1930.
[5] Adelmann, 1930; Stella, 1932.
[6] Dragomirow, 1932, 1933.   [7] Pasquini, 1927.

of the eye is introduced by the fact that the organiser, in the form of the primitive gut-roof, underlies the neural plate, and it now plays a part in the further determination of the eye-regions. The action of the primitive gut-roof in this respect has been tested by grafting portions of the eye-area without and with the primitive gut-roof.[1] It is found that the primitive gut-roof reinforces the eye-forming potencies of the lateral portions of the eye-area, and, further, it leads to the formation from median pieces of the eye-area of two eyes, with optic stalks, and separated by a region of the floor of the brain representing the optic chiasma; whereas similar pieces without gut-roof produce a single eye. The explanation of this "twinning" effect of the gut-roof on the eye-rudiment is still to seek.

The gut-roof or organiser further accelerates the processes leading to progressive subdivision of the eye-region into chemo-differentiated subregions. This is illustrated by experiments on *Rana esculenta*. If at the neurula stage in embryos of this species a rectangular piece is cut out from the neural fold region including part of the eye-area (together with the underlying primitive gut-roof), rotated through 180°, and grafted back again so that the original anterior edge of the piece is posterior, the rotated piece undergoes self-differentiation. The result in the later embryo is a reversal of the normal order of the structures of the brain: the diencephalon with the epiphysis lies behind the mesencephalon with its optic lobes. Anteriorly and posteriorly, these structures which have developed from the reversed piece are continuous with the parts of the brain which have developed undisturbed. It may be noticed, therefore, that the diencephalon, epiphysis, and optic lobes were already determined as subregions in the neural fold field at the time of operation,[2] and there is evidence that the infundibulum is also determined.[1]

But the most interesting feature of this experiment is concerned with the eyes. It is frequently found that there is a small pair of eye-cups in the normal position, and another pair farther back, situated either in front of or behind the ear-vesicles. The explanation of this result is that the anterior cut by which the rectangular piece was separated from its surroundings passed through the

[1] Adelmann, 1930.   [2] Spemann, 1912 A.

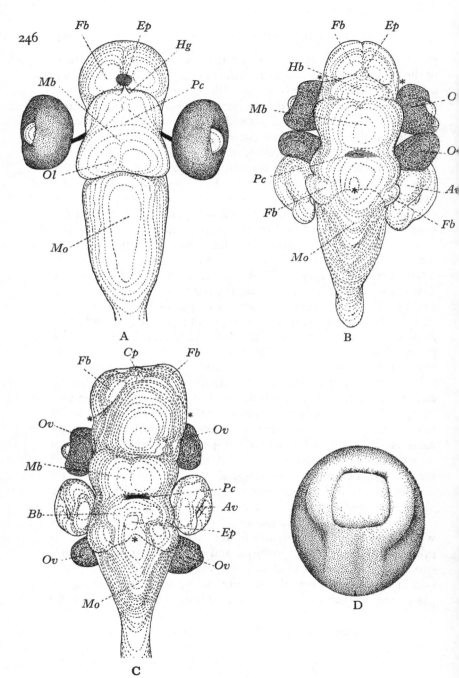

246

A

B

C

D

Fig. 120

THE MOSAIC STAGE OF DIFFERENTIATION 247

presumptive eye-rudiments: a portion of these rudiments was therefore left *in situ*, and another portion was included in the rectangular piece. The portion left *in situ* developed into the eye-cups in the normal position; the other portion after rotation developed into the eye-cups farther back. If the rectangular piece was short, its hinder edge was in front of the ear-vesicles, and the posterior pair of eyes then developed there. If on the other hand the rotated piece was longer, so that its hinder edge was situated behind the ear-vesicles, the posterior pair of eyes was behind the ear-vesicles also[1] (fig. 120).

These results show in the first place that the eye-rudiments were determined, since they could go on differentiating normally in abnormal surroundings. In some cases the four little eye-cups are not equal in size, but the sum of the sizes of the left front and right hind eye-cups is equal to that of the right front and left hind cups: the two eye-cups formed from one original rudiment divided by the cut are, together, equal to one normal eye. This means, therefore, that the eye-rudiment is not only qualitatively but also quantitatively determined, and that its topographical limits are now fixed. The actual time of onset of this determination cannot be stated, since the rudiment continues to be in contact with the underlying gut-roof, and progressive chemo-differentiation probably proceeds after the experiment. It is to be noted that in the experiment on

[1] Spemann, 1912 A.

Fig. 120

Mosaic development and self-differentiation of the brain and eyes in *Rana esculenta*. A, Dorsal view of the brain of a normal larva. B, Similar view of a larva in which at the neurula stage a square piece of neural plate with underlying gut-roof was rotated through 180°; the asterisks mark the line of junction between the rotated and non-rotated regions. C, Similar larva in which the piece rotated was longer. Note that the rotated regions of the brain have continued to develop by self-differentiation: the mid-brain lies in front of the epiphysis or fore-brain. There are four eyes, owing to the fact that the cuts went through the eye-region and, after rotation, parts of the eye-region found themselves behind the normal position (B) or, if the rotated piece was long, behind the ears. The sum of the sizes of the four eyes is equal to that of two normal eyes. In spite of their reduced size, the subdivided portions of the eye-rudiments have become rounded into cups approximating to the normal morphological differentiation. D, Neurula, showing the operation. *Av*, auditory vesicle; *Bb*, between-brain; *Cp*, choroid plexus; *Ep*, epiphysis; *Fb*, fore-brain; *Hb*, hind-brain; *Hg*, habenular ganglion; *Mb*, mid-brain; *Mo*, medulla oblongata; *Ol*, optic lobe; *Ov*. eye-cup; *Pc*, posterior commissure. (From Spemann, *Zool. Jahrb. Suppl.* xv (3), 1912.)

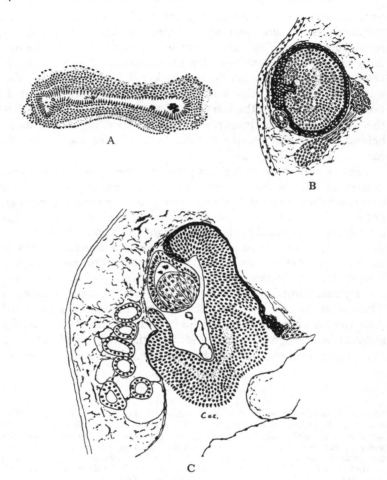

Fig. 121

Self-differentiation of the eye and of its constituent tissues. The optic vesicle was removed and grafted into the region of the ear in *Rana palustris*. The various tissues undergo self-differentiation regardless of their morphological differentiation, and of the proportions in which they are present. A, 19-day old graft, showing absence of pigment cells; rods and cones project into the cavity of a vesicle. B, 7-day old graft, showing excess of pigment cells. C, 5-day old graft, showing absence of part of pigment layer; rods and cones projecting into coelom of host. (From Lewis, *Amer. Journ. Anat.* VII, 1908, figs. 4, 5, 7.)

Amblystoma above-mentioned, the eye-region when not underlain by gut-roof is found to be still capable of regulation within itself at the neurula stage. The results of the experiments on *Pelobates* are apparently to be explained only on the supposition that chemo-differentiation of the eye-region in this form occurs more slowly. When these eye-cups are very small, it is frequently found that they are abnormally proportioned in that they may have too many or too few tapetum cells; too much or too little optic stalk or retina, compared with the proportions in which these constituents are found in the normal eye-cup. The explanation appears to be that the various constituents of the eye are now separately and individually determined. The cut, going haphazard through the eye-rudiment, will often separate parts which possess the prospective constituents of a normal eye in abnormal proportions. Similar results are obtained from experiments on *Rana palustris* in which incomplete eye-rudiments are grafted into various positions. Here, even the various retinal layers appear to be determined[1] (fig. 121).

§ 10

Yet another conclusion of importance emerges from the experiment mentioned above (p. 245). It will have been noticed that although some of the miniature eyes are abnormal in the proportions of their constituent parts, they nevertheless round themselves off into little spheres resembling normal optic vesicles, and some at any rate of these little vesicles become invaginated to form cups. In some cases, cups formed entirely of tapetum without any retinal tissue are produced. In other words, the processes of morphological differentiation, or production of form, are not dependent on the histological differentiation of the tissues which they are moulding. A similar conclusion can be drawn from the results of other experiments, in which it has been shown that portions of the neural fold region, or of the heart, or gut region, roll themselves up into tubes in spite of the fact that their histological differentiation may be abnormal.[2]

At the same time, other work, and on the most diverse groups, has shown that the histological differentiation of the tissues may

[1] Lewis, 1908.
[2] Roux, 1885; Ekman, 1924; Stöhr, 1925; Boerema, 1929; Holtfreter, 1931 B

250   THE MOSAIC STAGE OF DIFFERENTIATION

take place independently and in the absence of normal morphological differentiation. This may be seen in those experiments in which a portion of the embryonic area of the blastoderm of the chick is made to undergo development on the chorio-allantois of another egg;[1] or in embryos of Cephalopods, the normal development of which has been impeded by toxic agents.[2]

The same conclusion emerges from the results obtained by culturing *in vitro* various rudiments of the chick embryo, such as those of the eyes, fore-limbs, and ears. In these cases, histological differentiation may reach a high degree of perfection, while there may be little or no approach to the morphological differentiation of normal anatomy.[3] (See also Chap. XI, p. 375.)

A pretty example of abnormal morphogenesis is seen in the differentiation of reconstitution-masses of dissociated sponge cells which contain an excess of collar-cell tissue. In this case, partial spheres consisting of a single layer of collar-cells are produced with the collars directed outwards instead of inwards, as in the normal gastral lining.[4] (See p. 281.)

Purely morphological differentiation, then, seems to be in large part conditioned by physical and mechanical factors of available space, material, and pressure. Histological differentiation is in large part independent of these factors. While these two kinds of differentiation are sufficiently distinct during the later stages (i.e. after their initial determination) for the one to take place without the other, the question next arises as to what relation these two kinds of processes bear to one another at the start.

The first visible important steps in differentiation are concerned with the form-changes which result in gastrulation and neurulation. These may be held to constitute a phase of morphological differentiation, which, in development, is thus seen to precede histological differentiation. The question therefore arises as to whether the latter is dependent on the former in the initial stages. If it were, we should have another instance of the supposed effects of "dynamic determination", referred to on p. 163. The problem therefore presents itself as to whether it is possible to prevent a certain region

[1] Murray and Huxley, 1925; Hoadley, 1924, 1925, 1926 A.
[2] Ranzi, 1928.   [3] Strangeways and Fell, 1926; Fell, 1928.
[4] Huxley, 1911; de Beer, 1922.

of the embryo from passing through this early phase of morphological differentiation (mass movements at gastrulation and neurulation), and then to see whether it is capable of undergoing
histological differentiation.[1]

It will be remembered that during gastrulation in Amphibia, the
presumptive neural fold material undergoes a translocation in a
particular direction for each piece of tissue, so that the material is
brought into position for the formation of the neural folds in the
neurula (see p. 25). A piece of presumptive neural fold tissue may
be grafted into the dorso-lateral region of another embryo in the
gastrula stage and orientated in such a way that the movements
of the host tissues in which it becomes involved are either directed
parallel or perpendicular to the direction in which the tissue would
have moved had it been left intact *in situ*. It is found that the
tissue differentiates morphologically into neural folds regardless
of its orientation and of the direction of the movements which
it has undergone.[2]

It can be concluded from these experiments that specific form-
changes are not necessary for subsequent histological differentiation.
Other recent investigations of the histological differentiation of the
cell-regions in early embryonic stages of *Triton* have shown that
certain histological distinctions between presumptive epidermis and
presumptive neural fold are independent of form-changes. These
distinctions are already present at the earliest neurula stage. The
cells of the neural fold region are elongated, arranged in a single
layer, and have ellipsoidal nuclei; their pigment is concentrated at
the outer end of the cells. The cells of the epidermal region are
cubical, arranged in two layers; the nuclei are spherical, and the
pigment is distributed irregularly.

If at the neurula or late gastrula stages (i.e. after the organiser has
been invaginated and underlies the presumptive neural folds)
pieces of presumptive neural fold tissue or presumptive epidermis
are grafted into atypical positions, they develop, as we have already
seen, by self-differentiation, and undergo the histological differ-

[1] Goerttler, 1927. It may be noticed, however, that this experiment results
not only in forcing the piece of tissue to undergo abnormal movements, but it
also interferes with its polarity, which, as we have already seen (p. 243), plays
an important part in differentiation.
[2] Holtfreter, 1933 A.

entiation characteristic of their normal fate, but the morphological differentiation is not always achieved. If, on the other hand, such presumptive pieces are taken from an early gastrula (i.e. before the organiser has been invaginated) and grafted, they will undergo the morphological differentiation of their new surroundings: presumptive neural fold tissue in an epidermal area will remain flat, while presumptive epidermis in the neural fold area will become folded up into a neural tube. But, in spite of the morphological differentiation which these pieces are forced to undergo, they retain some of the histological characteristics of their normal prospective fates.[1]

In these experiments we have on the one hand the fact that histological differentiation can take place without morphological, and on the other, the fact that morphological differentiation when forced upon a piece of tissue does not entirely obliterate its presumptive histological characteristics. It is necessary to conclude, therefore, that in these cases, morphological and histological differentiation are independent of one another.

There are other facts which point to the same conclusion. In the larva of the sea-urchin, for instance, some histological differentiations (apical organ) take place without any antecedent form-changes of the tissue in question. In amphibian material, the results of experiments involving the culture of pieces of tissue *in vitro* likewise point to the independence of histological and morphological differentiation. We need only point to the instances mentioned above in Chapter III (p. 50), in which pieces of tissue taken from the blastula or early gastrula show far-reaching powers of histological differentiation without having undergone any specific form-changes, or any morphological differentiation.

Perhaps the most striking demonstration of the independence of morphological and histological differentiation is provided by those cases in which an insect egg (*Platycnemis*) gives rise to two embryos as a result of a transverse discontinuity in the blastoderm. The two embryos develop, each from its ventral surface in the normal way, and they are situated back to back. One is larger than the other, and when it folds up its sides to form its dorsal surface, it actually encloses its smaller brother within itself, and compels it to perform

[1] Lehmann, 1928 B, 1929.

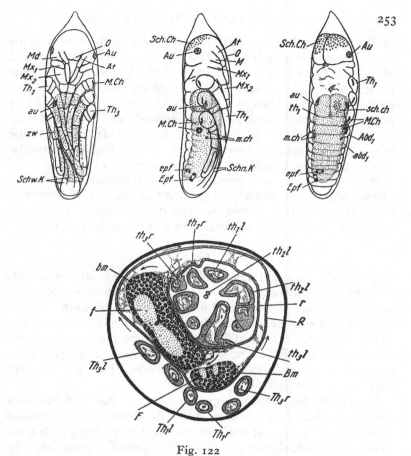

Fig. 122

Twin embryos in the insect *Platycnemis*. The embryonic rudiment was split
into two unequal portions at an early stage; the larger portion has produced an
apparently normal embryo, but within it (stippled) is the dwarf embryo produced
from the smaller portion, which is inside-out (see text). It has become enclosed
within the larger embryo as a result of the upgrowth of the sides of the latter; the
direction of upgrowth is shown by the dotted line arrows. This process was
too strong for the sides of the smaller embryo, which were forced to follow suit
and to fuse ventrally instead of dorsally, thus enclosing the limbs within a closed
cavity lined by the outer surface of the epidermis. The ventral nerve-cord of the
second abdominal segment of the larger embryo is in contact with that of the
first abdominal segment of the smaller embryo, which thus appears larger.
For explanation of lettering, see fig. 60: capital letters refer to the larger
embryo, small letters to the smaller embryo. In addition: *Abd*₁, 1st abdominal
segment; *Bm*, nerve cord; *F*, fibre-tracts; *R*, dorsal wall. (From Seidel, *Biol.
Zentralbl.* XLIX, 1929.)

similar movements. But, for this smaller embryo, these movements result in the folding and eventual fusion of its sides *ventrally* instead of dorsally, since it is back-to-back with the larger embryo. Thus the smaller embryo is inside-out: its limbs are contained in a closed cavity lined by its body-wall which is completely inverted; its organs and viscera lie outside its body-wall, and in contact with those of its larger brother, inside which it is.[1] In spite of these form-changes being the reverse of normal, histological differentiation continues as if nothing had happened (fig. 122).

While "dynamic determination", or the determinative effects of form-changes, may possibly be operative in the case of organisers (see Chap. VI, p. 163), it does not seem that histological differentiation in the mosaic stage of development is dependent on it.

§ 11

A special section may be devoted to the problem of the gonads and sex-differentiation, which present many interesting features. A full discussion of all aspects of the question has been given in recent books such as *The Development of Sex in Vertebrates*[2] and *Sex and Internal Secretions*;[3] accordingly here much controversial detail will be omitted. Here, only such points as bear upon morphogenesis and the problem of differentiation will be dealt with, and they in broad outline.

In general, the vertebrate gonad arises as what is doubtless a special gonad-field on the dorsal side of the coelom. It first consists of thickened coelomic epithelium (germinal epithelium) with some underlying mesenchyme, together with primordial germ-cells. In many vertebrates, these are undoubtedly differentiated precociously, in most cases in the endoderm, and then migrate into the gonad-rudiment. In other cases, especially among the higher forms, it seems equally clear that germ-cells arise directly from the germinal epithelium. It is possible that both these sources contribute to the formation of germ-cells in many vertebrates.

Later, the gonad-rudiment becomes differentiated into an external cortex and a central medulla, but the details vary considerably in different groups.

[1] Seidel, 1929.   [2] Brambell, 1930.
[3] E. Allen, 1932.

We may begin with the conditions in the Anura, which have been very thoroughly investigated.[1] The primordial germ-cells arise in the dorsal region of the gut-wall, and then become separated from the rest of the endoderm as a continuous ridge dorsal to the mesentery. This ridge later divides into two. In these two genital ridges, the germ cells are mixed with mesenchyme, and overlain by coelomic epithelium which becomes slightly thickened. Later the core of the ridges is invaded by the rete tissue, consisting of cords of cells which appear definitely to grow out from the rudiment of the mesonephros. The sexually undifferentiated gonad-rudiment is now completely constituted, and consists of two portions, a peripheral cortex composed of coelomic epithelium and primordial germ-cells with associated mesenchyme, and a central medulla derived primarily from the mesonephros. The cortex is broadly homologous with the germinal epithelium of Amniota.

Sexual differentiation now occurs. In the female the cortex enlarges, its contained germ-cells develop into oogonia and oocytes; meanwhile the medulla ceases growth and develops into epithelial ovarial sacs. In the male, the rete cords of the medulla continue to proliferate, and are invaded by the germ-cells, which leave the cortex and migrate inwards, then proceeding to differentiate into spermatogonia. Later the rete cords produce, among other structures, the non-germinal portions of the testis tubules. Meanwhile the cortex, after losing its germ-cells, becomes reduced to a thin peritoneal covering.

The evidence appears conclusive, first, that the type of sexual differentiation of the indifferent gonad-rudiment is normally dependent on its genetic sex-constitution, although, as we shall see later, this can be overridden by other agencies. The case is like that of any other mosaic differentiation, except that the gonad-field has one of two potentialities open to it, according to the sex-chromosomes which it contains. Secondly, the primordial germ-cells appear to be completely bi-potential as regards sex. What they shall become is determined by local influences emanating from the region in which they come to lie. In the cortex they become female, in the medulla, male. In other words, their differentiation is dependent.

Temperature exercises a differential effect upon the cortex and

[1] Full references in chapters by Willier and by Witschi in E. Allen, 1932.

medulla. Low temperature causes a differential inhibition of medullary development with consequent delay in males of the degeneration of the cortex and of the immigration of the primordial germ-cells from it to the medulla. As a result, the primordial germ-cells, exposed to cortical influences, become oogonia, and 40 mm. tadpoles are all somatically females. Later, however, in the genetic males among them, the delayed medulla succeeds in reaching the stage of development requisite to inhibit the cortex, upon which they become transformed into somatic males.

High temperature, on the other hand, has a deleterious effect upon the cortex, but not upon the medulla. As a result the sexes are early differentiated in the normal 1 : 1 ratio, but later the females show inhibition of the cortex. No further oocytes are differentiated, those already embarked on differentiation degenerate after a short period of further growth, the ovarian sacs derived from the medulla begin to proliferate and form cords, and any undifferentiated primordial germ-cells migrate inwards and join the medullary cords, where they differentiate into spermatogonia; thus the genetic females become transformed into somatic males. Essentially similar results have been obtained in Urodeles.

These experiments clearly demonstrate the existence of local sex-inductive agencies in cortex and medulla respectively. The nature and action of these factors is more fully revealed by a series of beautiful experiments on parabiotic twins.[1]

Amphibian embryos are united parabiotically either side by side (parallel pairs) or in series with the anterior end of one joined to the posterior end of the other (chains). This has been effected both homoplastically, between partners of the same species; or heteroplastically, between different species. Here we shall confine ourselves to homoplastic parabiosis, and to two-sexed pairs, in which the partners are genetically of opposite sex.

The most interesting experiments concern frogs (*Rana*). In these, no effect on sex-differentiation is exerted in chain pairs: sex-differentiation is normal both in the male and the female partner. In parallel pairs, however, the sex-differentiation of one member of the pair is modified. The affected partner is normally the female, and the modification consists in a certain degree of inhibition of

[1] Witschi, 1932.

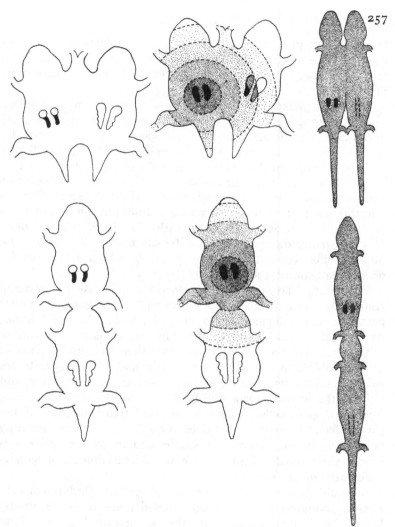

Fig. 123

Antagonistic sex-differentiation in Amphibia. Diagram of parabiotic twin pairs
of unlike sex in toads (left), frogs (centre) and Urodeles (right). Above, parallel
pairs; below, chain pairs. In the gonad: male differentiation, black; female
differentiation, white. In toads, there is no effect of one twin upon the other
(note Bidder's organ at the anterior end of the gonad in both sexes). In frogs,
there is no effect in chain pairs; but in parallel pairs, the male gonad affects the
gonads of the female partner, the effect diminishing with distance (degree of
shading of circles). In Urodeles (*Triturus*), the male gonad completely inhibits the
female gonad in both types of twin pairs. (Based upon Witschi, Chap. v in
E. Allen, *Sex and Internal Secretions*, London and Baltimore, 1932; modified.)

female-differentiation combined with a certain degree of encouragement of male-differentiation, resulting in progressive transformation away from the somatic female towards the somatic male-type of gonad.

The most unexpected result is that the effect always manifests itself first in the "inside" ovary, i.e. that nearer to the male partner, and always on the inside margin of this ovary; from here it gradually spreads, but with diminishing intensity, to the more distant regions. It appears clear that the medulla of the male produces a substance which not only promotes masculine sex-differentiation of germ-cells, but is also antagonistic to cortical development: further, that this substance is not strictly localised but can diffuse outwards in diminishing concentration and with diminishing effectiveness. In chain-pairs, the gonads of the female partner lie beyond the limit of effectiveness; in parallel pairs, they lie across a rapidly decreasing concentration-gradient (fig. 123).

Comparative studies on other forms provide further striking results. In toads, no effect is ever observed on the gonads of either partner, whether in parallel or chain-pairs. In the Urodele *Triturus*, however, the effect is equally marked in both kinds of combination. (The details here are slightly different: there is a long period of mutual inhibition, in which both male and female gonads are delayed and rendered nearly sterile. Later the male recovers, and reduces the female gonads still further to small rudiments almost free from germ-cells. There is no male transformation of the genetic females. In other Urodeles the effect is similar in affecting the female partner equally in chains and in parallel pairs, but neither the mutual antagonism nor the final inhibition of female-differentiation are so extreme.)

It would thus appear that the morphogenetic (inductive) substances produced by cortex and medulla are in toads strictly localised within the regions where they are produced, and wholly or almost incapable of diffusion. This is borne out by the existence in toads of Bidder's organ, an anterior section of the gonad of ovarian character, which develops from a portion of the gonad-rudiment consisting wholly of cortex.[1] This could not well develop, as it does, in males if the medullary substance could diffuse even a short dis-

[1] Witschi, 1933 B.

tance during ontogeny. In frogs, on the other hand, the inductive substances must be capable of moderate diffusion. The effect here recalls the graded distribution of limb-forming potencies in the limb-field of Urodeles (p. 223), and the probable diffusion of inductive substances from the presumptive dorsal lip region during cleavage (pp. 134, 311).

In Urodeles, on the other hand, diffusion is so complete that there is no evidence of any concentration-gradient. It is possible, though not demonstrated, that here the substance diffuses into the blood-stream and is transported by it. We have thus within the boundaries of one class of vertebrates either a complete or a nearly complete gradation between sharply-localised morphogenetic substances and freely-circulating hormones. It has indeed been suggested that the sex-hormones of the adult gonads are identical or homologous with these morphogenetic substances produced by the embryonic cortex and medulla, merely differing in being secreted into the blood-stream instead of soaking through the tissues.

In support of it we find indications in cases of hermaphroditism or asymmetrical development of gonads that the accessory sex characters (male and female ducts), whose differentiation is known to be dependent upon sex-hormones, are locally better developed in regions of greater development of the gonad of corresponding sex.[1] While no certainty can yet be reached on this point, it is a valuable suggestion to guide further research. In any event, it is clear that the physico-chemical conditions regulating diffusibility of morphogenetic substances are of great importance in ontogeny.

In Urodele parabiosis the failure of the medulla of genetic female gonads to differentiate in the male direction after regression of the cortex under the influence of the male partner is in marked contrast with the results in frogs. It appears to be general in the subclass.[2] No adequate explanation is yet forthcoming: in general, it appears to link up with the subject of metaplasia (p. 211). The medulla of all female Amphibia differentiates in a specifically female direction; that of the Anura retains its bisexual potency, and is capable of metaplasia and male histo-differentiation; that of the Urodela loses the original bisexual potency and is capable only of continued development or of regression within the limits of female-type

[1] Witschi, 1933 B.   [2] Witschi, 1933 A.

potency. The progressive restriction of potencies during ontogeny, and its variation between types of tissue and types of organism, remains a central problem for developmental physiology. Two further points may be mentioned. Occasionally in Anuran parabiotic twins, the female partner may obtain an unusually early start. In this case, the male-differentiation of the male partner is inhibited and a female phase may be passed through. This however is only to be seen in the parts of gonads nearest to the female partner, and is transitory, normal male-differentiation eventually gaining the upper hand again.

Secondly, in heteroplastic parabiosis between two species of frogs, a curious effect is visible in pairs in which both partners are genetic females, and in which one member belongs to the species *Rana sylvatica*. The *sylvatica* ovaries hypertrophy, those of the co-twin become reduced and degenerate. This can be explained on the assumption that some substance necessary for ovarian growth is present in limited quantity in the embryo, and that the faster-growing *sylvatica* ovaries obtain a disproportionate share of it.[1] In rare cases, the reduced ovary of the co-twin may even show some changes in the direction of male transformation. This may be explained on the hypothesis of antagonism between cortex and medulla; when the cortex is inhibited by being starved of the substance necessary for its growth, the medulla is released from inhibition.

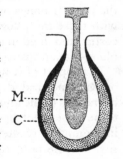

Fig. 124

Diagram of the two main components of the amphibian gonad-rudiment. *M*, medulla, responsible for male-differentiation; *C*, cortex, responsible for female-differentiation. (After Witschi, Chap. v in E. Allen, *Sex and Internal Secretions*, Baltimore and London, 1932. Modified.)

Thus we come to the general conclusion that sex-differentiation in Amphibia is under the control of substances provided by the gonad cortex and medulla respectively: that these substances are not species-specific: that they are mutually antagonistic: that they are capable of various

[1] The same hypothesis will account for the fact that removal of the gonad proper in either sex in toads leads to the hypertrophy of Bidder's organ to form a functional ovary. The substance in question is very possibly a hormone produced by the pituitary. See Witschi, 1933 B.

degrees of diffusion from the regions where they are produced: that in their action at a distance on another developing gonad the inhibitory effect of each is the primary or at least the stronger, and that the male-differentiating substance normally develops earlier and is more potent than the female-differentiating substance. It is also possible that they are or become converted into the sex-hormones of the adult.

Further highly interesting results have been obtained as the result of fertilising over-ripe eggs in frogs.[1] Below a certain degree of over-ripeness (about 3 days), no effects of any kind are to be noted in the resultant embryos. Beyond this critical point, the following main effects appear, all of them increasing with the degree of over-ripeness. First, a conversion of a certain proportion of genetical females into somatic males. The proportion is at first small, but finally, with eggs rather over 4 days over-ripe, all-male offspring are produced. Meanwhile a certain degree of delay in development is noticeable, and with increasing over-ripeness defects of development and abnormal mortality also appear, culminating in death of all embryos at an early stage when the eggs are about 5 days over-ripe. The defects of development have already been noted in Chap. v (p. 96); they arise mainly after 4-5 days' over-ripeness, and manifest themselves chiefly as a failure of coordination, leading to abnormal cleavage, production of double monsters, development of teratological outgrowths, and in extreme cases the production of abnormal tumour-like structures which partake of many of the characteristics of truly malignant growths. Comparable phenomena have been observed in trout.[2]

The sex-transformations are of peculiar interest, since the whole morphogenesis of the gonad-rudiment is modified. In extreme cases[3] the rod-like area of primordial germ-cells does not become detached from the endoderm, and the gonad-rudiment at its first appearance is a rudimentary fold containing no germ-cells (in such cases the germ-cells appear later to migrate into it, but how this occurs is not established). In other cases[4] the germ-cells while still in the endoderm become abnormally pigmented. The gonad-rudiment in highly affected specimens passes through a female

[1] Willier, 1932; Witschi, 1932.   [2] Mršić, 1923, 1930.
[3] Kuschakewitsch, 1910.   [4] Witschi, *loc. cit.*

phase followed by degeneration of the cortex and growth of the
medulla leading to male transformation as in the high temperature
experiments: in extreme cases, the cortex is inhibited from the
outset and sexual differentiation is male throughout.   In general,

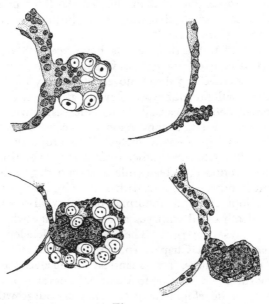

Fig. 125

The effect of late fertilisation on gonad-differentiation in frogs. Left, two stages
in the normal morphogenesis of the gonad. Above, genital ridge with primordial
germ-cells and mesenchyme. Below, later stage with peripheral cortex containing
primordial germ-cells and central medulla derived from the invading rete-cords
of nephrogenous origin. Right, corresponding stages in embryos from late
fertilised (over-ripe) eggs. Above, very small genital ridge with no primordial
germ-cells. Below, later stage with invading medulla (rete tissue) but rudimentary
cortex, with no primordial germ-cells. (Redrawn after Kuschakewitsch,
Festschr. f. R. Hertwig, 1910, vol. II.)

testis-differentiation is accelerated[1] (fig. 125). The thyroid of late-
fertilised frogs is also hypertrophied.[2]

We have thus a series of effects with progressive degrees of
over-ripeness. First, minor upsets of morphogenesis, notably in
regard to sex. Secondly, more general upsets of morphogenesis,

[1] Eidmann, 1922.   [2] Adler, 1917.

notably partial twinning. Thirdly, teratomorphic and malignant effects. It is interesting that, as Dr Waddington has pointed out in conversation, carcinogenic compounds are certainly related to oestrin, and are probably to the chemical substance responsible for organizing (p. 154). So our three effects, concerned with sex, with organiser abnormality (twinning), and with malignancy, may conceivably all be related to one fundamental process affecting substances of this type.

It is interesting that keeping the eggs for some time before fertilisation in conditions of relative lack of water (hypertonic salt solutions; keeping in air with a minimum of moisture) leads to a large preponderance of females.[1] Unfortunately no embryological study of this case has been made, but it too is evidence that conditions in the egg-cytoplasm during the earliest stages of development may modify morphogenesis at later stages: since sex-differentiation provides two alternative methods of morphogenesis in which the result is determined by a balance of two competing factors, we should expect to find in it the best indicator for such effects.

In Amniotes, the embryology of the gonad is not so simple. In general, however, we may say that the distinction between masculinising medulla and feminising cortex is maintained. The medulla is largely formed by the primary sex-cords which migrate inwards from the germinal epithelium; in the male these form the testis tubules, in the female they become inhibited and persist in modified form. In the male, only one set of sex-cords is formed. In the female, however, the cortex enlarges to produce a second set, which gives rise to the main ovarian structures, in association with which the female germ-cells differentiate. In the male, on the other hand, the cortex (germinal epithelium) becomes reduced to a mere peritoneal epithelium soon after the formation of the primary sex-cords.

The chick is here the best-investigated type. In the chick the germ-cells are formed in the extra-embryonic endoderm, in a crescent-shaped area of the blastoderm antero-lateral to the embryo. Embryos can be castrated by excision of this area, or by ultra-violet[2] or X-rays,[3] proving the mosaic determination of the primordial germ-cell tissue. Normally these cells appear to be attracted into the mesoderm when it invades the germ-cell field, and

[1] King, 1912.   [2] Reagan, 1916; Benoit, 1930.   [3] Danchakoff, 1933.

thence into the blood-stream. After being found in all parts of the body for a considerable period, they become localised during a few hours in the site of the future gonad-rudiment. Presumably the gonad-field, once determined, attracts the germ-cells chemically.

By various lines of evidence it has been shown that a gonad-field is determined and will differentiate into a gonad with typical sex-cords even in the total absence of primordial germ-cells, and that conversely in chorio-allantoic grafts germ-cells may be present in considerable numbers without giving rise to a gonad. There is, however, evidence that the germ-cells can induce some degree of early gonad-differentiation upon peritoneum which would normally never give rise to germinal epithelium. It is thus probable that the germ-cells are necessary for, or at least normally assist in, the process of gonad-differentiation, but that this effect can only be exerted within a "gonad-field" region of the dorsal coelomic epithelium, whose potencies are highest in the presumptive gonad-region.

In passing, it should be mentioned that, in some mammals at least, germ-cells appear to arise *in situ* at comparatively late stages in the already differentiated germinal epithelium. Here early gonad-differentiation cannot be dependent in any degree upon the presence of germ-cells.

The transition between a state of affairs in which an early determination of germ-cell tissue occurs in the endoderm and that in which a late determination occurs in the germinal epithelium is not easy to envisage, but we have at least analogies with the determination of other organs, i.e. the lens (p. 189), or the limbs of Amphibia, which in Urodeles are differentiated very early and are independent of thyroid action, while in Anura they differentiate later and will not display full growth in the absence of a certain concentration of thyroid hormone. Further work is needed to elucidate this point.

The ovaries of birds and of monotremes are of course asymmetrical, that on the left being large and functional, that on the right reduced, and non-functional. It is interesting to find in the bird that this difference is determined from the outset of differentiation of gonad-rudiment not as yet showing any sign of sexual differentiation. When grafted on to the chorio-allantois of other embryos, indifferent gonads of either side may differentiate into testes, presumably in the case of grafts from genetic males, since testes may develop on either male or female hosts. When they differentiate into ovaries, only a left rudiment will form a true ovary; the right rudiment will only develop the rudimentary ovarian structure typical of the right ovary. This difference must be determined in relation to the original asymmetry gradient of the embryo.

The different initial determination of right and left gonads is further shown by another experiment. If all the germ-cells are destroyed during the first half of the incubation period (which can be accomplished by X-rays, owing to the high susceptibility of the germ-cells to this agency), the subsequent complicated differentiation of the non-germinal portions of the gonad will continue, leading to the formation, shortly before hatching, of testes, functional (left) ovaries, and non-functional (right) ovaries, which are of typical structure except for being sterile.[1]

Grafting and other experiments have also elicited other interesting facts. In the first place, the grafting operation, and, still more, brief exposures to low temperatures soon after visible sex-differentiation has begun, favour the persistence of structures which normally atrophy during development, such as the right Müllerian duct (oviduct) of females, and both Müllerian ducts in males. The percentage of survival of such structures is raised by low temperature from about 18 per cent. found in controls to over 70 per cent.[2]

Low temperature is known to inhibit or retard many developmental processes: it would appear either that it has a specially strong effect on processes leading to the reduction of organs, or that since these processes, as shown by the 18 per cent. of persistence in controls, are unusually labile, slight alterations in conditions will cause large changes in their results (see also Huxley, 1932, Chap. VI, 8).

Another interesting fact is that the capacity of grafted portions of the gonad-field for differentiation increases with their age when grafted. Before the time of visible differentiation of germinal epithelium, few or no grafts of this area give a gonad at all. A little later, gonad-like bodies of uncertain sex are produced. If the graft is taken still later, when it includes gonads which are well-formed but still microscopically undifferentiated as regards sex, sex-different-

---

[1] Danchakoff, 1932.   [2] Willier, 1932.

iation either in the male or female direction occurs in the grafts, and its completeness and frequency (as well as the size of the resultant gonad) increases with the age of the grafts. It would thus appear that, as a result of processes occurring in the embryo as a whole, gonad-determining substances tend to accumulate in higher concentration in the presumptive gonad-field during, and presumably for a short time before, the stage of its early differentiation. This is perhaps somewhat parallel with the intensification of organiser potencies in the grey crescent area during the time from fertilisation to gastrulation (p. 68) and with the progressive capacity of the lens of the bull-frog for self-differentiation after determination (p. 189). It would be of great interest to discover whether a similar state of affairs can be detected for other rudiments in other groups. A parallel increase of differential potency in whole embryos and large fragments will be discussed in the next section.

The remarkable condition of the avian right ovary is correlated with a marked regression of the cortex after its first formation. The capacity for this must, as we have seen, be determined intrinsically within the right gonad-rudiment; but one of the results of this primary differentiation as a right ovary appears to be sensitivity to substances emanating from the left ovary; for when the left ovary is removed, the right hypertrophies. What it shall then produce is determined by the degree of degeneration which its cortex has previously undergone. If considerable cortex is still left, this dominates and it becomes a gonad of true ovarian type; if less cortex is left, both it and the medulla participate in the hypertrophy, forming an ovo-testis. If the cortex had completely regressed, the medulla hypertrophies and it forms a testis. The germ-cells normally disappear from the right ovary in the first month after hatching. If the hypertrophy takes place later, the resultant gonad is sterile; if earlier, as in very early left ovariotomy, spermatogonia may be formed.

In conclusion, the special interest of the gonads for our purpose lies in the fact that two alternative paths of differentiation lie open to them, the actual path taken being normally first decided by the sex-chromosome mechanism, and implemented by a quantitative balance between two mutually antagonistic male-differentiating and female-differentiating substances, locally produced by the two main regions of the gonad-rudiment, formed in different quantities, at different rates, and with different capacities for diffusion. The result is a bipotentiality, and therefore a lability of differentiation which makes the gonad especially interesting for a study of the effects of external agencies upon morphogenesis.

§ 12

Problems of possibly a special nature are presented by experiments of grafting portions of chick blastoderms on to the chorio-allantois of other embryos. In general, it seems that the power of histological differentiation of a piece of blastoderm is conditioned by its size, and by its age at the time of its isolation from the rest of the blastoderm. An entire unincubated blastoderm, when grafted, will show a degree of histological differentiation approximating to that found in normal chick embryos which have been developing for the same length of time. As mentioned above (p. 250), the morphological differentiation of such a piece may be very abnormal indeed.[1]

Small pieces, representing about one-fifth of the area of a whole blastoderm, are restricted in their powers of histological differentiation. Pieces of unincubated blastoderms grafted on to the chorio-allantoic membrane will differentiate only into epidermis and gut. Pieces cut from blastoderms incubated for 2 hours will, in addition, differentiate into nervous tissue. After 4 hours' incubation of the blastoderm, pieces cut from it will produce brain, eye, cartilage, and muscle. After 10 hours, corium and feather-buds are formed.[2] It may also be mentioned that the earlier a piece is isolated, the smaller are the organs which it forms.[3]

It is thus apparent that the older a piece of tissue is at the time of grafting, the better it will differentiate. This is especially well shown in the case of the eye. A 4-hour piece will produce an eye consisting of pigment cells only; a 6-hour piece gives an eye differentiated into pigment cells and retinal cells. After 8 hours, the various layers of the retina are differentiated, while complete self-differentiation of the eye is obtained from pieces cut from blastoderms that have been incubated for 33 hours. In the case of the mesonephros, 4-hour pieces give secretory tubules, 6-hour pieces

[1] Murray and Selby, 1930.   [2] Hoadley, 1926 A.
[3] Hoadley, 1929.

also produce glomeruli, 10-hour pieces differentiate a Wolffian duct; older pieces give complete self-differentiation of the mesonephros. Feather-buds present the same picture.[1]

Other experiments have been performed in which the pieces cut from the blastoderm were larger, representing one-third instead of one-fifth of the whole area. While further results are desirable, it seems from those already obtained that these larger pieces differentiate more fully than the smaller ones, thus indicating that the size of the piece is also a factor in the capacity of tissues to undergo histological differentiation.[2]

It will be remembered (see Chap. VI, p. 160) that in the chick blastoderm there is a gradient of developmental potencies at different levels along the axis, and this must be taken into account in interpreting the results of grafts of portions of blastoderms.[3]

A similar gradient of potency for differentiation has been noted in the case of portions (thirds) of 11-day rat embryos grafted on to the chorio-allantoic membrane of chicks, and can be stated in tabular form.[4]

Tissue differentiated	In grafts from		
	Anterior one-third	Middle one-third	Posterior one-third
Nasal sac	Present	Absent	Absent
Brain tissue	Present	Absent	Absent
Hair follicles	Present	Present	Absent
Epidermis	Present	Present	Present
Cartilage	Present	Present	Present
Bone	Present	Present	Present
Mesonephros	Absent	Present	Present
Gut	Absent	Absent	Present

These cases in the rat are perhaps hardly comparable to those in the chick, owing to the difference in age of the fragments tested. Until further information is obtainable concerning the existence and possible spread of a labile determination in these forms, it is hazardous to attempt an interpretation of these cases.

[1] Hoadley, 1924, 1925.
[2] Murray and Selby, 1930.
[3] Willier and Rawles, 1931 B; Hunt, 1932.
[4] Hiraiwa, 1927.

§ 13

A word may be said as to certain problems of determination which occur in later stages. As an example, we may take the case of the spurs of fowls, grafted into hosts of the same or opposite sex, a few days after hatching.[1] As might be expected, grafts into hosts of the same sex as the donor develop in the way characteristic for that sex,

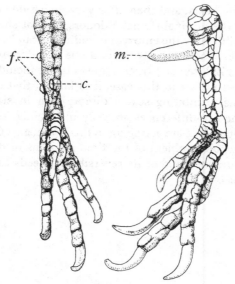

Fig. 126

Differential behaviour of juvenile male and female spurs grafted into young female fowls. The two legs of a hen 18 months old into which, when newly hatched, two female spurs and one male spur had been grafted from day-old donors; *f.* the two grafted female spurs have remained the same size as the control spur (*c.*) which has developed on the host; *m.* the grafted male spur has enlarged to the dimensions characteristic of a spur on a normal cock. (Redrawn after photo in Kozelka, *Journ. Exp. Zool.* LXI, 1932.)

remaining rudimentary in the female, but attaining a large size in the male. Whereas, however, female spurs in a male host are capable of male-type development (although, owing to an inhibitory effect of male environment on female tissues, this is not universal), male spurs on a female host regularly develop masculine

[1] Kozelka, 1932, 1933.

size and other male-type characteristics. It would thus appear that male-type development has been already determined in the spur-rudiments of the young male chick, although these are still very small. The spur-rudiments of the female chick, on the other hand, are in a labile, undetermined state. Whether the determination in the male has been effected by the testis hormone acting on the rudiment is not known. Against this is the fact that female spurs grafted to male hosts, and then after varying periods up to 24 days replanted in the original (female) donors, do not show male-type growth. Possibly only the embryonic rudiment can be sensitised by male hormone. The alternative hypothesis of different reactivity of ZZ (male) and ZO (female) tissues must also be included (fig. 126).

In contradistinction to this case, it is known that in many vertebrate organs, exhibiting sexual dimorphism in size and other characters, the sex-differences are only maintained so long as the hormones responsible are acting upon them (see, e.g., Goldschmidt, 1923). The whole problem of the time-relations of determination exerted by hormones, and of its reversibility, needs careful experimental analysis.

Chapter VIII

FIELDS AND GRADIENTS

§ 1

If a simple animal such as a Planarian is cut transversely into two pieces, normally the front piece will form a tail at its hind end, and the hind piece will form a head at its front end. But if the transverse cut had been made a short distance farther back in the body, those cells which in the previous experiment belonged to the hind piece and proliferated to form a head, will now belong to the front piece and will proliferate to form a tail. Therefore the determination of the quality of the structure which is formed cannot be based on any localisation of specifically different materials or substances, for, if so, it would be impossible to understand how either a head or a tail can be formed from identically the same tissues. How, then, is the quality of the structure which will be formed determined?[1]

A situation in some respects comparable with that just described occurs in the regeneration of the limbs of newts. An amputated limb gives rise to a regeneration-bud, from which an arm or a leg, as the case may be, is eventually formed. These structures can be easily distinguished by the number of digits and other criteria. But at the outset of this process of regeneration there is no qualitative determination of arm-forming as opposed to leg-forming material in the regeneration-bud, for an arm regeneration-bud can be grafted on to the stump of an amputated leg, where it will develop into a leg, provided that the operation is performed soon enough after the amputation of the arm and the formation of the arm regeneration-bud. The converse experiment of grafting a leg regeneration-bud on to the stump of an arm leads to the formation of an arm under the same conditions.[2]

The tissue regenerated by an arm or a leg is at the outset not even determined to produce a limb. The early regeneration-bud of a

---

[1] See also J. Loeb, 1912. [2] Milojević, 1924.

limb grafted on to the base of a tail actually produces a little super-numerary tail (fig. 127).

The undetermined stage of the regeneration-bud is of limited duration. Whereas a bud of hemispherical form is still undetermined, by the time a markedly conical shape has been attained, the bud is determined, and if grafted elsewhere will now continue to differentiate in accordance with its place of origin instead of in accordance with its new situation. In this respect regenerated tissue behaves just as do the various regions of the amphibian egg, which also pass from a plastic to a determined phase.[1]

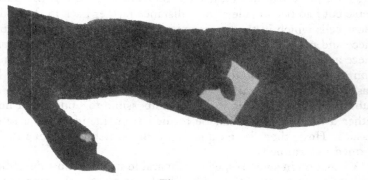

Fig. 127

Lack of determination in early regeneration-buds. *Triton* larva showing a tail (against a square of white paper) developed from an early limb regeneration-bud grafted into the tail-field. (From Guyénot, *Rev. Suisse de Zool.* xxxiv, 1927.)

The success of the converse experiment in which the early re-generation-bud of a tail is grafted on to the stump of an amputated hind-limb, or into the fore-limb field, close to the base of the (un-operated) host-limb, and then produces a limb,[2] has also been reported, but this, though highly probable, cannot be regarded as conclusively proved[3] (fig. 128).

[1] Guyénot, 1927; Guyénot and Ponse, 1930.

[2] Weiss, 1927 B.

[3] The experiment was done with regeneration-bud and host belonging to the same species, and it is difficult therefore to be absolutely certain that the limb developed from the grafted cells. Further, as pointed out by Guyénot, the graft may have come under the influence of the endings of the brachial nerve, which are known to be able to produce the formation of a limb (see p. 362). However, the presumption is that Weiss' interpretation is correct.

The results of the regeneration experiments, as well as those conducted on embryos undergoing embryonic development, agree in
demonstrating two important points. The first is that tissue which
is about to differentiate into a given structure is at the outset undetermined, and therefore capable of differentiating into other
structures, of wholly different type. The second is that the actual
decision as to the fate of such undetermined tissue rests with its
position relative to some major system. In the amphibian egg, the
determining factors are the level of the tissue along the main egg-
axis, and its distance from the organiser region (see p. 139). In the

Fig. 128

Lack of determination in early regeneration-buds. The smaller limb here shown
(right) was produced from the early regeneration-bud of a *Triton* tail grafted on to
the stump of a hind-limb. Left, unoperated hind-limb of other side. (From
Wells, Huxley and Wells, *The Science of Life*, London, 1929, after Weiss.)

case of the Planarian cut transversely, the determination of the
pieces of freshly regenerated tissue are controlled in relation to the
polarity of the whole organism: front edges of hind halves produce
heads, hind edges of front halves produce tails. In regeneration in
newts the type of differentiation is controlled by the local environment of the regeneration-bud; this exerts qualitatively different
effects in different regions of the body (e.g. region of leg as against
region of tail). The material of the early regeneration-bud is indifferent. So far, it has been shown that its capacities of differentiation include organs of such different type as limb and tail; it would
be of great interest to determine whether it was so completely

undetermined as to be able to produce any structures, internal or external.

§ 2

On pushing analysis further, it is found that the original control of differentiation in all cases appears to be exerted in relation to what may be called a biological or morphogenetic *field*. Within these

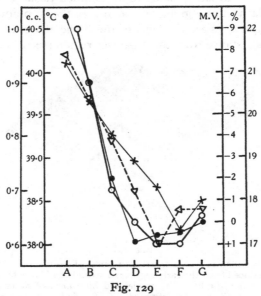

Fig. 129

Gradients of various kinds in the earthworm *Pheretima*. O——O and outer left-hand scale, oxidisable substance as determined by the Manoilov reaction. ×—— × and outer right-hand scale, solid content, per cent. ∇——∇ and inner left-hand scale, temperature at which heat-shortening occurs. ●——● and inner right-hand scale, electrical potential, millivolts. (Redrawn after Watanabe, *Sci. Rep. Tohoku Imp. Univ.* VI, 1931.)

fields, various processes concerned with morphogenesis appear to be quantitatively graded, so that the most suitable name for them is field-gradient systems, or simply *gradient-fields*.[1]

[1] Historically, concepts of this type were first introduced into embryology by Boveri (1901, 1910); later Child (1915 A) generalised a large number of observations in the form of his theory of axial gradients; the term *field*, however, was only introduced in the last few years, notably by Spemann ("Organisationsfeld") (1921), Gurwitsch (1922, 1927), Weiss (1927 C), Bertalanffy (1928), de Beer (1927).

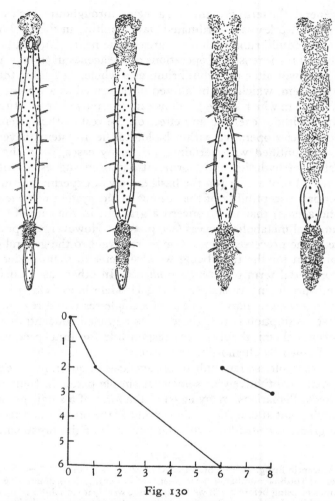

Fig. 130

Differential susceptibility in the primitive oligochaete *Aeolosoma*. Above, four stages in the disintegration of a worm with a well-developed posterior zooid nearly ready for detachment, exposed to $N/100$ KCN. Below, graph of the axial gradient of a similar specimen. The abscissae represent the ordinal number of the segments, the ordinate the time to death in the toxic solution, in minutes. (From Hyman, *Journ. Exp. Zool.* xx, 1916.)

In general, the term *field* implies a region throughout which some agency is at work in a co-ordinated way, resulting in the establishment of an equilibrium within the area of the field. A quantitative alteration in the intensity of operations of the agency in any one part of the field will alter the equilibrium as a whole. A field is thus a unitary system, which can be altered or deformed as a whole; it is not a mosaic in which single portions can be removed or substituted by others without exerting any effect on the rest of the system.

The agencies operative within biological field-systems have not yet been identified with certainty. In many cases, as in the regeneration of hydroids and worms, it has been suggested with a good deal of probability (on the basis largely of experiments on the differential susceptibility of the regions of the system to toxic and narcotic agents) that they concern a gradient in the rate of some fundamental metabolic process (see p. 301). However, the precise nature of the processes in question is irrelevant to the general discussion, and for the time being we shall refer to them under the non-committal term of *activity-gradients*. In other cases, such as the limb-producing capacities of the Urodele limb-field (p. 222) which concern the morphogenesis of a single restricted region, the simplest assumption is that there exists a graded concentration of the specific chemical substances responsible for limb-production and laid down by chemo-differentiation.

In all examples so far studied, the agencies in question appear to be graded quantitatively in somewhat simple patterns, frequently (Hydroids, Planarians, many eggs) in the form of a single gradient with high point at one end and low point at the other, the direction of the gradient coinciding with the long axis of the organism. It

---

Fig. 131

Axial susceptibility-gradients of various oligochaete worms. The abscissae represent the ordinal number of the segments of the worm, the ordinates the time in minutes elapsing before death when exposed to weakly toxic solutions of KCN ($N/100$ to $N/500$). Above, left, susceptibility-gradient of a mature *Aeolosoma* in which secondary zooid formation has not begun. Above, right, the same for an *Aeolosoma* in which the shape of the gradient indicates that the processes leading to the formation of a posterior zooid have been initiated (compare also fig. 130, in which a posterior zooid is visibly differentiated). Centre, susceptibility-gradient of an individual of *Dero* without visible fission-planes. The posterior rise in susceptibility, characteristic of most oligochaetes and associated with the subterminal growth-zone, is well shown. Below, the same in *Lumbriculus*. The posterior rise is more marked, and concerns a larger proportion of the body-length. (From Hyman, *Journ. Exp. Zool.* xx, 1916.)

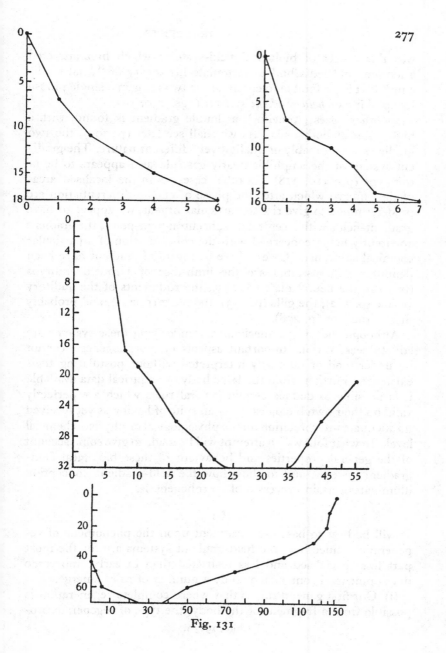

Fig. 131

was this aspect of biological field-systems which first attracted attention, and led Child to formulate his theory of "axial gradients".[1] It is preferable to combine the two ideas in a single phrase by speaking of *field-gradient systems* (figs. 130, 131).

In other cases (Annelids), a double gradient is found, with a high point at both ends. As we shall see later (p. 309), the two gradients are probably of qualitatively different nature. The gradient-system of the amphibian early gastrula also appears to be of this type (pp. 310, 318). In other cases, as in the localised areas of the embryo which after the phase of chemo-differentiation are predetermined to give rise to particular organs, we appear to have gradient-fields with a central or subcentral high-point, the gradient apparently being concerned with the concentration of a particular chemical substance. Cases where this form of gradient have been definitely demonstrated are the limb-disc of Urodele embryos (p. 222), the neural plate (p. 243), the rudiments of the auditory vesicle (p. 232), the gills (p. 233), the heart (p. 233), and probably that of the lens (p. 238).

Although the precise mechanisms underlying these systems are still to seek, various important aspects of morphogenesis cannot be understood or rationally interpreted without postulating their existence. Further, from the large body of empirical data available it is possible to deduce certain general rules which are perfectly valid on their own biological level, in spite of having as yet received no adequate interpretation on the physiological or physico-chemical level. In what follows, an attempt will be made to give some account of the general properties and behaviour of these biological field-gradient systems, and to show how the field-gradient conception illuminates certain processes of morphogenesis.

§ 3

It will be best to base our treatment upon the phenomena of regeneration, since here the field-gradient systems are for the most part less specialised and less restricted than in early embryonic development. From such a study a number of rules emerge.

(i) Our first general rule is that where complete regeneration is possible from a fragment of the body, the type of regenerate pro-

[1] Child, 1915 A.

duced is normally controlled in relation to the polarity of the fragment.

It is well known that the bodies of Coelenterates, Planarians and Annelids are polarised. A differential of some sort exists between different levels, so that in cut pieces the end nearest the apical region usually regenerates a new apical region (see p. 271), while the other cut surface usually regenerates a posterior end. (The exceptions to this statement are treated later (p. 296), and it will be found that they can all be formulated in terms of another general rule.)

In these cases, the cut piece contains a portion of the general gradient-system of the entire individual, which piece by the fact of cutting becomes isolated as a separate field-system in which the factors determining polarity are still graded from apical to basal end.

This rule, however, needs some amplification. In *Planaria*, if a transverse fragment is divided in the middle line into two halves, both will form a head at the anterior end. But if it is divided into three pieces, the central of which includes the main longitudinal nerve-trunks, the two outer pieces will, if below a certain length, form heads either obliquely at the medio-anterior corner, or at right angles to the original main polarity, on the median cut surface: the percentage of medianly directed heads increases with decrease in the length of the piece.[1] It would appear probable that in this case the new heads are determined in relation to the cut ends of the lateral nerves, which come off transversely from the main longitudinal nerve-trunks, and are of the same essential structure, containing cells as well as fibres. There is of course also a secondary medio-lateral susceptibility gradient in the intact animal, and this is presumably correlated both with the course of the lateral nerves and with the determination of medianly directed heads.

(ii) The second general rule is that the origin of polarity is to be sought in external factors. Either the polarity of the regenerating fragment is taken over from that of the whole organism, which is derived from that of the embryo, which in turn is due to factors external to the egg (pp. 36, 60); or the regenerating fragment acquires a new polarity under the influence of the external agencies acting upon it after its isolation. In some cases, although the frag-

[1] Beyer and Child, 1930.

ment is originally polarised by virtue of possessing part of the general gradient system of the organiser from which it has been isolated, it is possible to abolish this original field and to substitute another for it. This can be done, for example, with pieces of the stem of the hydroid *Corymorpha*, by placing them in dilute solutions of various narcotics (see p. 63). In these conditions the pieces round themselves off, and dedifferentiate. If sea-water is now substituted, they redifferentiate, but with a new polarity, at right angles to the substratum.[1] This is probably to be explained by the greater oxygen-concentration away from the substratum. Normally, the differential established by this means, at right angles to the original polarity and to the long axis of the piece of the stem, is less powerful than the already existing differential due to the physiological gradient between the two ends of the piece. But when this latter has been abolished by narcotics, the other comes into play, and establishes a new physiological gradient.

It is to be noted that although this differential is smaller than that constituting the original polarity of the piece, the polyp eventually formed is normal. Once the gradient has been established, it acts as a realisation-factor for the production of an apical region. If the conditions permit of this developing normally, then, as will be seen later, the rest of the reconstituted organism will be normal, provided that the piece is not too small (see p. 285). This is clearly similar to the processes leading to the establishment of the plane of bilateral symmetry and the grey crescent in Amphibia, described in Chap. IV. Numerous differentials, of very varying intensity, can lead to the establishment of bilateral symmetry: if conditions are normal, the bilaterality of the embryo is always normal, whatever the intensity of the trigger action which has released the processes leading to its formation.

Reversal of polarity has also been obtained by appropriate methods of grafting in *Hydra* and other forms. The reversal may occur in the small engrafted fragment or in the major "host" portion.[2]

In other cases, the regenerating portion of tissue is not isolated in such a way as to take over a part of the original field-system of the organism from which it is derived, and therefore possesses no

[1] Child, 1925 B, 1927.   [2] Goetsch, 1929.

original polarity at all. This is seen in the reconstitution-masses formed from pieces of sponges or hydroids after being strained through bolting-silk.[1] Yet here, too, axiation, or the development of polarity, later appears, presumably in response to external differentials in such factors as oxygen supply. This appears to be comparable with the determination of the main axis of polarity in the oocyte (pp. 36, 65; figs. 27, 132). In passing, it may be mentioned that in the sponge reconstitution-masses, one important step in differentiation, namely the attainment of the two-layered condition, appears to be caused by the migration outwards of the dermal and inwards of the collar-cells from their original scattered positions. Here the fate of the cells is not determined by their position, as is the case with undifferentiated cells (e.g. blastomeres of regulation-eggs, early regeneration-buds), but the already acquired differentiation determines the position taken up. When an excess of collar-cells is present, these cannot be overgrown by dermal cells, and they form spheres or vesicles with the collars directed outwards instead of inwards (see p. 250).

(iii) Our third general rule is that in regeneration the apical region or head is the first to be formed; and that its formation, once initiated, is an autonomous process, independent of the level of the cut, and also of the formation of other regions, whether in the regenerated material or within the old tissues of the piece.

The autonomy of a limited apical region is most clearly seen in the regeneration of Annelid worms. In many of these, the tissue actually regenerated at an anterior cut surface, whatever its level in the body of the worm, never forms more than a restricted head region, composed of a definite number of segments (the precise number varying with the species: it may be as low as two).[2] In Planarians the autonomous apical region is the head; its posterior limits, however, seem not to be quite so sharply fixed as in Annelids. The formation of the cephalic ganglion appears here to

[1] H. V. Wilson, 1907; Huxley, 1911, 1921 A.
[2] In some Annelids, anterior regeneration is complete; i.e. the regenerated tissue produces just those segments needed to complete the front end of the worm, and not a fixed number of segments only (see Berrill, 1931; E. J. Allen, 1921). This appears to depend on the power of growth in the regenerated tissue. However, the extreme anterior end would here too be the dominant region (p. 285).

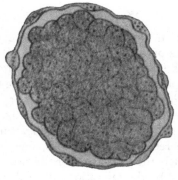

A

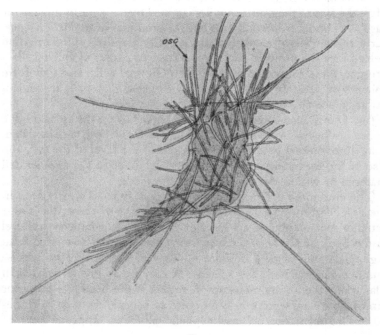

B

Fig. 132

Reorganisation of cell-masses from dissociated cells in the sponge *Sycon*.
A, 2-layered stage attained after 5 days, probably by migration of the dermal cells
to the exterior and the gastral cells to the interior. B, 34 days, a typical *Ascon*
stage has been reached, with open osculum (*osc*) and uniradiate and triradiate
spicules. See also fig. 27, p. 66. (From Huxley, *Phil. Trans. Roy. Soc.* B, CCII,
1911.)

be the most essential feature in the production of a new apical region.

The autonomy of regenerated apical regions in these forms is in striking contrast to the dependence of regenerated basal (posterior)

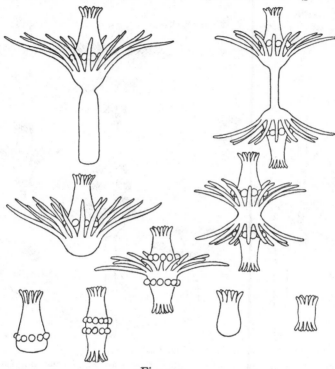

Fig. 133

The independence of the apical region. Partial regeneration in short stem-fragments of *Tubularia*, whether the result is uniaxial or biaxial, gives rise to apical regions, together with as much of the rest of the organism as can be formed from the material available. (From Child, *Individuality in Organisms*, Chicago, 1915.)

regions. In all Annelids and Planarians, the tail region regenerated at a posterior cut surface is formed as a direct continuation of the fragment, and completes the missing parts of the animal. No remodelling is needed, either in the new tissues or in the original fragment; whereas after the formation of a head region of limited extent, a complete animal can only be produced by a remodelling of the organisation of the original fragment (see below).

In Hydroids such as *Tubularia* and *Corymorpha*, the new hydranth is produced entirely by reconstitution within the old

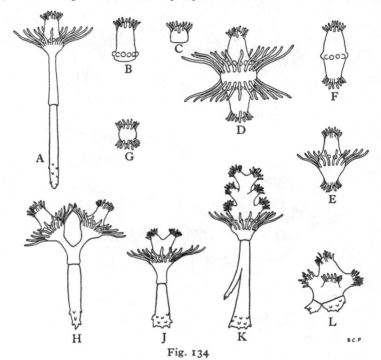

Fig. 134

Reconstitution from pieces of stem in *Corymorpha*. A, Normal unipolar form showing hydranth and base with holdfasts. B–G, Reconstitution of very short pieces to form partial structures, either unipolar (B, C) or bipolar (D, E, F, G). The extreme apical region is always present. In E, the original apical end has formed more than the basal end. H, J, Total and partial twinning of hydranth. K, L, Formation of numerous apical and basal regions in relation to a single hydranth (K) or independently (L). (Redrawn after Child, *Biol. Gen.* II, 1926.)

tissue, not by regeneration from the cut surface, so that here the delimitation of the apical region is less clear-cut. The independence of extreme apical regions is, however, very well shown in these forms (fig. 134).

Extremely small fragments of their stems do not become reconstituted into miniature whole polyps; they produce only apical portions of polyps, but these are of normal size. Such small fragments frequently form an apical region at both ends, for reasons to be discussed later: in such cases, two sets of apical structures are produced, without any basal portion. Similar phenomena occur in the regeneration of very short pieces of *Planaria*. In *Corymorpha*, reconstitution-masses produced from the aggregation of dissociated cells may produce only apical portions of hydranths[1] (see p. 65).

In all cases, what is determined in the first instance is, in fact, the formation of an extreme apical region of a certain standard size, this varying with the size of the piece and also with external conditions. Once this extreme apical region is determined, the region next more basal is determined, and so on, until all the available material is used up. This process may be initiated either at one or at both ends of the piece.

Abnormal external conditions influence the size of the apical region produced. In Planarians, for instance, cold and narcotics reduce its size, while heat up to a certain degree increases it. Beyond a certain degree of cold or concentration of narcotics, no apical region will be formed at all (fig. 135; see also p. 301).

In the most general terms, it appears that the relative size and the degree of differentiation of the apical region depend in some way upon the physiological activity of the regenerated tissue. If this is depressed by cold or narcotics, the development of the apical region is subnormal.

(iv) Our fourth rule is that, once an apical region is produced, it then exerts an influence on other organs and regions within the old tissues of the fragment: this influence is, however, limited in extent. Accordingly, the apical region has been called by Child the "dominant" region. In terms of the field-concept, the apical region establishes a field of a certain extent, which it dominates so as to control the morphogenetic processes of the other regions of the field. The control is exerted in such a way that the various morphogenetic processes occur in harmonious relation with each other: this is because it exerts its control through the establishment of a field.

If the range of dominance is artificially reduced, as by removal of

[1] Child, 1928 B.

some of the more basally situated tissue, the gradient-field set
up by the dominant region is in relation to the reduced size of its

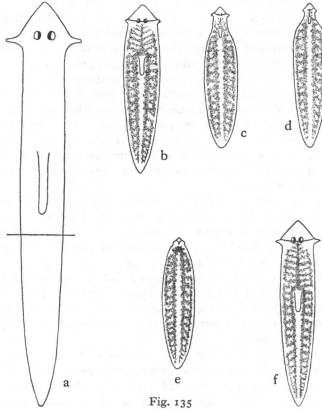

Fig. 135

Correlation between size and degree of development of a regenerated apical region
in *Planaria*, and the extent of its inductive capacity. Posterior fragments are
isolated as shown in *a*. *b*, Regeneration in standard conditions. *c–e*, Regeneration in increasing concentrations of narcotics, showing decreasing size and increasing abnormality of the regenerated head. Correlated with this, the pharynx
induced in the old tissues becomes smaller and less remote from the apical
region. *f*, Regeneration at high (optimal) temperature. The head and eyes are
larger, the induced pharynx farther away and of greater size. (From Child,
Individuality in Organisms, Chicago, 1915.)

possible range. For instance, the reconstitution of a polyp in a
portion of stem of *Tubularia* of a certain length normally results in

the formation of rudiments of distal and proximal tentacles of a certain size, distance apart, and distance from the apical point. But these values are smaller if the piece of stem is shorter[1] (see also p. 318). Similarly, when the Ascidian *Clavellina* undergoes dedifferentiation into a small mass of cells, and subsequently redifferentiates into a well-proportioned *Clavellina* of reduced size,[2]

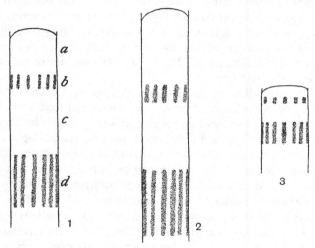

Fig. 136

Modification of the scale of organisation in reconstitution in stem-pieces of *Tubularia*. *a*, Future mouth region; *b*, primordia of apical tentacles; *c*, future hypostome; *d*, primordia of main (basal) tentacles. 1, Under standard conditions. 2, In optimal conditions: the scale of organisation is enlarged. 3, In subnormal conditions: the scale is decreased. (From Child, *Individuality in Organisms*, Chicago, 1915.)

one might say that the various fields are localised in terms of relative quantitative positions along the main gradients: and these relations holding for different total sizes, the control exerted by the dominant region will be harmonic.

As already mentioned (p. 165), the extent of the field dominated by an apical region can be experimentally modified. Narcotics reduce the size of the regenerated head in pieces of Planarians; the size of the reconstituted pharynx, as also its distance from the

[1] Driesch, 1899; Child, 1931.   [2] Huxley, 1926.

288 FIELDS AND GRADIENTS

anterior cut surface, is then a function of the size and degree of differentiation of the head, which in turn appears to be a function of the activity of the regenerating tissue from which it was formed.[1] A similar state of affairs is found in the reconstitution of pieces of the stem of *Tubularia*. Here the apical tentacles constitute the dominant region. The size of the rudiments of these determines the distance between these and the rudiments of the basal tentacles, and can be modified experimentally. In one series of experiments on stem fragments of a definite length, the average length of the primordia of the two sets of tentacles was reduced by 12 per cent. by immersion in $M/150,000$ KCN, and 23 per cent. by immersion in $M/50,000$ solution[2] (fig. 136).

In reconstruction from dissociated cells in *Corymorpha*, the frequency of complete hydranths was much reduced and that of partial forms, consisting of apical portions only, much increased by moving the undifferentiated cell-aggregates about during a certain critical time after their formation instead of leaving them attached to the substratum. The interpretation advanced is that when attached, the differential established between well-oxygenated upper surface and poorly oxygenated lower surface will be large, the resultant gradient steep; when moved, the gradient between apical and basal regions will be less steep, and the structure can therefore differentiate on a larger scale, whereas with a steep gradient it is more compressed.[3] In other words, the morphogenetic field of the polyp in process of reconstitution can be altered as a whole by altering the differentiation of the apical region.[4]

Interesting results have also been obtained in *Sabella* (p. 165). In abdominal fragments of this worm, the number of segments of abdominal type which are transformed into segments of thoracic type by a regenerated head varies from 0 to 75 (the number produced in normal ontogeny is 5 to 11). Here, the agencies responsible for the wide range in the extent of the region morphogenetically affected by the new head in this case appear to reside chiefly in the old tissues[5] (fig. 137).

The conversion of abdominal into thoracic segments, obtained experimentally in *Sabella*, occurs as a normal process in the develop-

[1] Child, 1915 A.   [2] Child, 1931.   [3] Child, 1928 B.
[4] Child, 1915 A, p. 128.   [5] Berrill, 1931.

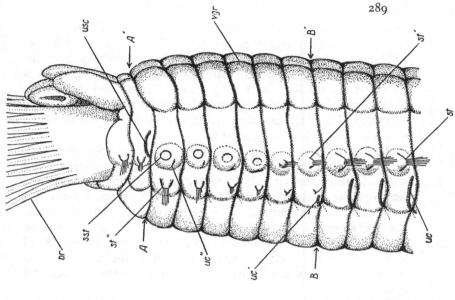

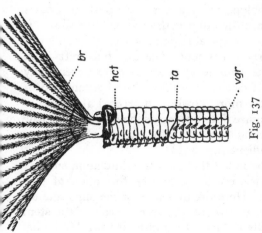

Fig. 137

Induction by a regenerated head in the polychaete worm *Sabella*. Left, anterior end of intact worm, showing branchial palps (*br*); *hct*, posterior limit of collar region, followed by thorax bearing parapodia with dorsal setigerous and ventral uncinigerous lobes; *ta*, abdominal setigerous and uncinigerous lobes reversed in position; *vgr*, ventral ciliated groove. Right, an abdominal fragment with advanced regeneration. *A–A'*, posterior limit of regenerated portion. Between *A–A'* and *B–B'* region with progressive antero-posterior degeneration of old (abdominal) parapodial lobes and regeneration of new lobes in reversed (thoracic) position. Beyond *B–B'* unmodified abdominal segments. *st*, old setigerous lobe, unchanged; *st'*, *sst*, old setigerous lobes commencing and terminating regression; *st''*, newly formed setigerous lobe in thoracic position; *uc*, *uc'*, *uc''*, corresponding phases of uncinigerous lobe; *usc*, uncinigerous lobe of regenerated segment; *br*, branchial palps. (From Berrill, *Journ. Exp. Zool.* LVIII, 1931.)

ment of *Filigrana* and *Salmacina*. The young forms of these worms have three thoracic segments, and the budding zone at the hinder end adds a number of segments of abdominal type behind them. In subsequent development, the number of abdominal segments is increased as a result of the activity of the budding zone, but such a method is of course out of the question in the case of the thoracic segments. These increase their number to ten by conversion of the most anterior abdominal segments.[1]

(v) This fourth rule is really a special case of a more general fifth rule, which is that, within a given field, the differentiation of all regions, other than an apical region, is dependent on influences which proceed from more apical levels. For instance, a piece of a Planarian can regenerate a tail posteriorly even if it fails to regenerate a head. Similarly, whereas a piece of a Planarian from the post-pharyngeal region will not form a new pharynx unless a head is regenerated at its anterior end, a piece from the prepharyngeal region is capable of producing a pharynx even in the absence of a head.[2]

Corymorpha also provides a good example of this. In this hydroid, grafts of a portion of the stem of one polyp inserted laterally in the stems of other polyps will in a certain proportion of cases act as organisers and induce the outgrowth of a new hydranth. It was found that grafts from the apical region inserted at basal levels induced hydranths in nearly 85 per cent. of cases, while grafts from basal levels inserted at the same level in another stem were only effective in 45 per cent. of cases; in addition, the hydranths produced by basal grafts grew more slowly and arrived at a smaller size.[3] The capacity to organise does not reside in any specific tissue but is a physiological condition, the efficacy of which varies quantitatively down a gradient (fig. 138).

In addition, it should be noted that general stimulation such as that produced by an incision will induce the formation of new hydranths in *Corymorpha*. Here the influence of the substrate on the result emerges clearly: for whereas at apical levels of the stem a single incision will usually induce a hydranth, at basal levels this is ineffective, and lacerated incisions are required for induction.

[1] Malaquin, 1919.   [2] Child, 1915 A, p. 102.
[3] Child, 1929 B.

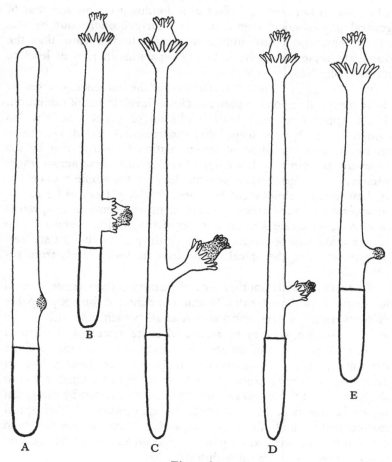

Fig. 138

Induction by grafts in stems of the hydroid *Corymorpha*. The grafted fragment
is shown stippled. A–C, Distal fragments of stem engrafted at proximal levels in
the host stem. A, An early stage. B, C, Two specimens after 48 hours' development. The graft induces an outgrowth, which it organises to form a complete
large hydranth. D–E, Proximal fragments of stem engrafted at proximal levels
in the host stem, after the same length of time as B and C. The resultant hydranth is smaller (D) or subnormal and delayed (E). (From Child, *Physiol. Zool.*
II, 1929.)

The relation between the effect of a dominant region and that of general physiological stimulation is clearly brought out by this experiment, and lends additional weight to the view that the dominant region owes its inducing capacities in part at least to its high physiological activity.

(vi) The sixth rule is that one at least of the influences exerted by the more apical regions on regions at lower levels is that of inhibition. There appears to exist both inhibition of general activity (as evidenced chiefly by susceptibility experiments), and also of differentiation. The inhibition of differentiation is well shown by the following experiment. If a polyp of *Haliclystus* be cut across transversely, it will regenerate new tentacles over the whole cut surface. If, however, an oblique cut be made, reaching down as far as the transverse cut in the previous experiment, and continuing upwards so as to leave intact a small portion of the original distal rim, no regeneration will occur on the less apical part of the cut surface. The presence of the apical region inhibits lower levels from regenerating.[1]

This rule is really another way of putting certain consequences of our third and fourth rules. Within the region of the body capable of regenerating a new apical region at all (which may include the whole organism, or may be restricted to its more apical portion: see p. 297), any piece of tissue, if by reason of an operation it finds itself at the front cut surface of a fragment, can develop into an apical region. That it does not do so in the intact animal is due to the presence of the apical region. The control exerted by the apical region is thus twofold: it inhibits the appearance of other apical regions within the limit of its field, and it influences the tissues to develop into subordinate organs in relation to the morphogenetic gradients which it sets up within its field.

The inhibition set up is not merely morphogenetic; it is also trophic. In portions of Hydroid colonies kept in suboptimal conditions, the stolons that are formed frequently detach themselves from the stock and move slowly across the substratum, their original tip leading the way. This appears to be due to the tip being the dominant region within the subsidiary gradient-field of the stolon: it is able to grow by abstracting material from the proximal, sub-

[1] Child, *Sci. Rep. Tôhoku Imp. Univ.* 4th Ser. Biol. VIII, 1933, p. 75.

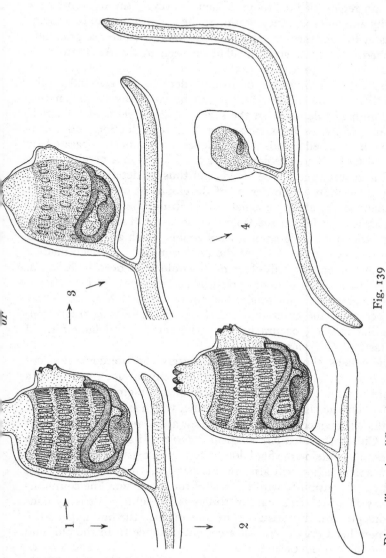

Fig. 139

Diagram illustrating differential susceptibility and resorption the Ascidian *Perophora*. 1, Normal zooid isolated from the colony together with a short length of stolon. 2, In favourable conditions, maintenance and trophic dominance of the zooid, resorption of the stolon tissue. 3, 4, In slightly unfavourable conditions, dedifferentiation and resorption of the zooid, trophic dominance and growth of the stolon. (Original, based on Huxley *Quart. Journ. Micr. Sci.* LXV, 1921.)

ordinate region of the stolon. When, however, the hydranths are healthy and vigorous, they dominate the stolon and maintain themselves at the expense of any attached stolons, which are gradually resorbed. A similar state of affairs is seen in the Ascidian *Perophora* (see fig. 139 and p. 425).

The most striking case of trophic dominance is found in the flatworm *Stenostomum*.[1] Here it can be conclusively shown that the dominance depends on the *degree of development* of the apical region. *Stenostomum* possesses asexual reproduction and forms chains of attached zooids (up to eleven in number), separated by fission-planes. These fission-planes are formed in a regular order, and the relative age of the zooids can thus be determined, as well as by inspection of the degree of development of the head. If a fragment of the chain be isolated by cutting, the zooid possessing the oldest head left in the fragment normally resorbs all younger zooids and any headless portions of zooids which are anterior to it. This is shown in fig. 140. If the posterior cut had been made a little farther back, a still older head would have been included in the fragment, and would have resorbed all regions anterior to itself. If the fragment is made so short as not to contain a head, regeneration occurs at the anterior cut surface, and there is no resorption. A similar relation occurs between the earlier- and later-formed holdfasts of *Corymorpha*.[2]

These facts show that it is not merely the presence of a cut surface which leads to regeneration: the cut surface must be in a certain relation to the gradient-system of the fragment as a unit.[3]

(vii) This leads on to a seventh rule, which is a corollary of the fourth. This concerns the origin of new apical regions as a result of what Child has called *physiological isolation*. If a portion of tissue comes to lie outside the field dominated by an existing apical region, a new apical region will arise in this portion, even though it is still in physical continuity with the rest of the organism. The commonest way in which this state of things is brought about is by continuous growth. For instance, in *Stenostomum* the first appearance of a new head only occurs at a certain distance from the old, and

[1] van Cleave, 1929; Child, 1929 A.   [2] Child, 1928 A.
[3] It is possible that the phenomena of the graded distribution of growth-potency in the animal body (see p. 366) is correlated with this trophic effect of one part of a morphogenetic system upon another.

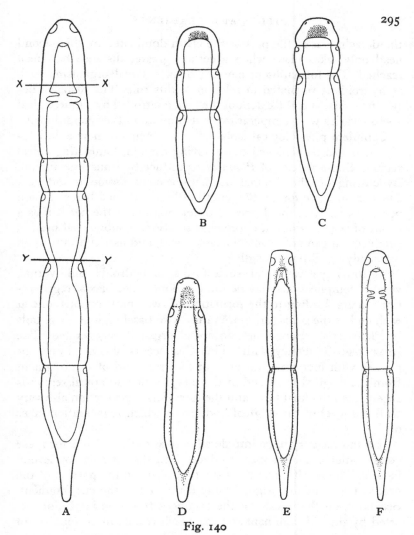

Fig. 140

A chain-forming flatworm, *Stenostomum grandis*. Physiological dominance of zooids with older head-regions over those anterior to them which are headless or have less advanced head-regions. A, Chain of five zooids, showing piece isolated, between X–X and Y–Y. B, The headless anterior zooid-fragment is partly resorbed. C, It is further resorbed but is attempting to differentiate a head. D, It has been totally resorbed, and the next zooid is undergoing resorption. E, F, The original posterior zooid, with the oldest head in the fragment, has resorbed all the material anterior to it, and has divided to form a younger, more posterior zooid. (From Child, *Arch. Entwmech.* cxvii, 1929.)

the detachment of the part of the chain dominated by this second head only takes place when a certain greater distance has been reached. The formation of new zooids in colonial organisms such as hydroids is regulated in relation to this rule. The distance between zooids—i.e. the extent of the field controlled by a more apical zooid—varies with temperature, nutrition and other conditions.[1]

Complete physiological isolation of an incipient new apical region can also be achieved by removing the old dominating apical region. Some species of *Planaria* reproduce by transverse fission. By cutting off the original head, precocious fission is induced. Further, in these forms, the length of body attained before fission occurs varies with the degree of differentiation of the head: if as a result of regeneration in depressant solutions a subnormal head is produced, it can only control a small field, and fission occurs at an unusually small body-length.[2]

In the regeneration of such forms as hydroids and worms, various complications may be found. Sometimes biaxial regeneration occurs, leading to the formation of two apical regions, one at each end of the piece (or, more rarely, two basal regions—e.g. tails in Planarians). Sometimes no apical organ is regenerated. The percentage frequency with which this occurs almost always increases with increasing distance of the front end of the fragment from the original front end of the body. When an apical region is regenerated, its final form and the rate of its regeneration also vary with the level of the original body from which regeneration takes place.[3]

It is unnecessary to go into detail here as to the reasons for these complications. They appear to depend on the interplay of several factors. The result depends in the first place on the portion of the gradient-field of the original body contained in the cut fragment. Secondly, on the release of the fragment from the inhibition exerted by the old dominant region, which results in an increase of

---

[1] Child, 1929 A.

[2] In plants, physiological isolation has been obtained by exposing to low temperature a portion of the region (e.g. a runner) connecting dominant and subordinate parts. Even though under these conditions the runner continues to grow, the field is interrupted, and a new plant is precociously formed at the free tip of the runner. Similar experiments have not yet been successfully carried out on animals. See Child, 1915 A.

[3] Šivickis, 1931 A and B.

physiological activity throughout the piece; the extent of this increase will vary with the age and position of the fragment. Thirdly, the operation of cutting also results in an increase of activity: this is intense close to the cut, and then appears to grade away rapidly. Fourthly, external conditions influence the activity both of the old tissues and still more of any new tissue proliferated at the cut surface. This question has been discussed at some length by Child.[1] Comparable results have been shown to occur in the regeneration of fragments of certain plant tissues, such as seakale roots.[2]

(viii) So far as the facts are relevant here, we may sum them up in the form of the following rule: The frequency or absence of regeneration, and the type of structure regenerated appear to depend (a) on the level of the cut surface within the original gradient-field, and (b) upon the form and steepness of the gradient eventually established between the proliferating tissues at the cut surface and the rest of the piece.

As Child has epigrammatically put it, when a new apical region is regenerated, it arises not because of the activities of the rest of the fragment, but in spite of them.

As a corollary of these various rules with regard to the establishment of polarity, at the autonomy and subsequent dominance of the apical region, the facts concerning the varying number of structures produced by a given piece of tissue may be satisfactorily explained. A given length of *Tubularia* stem normally possesses but a single hydranth, whereas regeneration experiments show that it is capable of producing dozens. The limb-disc of a Urodele, if cut up and the pieces grafted, can produce several fully developed limbs: why in normal development does it only produce one? In the regeneration of a fragment of *Corymorpha* stem, sometimes one new dominant region is produced, sometimes two, sometimes several: why is this?

The reason that a given field normally gives rise only to one of the structures characteristic for it is due to the inhibiting effect of a dominant region, once initiated, upon the development of other dominant regions. Normally the gradients within the field are such as to give one region a start; this becomes the dominant region and inhibits the potentialities of other regions. This is well seen in

[1] Child, 1915 A.   [2] Jones, 1925.

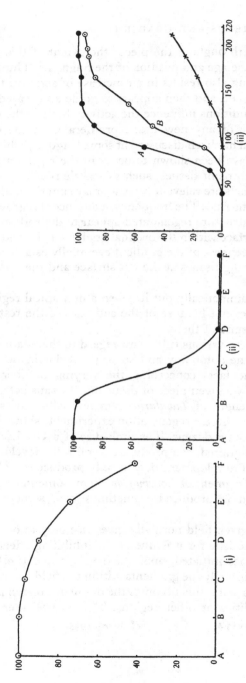

Fig. 141

(i) and (ii), Gradients in regenerative capacity in the triclad flatworms, *Phagocata* (i) and *Dendrocoelum* (ii). The ordinates represent the percentage of heads with eyes, which are regenerated; the abscissae the equal portions into which the body was cut, *A* being the most anterior. (iii) Gradient in regeneration-rate of pieces of the body from different levels (as in (ii)) in *Dendrocoelum*. Ordinates as in (i) and (ii); abscissae, time in hours. (After Šivickis, *Arch. Zool. Ital.* XVI, 1931, and *Arb. 2. Abt. Ung. Biol. Forsch. Inst.* IV, 1931; modified.)

Urodele limb-buds. If a limb-disc is removed and grafted on to the flank of the same animal, it will develop into an independent limb if sufficiently far from its original position. But if the site of grafting is within three segments of its original position, it is within the sphere of dominance of the limb developing from the portion of limb-disc left *in situ* and becomes resorbed.[1] Experiments on Anura have had similar results.[2]

If an entire limb-disc be grafted, it often develops into two or three limbs. In this case the operation has upset the normal gradient system, and permitted supernumerary centres of activity to develop. This fact is of great interest in its bearing upon dichotomous growth; for it shows that a field which normally gives rise to a single set of structures can under slightly altered conditions be made to give rise to two.[3]

Similar agencies are at work in a Hydroid or a Planarian. The various regions of the body, though each capable of producing a new apical region, are all held in check by the existing head or hydranth. When, however, growth has removed them to a sufficient distance, the inhibition can no longer act on them, and they do develop into apical regions.

The double multiple forms are of great interest. Two-headed Planarians can be produced by splitting the anterior end and preventing the two halves from reuniting; and types with doubled apical region can be produced by similar means in sea-anemones.[4] In cases of normal dichotomy of branching organisms, the duplication of the axis and main structures is brought about by growth; however, what initiates the division of the growing-point in these forms is not yet known.

In fragments of hydroid stems, biaxial hydranths are formed when conditions at the two ends are such as to produce two positive gradients of sufficient intensity to initiate the formation of an apical region. Neither region has a sufficient advantage to inhibit the development of the other. If either cut surface is handicapped by being enclosed in paraffin or stuck in the sand, only the other end produces a new hydranth. In *Tubularia* the stem is enclosed in

---

[1] Detwiler, 1918.　　　　　　　　　　　[2] Hellmich, 1930.
[3] The symmetry relations of the supernumerary limbs are of much interest: this problem is considered on p. 224.
[4] Child, 1924, p. 161. See also below, p. 327.

a perisarc which acts as a handicap to all regions within it, and permits of regeneration only at the ends. In *Corymorpha*, however, the perisarc is absent over most of the stem-length in large specimens. As a result, multipolar forms often arise, especially from short fragments[1] (fig. 142).

A very striking example of the multiple production of dominant regions is seen in the sea-anemone *Harenactis*. When portions of the body are isolated, they roll up to produce hollow structures like a tyre, the original distal and proximal cut surfaces meeting and growing together. It is then found that regeneration is initiated at a number of places along the line of suture. Regeneration is found especially at places where, owing to irregularities of the cut surfaces, union has not been smooth. At each of these spots, conditions are favourable for new growth, and therefore the establishment of new apical organs; and the various regions of new growth are isolated from each other by other regions in which smooth union, leaving no free cut surface, has taken place.[2]

As will be seen in the next chapter, these phenomena of double or multiple organisations arising from a single portion of tissue are of great interest in the interpretation of various facts in ontogeny.

Fig. 142

Single and multiple regeneration in the sea-anemone *Harenactis*. Left, diagram of *Harenactis* to show (*a* and *b*) sections isolated for regeneration. These roll in to form hollow tubes, as shown (centre) in section: the distal and proximal cut surfaces unite in a suture, here shown centrally. Right, regeneration of apical regions (whole or partial tentacle groups) from the suture. When the suture is irregular, with considerable proliferation, a number of apical regions can arise (above); when the union is smooth, one regenerate dominates and inhibits the development of others (below). (After Child, *Physiological Foundations of Behavior*, New York, 1924; modified.)

Before passing to our next section, the views of Goetsch[3] should be mentioned. He finds that regeneration is frequently accompanied by a polarised migration of cells. In some cases, certain types of cells have the tendency to migrate apicalwards, other types

[1] Child, 1926. [2] Child, 1924, p. 119.
[3] Goetsch, 1929.

basalwards. In other cases, indifferent cells migrate in both directions, but become progressively differentiated in different ways according as they are moving apicalwards or basalwards. The presence of a lateral graft, e.g. in *Hydra*, will induce a flow of cells towards the graft; as is shown by heteroplastic experiments in which the two types of tissue can be distinguished, some of these cells grow out to form a base below the graft.

There is thus a form of "dynamic determination" (see p. 163), and although the graft acts in a way resembling an amphibian organiser, it does so largely by a different method, namely, by stimulating directive growth-processes. Something of the sort occurs as part of the induction of new hydranths by stem grafts in *Corymorpha* (p. 164), and in the case of grafted amphibian limbs (p. 364). On the basis of these and numerous other experiments he comes to conclusions rather different from those of Child. However, although it is probable that the further analysis of these directive movements of migration and growth will throw much light on regeneration and differentiation, they cannot explain a number of the facts previously cited in this chapter, for which some form of field-theory is indispensable.

(ix) Next we come to an extremely important rule, which is that the action of external conditions upon gradient-fields and the morphogenetic processes associated with them is always differential. This appears to be a consequence of the quantitatively graded nature of the fields. The differential action is revealed under three main heads, (*a*) differential inhibition, (*b*) differential stimulation or acceleration, (*c*) differential acclimatisation and recovery.

(*a*) When depressant agents are used in concentrations not permitting acclimatisation, the most active regions are the most susceptible, and suffer most. This is well seen when regenerating Planarian fragments are exposed to narcotics. The heads regenerated under such conditions are not only subnormal in size, but abnormal in form, in that certain regions are missing. The process takes place progressively as the toxicity increases: first of all the eyes become approximated, and then fused (absence of interocular region); then the median part of the pre-ocular region fails to form. Higher concentrations affect more lateral and more basal parts of the head, until finally only a small basal head-rudiment, eyeless but with rudimentary ganglia, is produced. Higher concentrations inhibit head-formation altogether, and only healing occurs (fig. 143).

(b) Differential acceleration occurs in response to exceptionally favourable conditions. The effects are the exact reverse of those obtained by differential inhibition, though such extreme departures from the normal are not seen, since there are no regions which fail to form. As illustration we may take the fact that optimal high temperature applied during regeneration of Planarian fragments leads to the formation of heads which are not only relatively large, but have widely separated eyes and an unusually large pre-ocular region.

(c) Differential acclimatisation occurs in certain low concentrations of depressants. In these it appears that the most active regions, although the most susceptible, have the greatest power of acclimatisation, and after a time show differential development. For instance, intact normal Planarians placed in weak alcohol or ether first show a differential reduction in size of head, the whole pre-ocular region disappearing. Later, new growth sets in, and this is abnormally high in the most median and most anterior regions, leading to "snouted" forms[1] (fig. 143).

In slightly stronger concentrations, this differential action will not take place during exposure to the solution, but occurs on replacement in water. In such cases the process is strictly speaking one of differential recovery instead of differential acclimatisation, but the results are in most respects similar.

(x) In addition to these statements, applicable to regeneration of the complete type, within total fields permeating the whole body, there must be mentioned another very important rule derived from a study of partial regeneration in a local field. This is that the various tissues of the regenerated region need not be proliferated from corresponding tissues in the old region, but are determined in relation to a gradient-system which extends out from the old region into the proliferated material. Total regeneration appears normally to take place in two phases—first the formation of a new apical region, and secondly the remodelling of the old tissues under the influence of this apical region. However, in partial regeneration, e.g. of an amputated limb or tail, the new tissues are not known to

[1] Child, 1921 A.

exert any morphogenetic effect on the old tissues of the stump: if,
as usually occurs, a complete appendage is restored, this is effected
entirely by means of new growth.

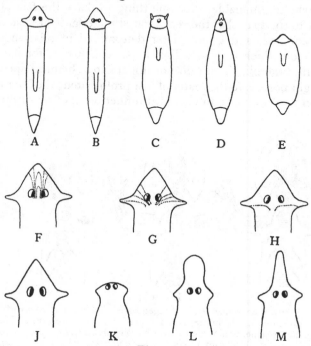

Fig. 143

Differential susceptibility in *Planaria dorotocephala*. A–E, Various grades of
head differentiation after regeneration. A, Normal. B, Teratophthalmic (eyes
approximated or partially fused, head form nearly normal). C, D, Teratomorphic
(single median eye, lateral sensory projections approximated or fused anteriorly).
E, Anophthalmic (no eye, median or no sensory projection, rudimentary cephalic
ganglion). F–H, Diagrams showing, between the dotted lines, the regions
missing in hypotypic heads. F, In teratophthalmic forms (cf. B). G, In teratomorphic forms (cf. C, D). H, In anophthalmic forms (cf. E). J–M, Differential
acclimatisation. J, Normal head. K, Reduction of apical region after 2–3 weeks
in dilute anaesthetics. L, M, Subsequent hypertrophy of the apical region after
1½–2 weeks more in the solution. (Redrawn after Child, *Individuality in Organisms*, Chicago, 1915 (A–E), and *Journ. Exp. Zool.* XXXIII, 1921 (F–K).)

For this to occur, it is clear that exactly those regions removed
by the operation must be restored by the new growth, a phenomenon abundantly confirmed in limb-regeneration in Arthropods and

Amphibia. It has always been difficult to connect this with any purely chemical specificity of the regenerating tissues at one level as against another level of the limb, and recent work has made such a view wholly untenable. For one thing we have the fact already referred to (p. 271) that the regenerated material is at first wholly undifferentiated, and is only later determined in relation to the substrate on which it grows. This is not conclusive, for it merely proves that the old tissues do not impart any chemical specificity they might possess to the material just proliferated; the later determination might be due to chemical influences specific to the level

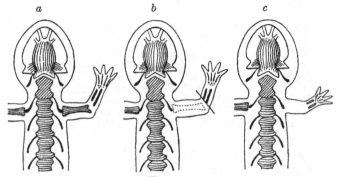

Fig. 144

Diagram to show the independence of regenerated tissues. *a, Triton* with normal fore-limb skeleton. *b,* The humerus is removed, and the fore-arm and hand removed. *c,* The regenerated fore-arm and hand contains the normal complement of skeletal elements. (Przibram, in *Handb. norm. u. path. Physiol.* XIV (1) (i), 1926.)

of the cut. However, it has now been shown that total absence of one kind of tissue, or the substitution of one kind of tissue by another in the regenerating base of the limb, does not interfere with normal regeneration. If the skeleton be removed from the upper arm or thigh region of a Urodele limb, and the limb later cut across in this region, the distal regenerated portion possesses a normal skeleton, whereas no regeneration of the missing parts occurs in the stump.[1] Similarly, if the skin is removed from a limb, an envelope of lung tissue grafted on, and the limb cut across after healing has occurred, the regenerated portion is found to possess normal epidermis, in spite of the absence of such tissue in the stump.[2]

[1] Weiss, 1925; Bischler, 1926.   [2] Weiss, 1927 A.

Such facts can only be interpreted in terms of a field theory. Some general activity must be distributed in a graded way through the limb so as to constitute a gradient-field. The fate of the regenerated tissue is determined in relation to the *level* of the gradient at which regeneration is made to occur, not to the specific tissues present on the cut surface. Further, the determination of the regenerated portion is a unitary process. The regenerated portion is determined as a field, the morphogenetic agencies in which are in equilibrium with those operative in the stump, so that the fractional field of the regenerated portion and that of the stump together make a whole (see also Chap. x, p. 362). Both the products of undifferentiated cells and also certain types of already specialised cells contribute to the regenerated material.[1]

Presumably the morphogenetic gradients in the stump extend as it were by extrapolation into the new tissue, so that it comes to be permeated by the missing portion of the total field: when this occurs, the gradient activities of the whole field are in equilibrium. As regards its gradients, the regenerated portion then constitutes a fraction of a field: but since it alone contains undifferentiated tissue, in its subsequent morphogenesis it behaves as an autonomous field system with basal boundary set by the level of the cut.

The same type of behaviour is seen in the regeneration of a tail in Planarians; the new tissue from the start is determined in relation to the existing gradient-stem of the old piece. It would thus appear that the basalmost regions of a limb are dominant, and correspond, as regards their activities in the gradient-system, to the anterior (apical) region of the whole body in animals capable of total regeneration.

It should be noted that in such cases quite a small fraction of the field (e.g. a short disc cut from a limb) will be able to exert this morphogenetic effect on material proliferated from its cut surface, even when grafted into another region of the body altogether (e.g. a short section of fore-limb stump taken with an indifferent regeneration-bud that has been proliferated from it, and grafted into the hind-limb field; see p. 273). It is also important to find that when a section of a limb is cut out and engrafted elsewhere in reversed orientation, with original proximal cut surface away from the body,

[1] Hellmich, 1930.

306

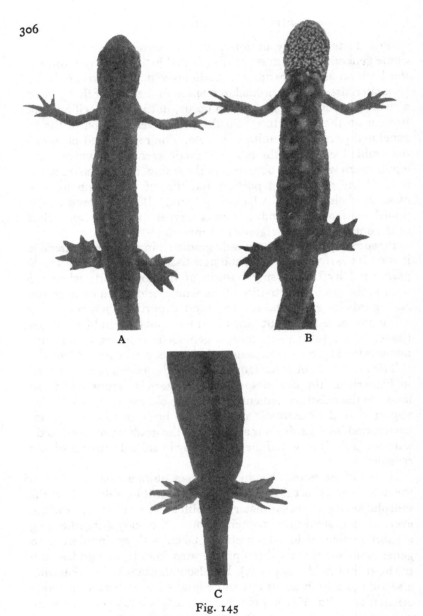

A

B

C

Fig. 145

Regeneration is determined by the level of amputation within the limb-field, and not with reference to the organism as a whole. A, Dorsal, and B, Ventral, views of a newt (*Triton*) in which the legs were amputated above the thigh, and shanks were grafted in their place and subsequently amputated. C, Radiogram, showing that only the tarsus and foot have been regenerated: i.e. structures distal to the graft. (From Guyénot, *Rev. Suisse de Zool.* XXXIV, 1927.)

this free cut surface does not regenerate the missing (i.e. proximal) regions, but produces a structure representing the parts of the limb distal to the level of the cut, although this duplicates regions of the stump.[1] The same is true of tail-fragments.[2] These results show that

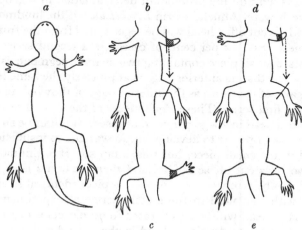

Fig. 146

Diagram showing the morphogenetic effect of the limb-field in regeneration. In *Triton*, an early regenerate bud from a fore-limb cut as in (*a*) is taken and grafted on to a hind-limb stump. If (*b*) grafted with a portion of the original stump, it produces (*c*) a fore-foot; if (*d*) grafted alone, it produces (*e*) a hind-foot. (Przibram, in *Handb. norm. u. path. Physiol.* XIV (1) (i), 1926.)

the explanation given above needs modification. The field is not active within its differentiated regions: the morphogenetic influence is exerted in relation to the character of the differentiated tissue *at the cut surface*.

§ 4

Little is known as regards the precise time-relations of some of the processes, e.g. whether the new morphogenetic gradient-field is established immediately the new head is determined, or not until it has reached some degree of morphological development, such as the formation of a brain, or the penetration of nerves from the new brain into the old tissues. The general sequence, however, is clear.

[1] References in Milojević and Grbić, 1925.
[2] Milojević and Burian, 1926.

But the precise method by which the dominant region exerts its morphogenetic control over the rest of the field is still unknown.

However, an experiment may be described here which not only illustrates the importance of quantitative potential difference, but also throws light on the problem of determination in regeneration. In the fresh-water Annelid worm *Lumbriculus*, if the hindmost fifth of the body is cut off, a head will be regenerated from the front edge of this piece in 90–95 per cent. of cases. In a second series of experiments, a small piece containing two or three segments is cut off in such a way that its anterior edge is at precisely the same level on the long axis of the worm as the anterior edge of the whole hindmost fifth in the first series. These small pieces of the second series only regenerate a head in 20–30 per cent. of cases. It might be supposed that this lack of power to develop a head was due to insufficiency of material in the small piece, but this is not so. If a hindmost fifth of the worm is cut off as before, and then, 20 hours later, a large piece of this be removed so as to leave a piece identical in size and in level with that used in the second series of experiments, it is found that a head will be regenerated in 70 per cent. of cases.[1]

Lack of power to develop a head in the second series of experiments is therefore not due to lack of material, for the pieces of the third series are of the same size as those of the second, but can regenerate a head almost as well as those of the first series. The only difference between the pieces of the third and second series is that for 20 hours the anterior end of the pieces of the third series has been in continuity with the whole hindmost fifth of the worm, and this period of time is apparently long enough for the qualitative determination of a head to be effected, as in the first series. After this determination, reduction in size of the piece does not hinder head-production. The fact that the act of cutting raises the activity of the old tissue in small pieces more than in large pieces where the cuts are farther apart and the stimulation consequent upon them has to act on a much larger mass of material. The anterior edge of small pieces will therefore have more difficulty in obtaining the necessary threshold potential difference for head-determination.

Experiments in every way analogous to those just described on *Lumbriculus* have been performed on *Planaria*, and with similar

[1] Hyman, 1916.

results.[1] It appears that at room temperature the formation of a head is determined in about 6 hours from the time of operation.

§ 5

There is another fact concerning the gradient-systems of adult lower invertebrates which requires consideration, for it throws light on certain processes of embryology. This is the double gradient analysed by Child and his school in Annelid worms. In these animals, as is well known, new segments are added from a growing zone in the penultimate segment of the body. Experiments with dilute toxic solutions show that there is a region of high susceptibility at both ends of the worm, with a minimum at an intermediate point. Child and his school have always attempted to reduce all gradient-phenomena to variations in a single variable, which they have tried to identify with oxidative metabolism, but which, theoretically, might be any general activity of protoplasm. This conception, however, seems definitely to break down in face of the facts in Annelids. Here, two distinct processes appear to be at work. One is the formation of new segments at the hind end associated with the presence of undifferentiated, physiologically young tissue; the other is the controlling and morphogenetic activity of the front end, associated with old tissue and a high grade of differentiation. It is worth recalling that the conditions of formation and the morphogenetic effects of the dominant region in Annelids are similar to what is found in Planarians (see p. 279). If regeneration occurs at all at an anterior cut surface, the normal result is a new dominant region, which never consists of more than a small number of segments, constant for each species; and this, once produced, causes morphogenetic changes in the old tissues, such as the production of a new genital region at the correct distance behind the head, or the transformation of a certain length of intestine into crop and oesophagus, or the conversion of abdominal segments into thoracic segments.[2]

Though both head and tail in Annelids are regions of high susceptibility, the processes at work in the two are entirely distinct. There are therefore two qualitatively different gradients in the organism, and there is every right to believe that the effects of the

[1] Child, 1914. See also Abeloos, 1932.
[2] Harper, 1904; Berrill, 1931.

two will interact—e.g. that the morphogenetic effect of a head of given activity will differ according to the tail-gradient and the effects which this exerts on the old tissues, just as in regeneration from a posterior cut surface, with a given tail-gradient, the morphogenetic results will vary according to the size and activity of the head.

This leads on to a point which may prove to be of great theoretical importance, although so far only limited discussion of it has taken place.[1] It concerns the classification of gradient-fields into two types. The first constitutes what Waddington refers to as an *individuation-field*, in which there exists some form of dynamic equilibrium controlling morphogenetic processes. Removal of one part of the system will, if growth is still possible, lead to the regeneration of what is missing, as above pointed out (p. 276). Further, the induction effect of a dominant region is exerted not by contact as with the amphibian organiser, but apparently at a distance, as with regenerating *Sabella* or *Planaria*; this is because the essential effect of the dominant region is to establish a total field. Another term for these would be gradient-fields of direct effect.

In contradistinction to this we find what may be called gradient-fields of secondary effect. A gradient-system exists, and exerts its effects, not directly, but by giving rise to a graded concentration of some chemical substance which is then responsible for certain morphogenetic effects. It appears that in amphibian eggs the dorso-ventral gradient with the organiser at its high point is of this type. The reasons for this assertion are in the first place that induction is exerted mainly by contact (see p. 135); secondly, that dead organisers may continue to exert their inductive effect (p. 153); thirdly, that there is no evidence of equilibrium or saturation being obtained in the organiser region. This last point requires elucidation. In the bird, a complete "organiser-field", i.e. a sheet of epiblast (ectomesoderm) containing the whole primitive streak and an extensive area around it, is still capable of inducing a neural tube and other organs in a sheet of epiblast from another embryo.[1] The organising capacities of the primitive streak have not been "saturated" in the formation of its own field, as would be expected if

[1] Waddington and Schmidt, 1933.

organisation were an affair of equilibrium between the morphogenetic capacities of the organiser and the neighbouring tissues, as
is clearly the case with a regenerating Planarian (see p. 287).
These two types of gradient are of course not mutually exclusive.
The primary gradient-field of the amphibian egg is an individuation-field; but as a result of its existence, graded accumulations of
yolk and other substances occur, which then exert effects upon
development. In this case, the substances accumulated are mere
raw materials, but in other primary gradients, doubtless, true organ-
forming substances are formed in this way. It is, of course, also
possible to conceive of the graded formation within a gradient-field
of some substance which has no further effect on development, so
that there is no secondary action of the field as occurs with the
organiser. Such fields we may if we like distinguish as fields of indirect action but without secondary effect. Many cases of graded
distribution of pigment within organs are doubtless of this type.
The gradient-field of the amphibian organiser appears to be
essentially one of secondary effect; but it very possibly acts as a
weak individuation-field in the stages before gastrulation. In birds,
as already mentioned (p. 160), portions of the organiser (primitive
streak) when isolated regularly produce more than their presumptive fates, thus showing a tendency to individuation.
Partial fields such as the limb-field in Amphibia appear to partake of both these aspects of field-action. They seem undoubtedly
to be areas in which there has resulted a graded concentration of a
specific chemical substance which is capable of producing limb-
formation: but they also have their own individuation-field, which
sees to it that what is produced is normally neither a partial nor a
multiple structure but one whole organ.
As with any new concept, considerable analysis, both experimental and theoretical, will be needed before the different rôles of
field-systems in ontogeny can be properly understood. Meanwhile,
however, this distinction between fields of direct and indirect action
is a first important step, helping considerably to clarify amphibian
development (see Chap. IX, p. 318).

Chapter IX

FIELDS AND GRADIENTS IN NORMAL ONTOGENY

§ 1. *Polarity in ontogeny*

As already mentioned, the conclusions reached in the preceding chapter are derived from experiments on regeneration and grafting in adult animals. They are also, however, relevant in the normal ontogeny of higher forms, though the conditions here are often more complex and more specialised. In the present chapter it is proposed to illustrate the various principles, so far as possible, from early development.

(i) Polarity and the main axis of the resultant organism

The first rule mentioned in Chap. VIII was that the inherent polarity of a fragment normally determined the polarity of the organism which arose from it. This obviously holds good in normal ontogeny. The egg is a fragment of the mother, in which a well-marked polarity has been set up before it is detached. In the great majority of cases, the main animal-vegetative axis of the egg gives rise to the definitive antero-posterior axis of the resulting organism, with the head or apical region arising at the animal end. In various Echinoderms the main axis of the egg persists as that of the larva, but later a new axis in a different direction is established in the rudiment of the adult.

(ii) Polarity determined by external agencies

We next come to the point that the polarity of an organised portion of living matter has in the long run been determined by agencies external to it; and that in certain cases the existing polarity can be overridden and a new polarity imposed by external conditions. Examples have already been given of how the polarity of the developing oocyte or egg may be determined by factors external to itself, either by conditions within the ovary, or external agencies acting after fertilisation (*Fucus*). Cases have also been adduced in which the axis of bilateral symmetry is determined from without

(p. 60). The rule appears to be of general application for the developing egg.

The overriding or abolition of the original polarity by external agencies appears seldom to be obtainable with eggs; but some remarkable cases are known from Echinoderms. We have already referred (p. 83) to the fact that in developing fragments of *Lytechinus* and *Patiria* eggs, which have been obtained by cutting before fertilisation and subsequently inseminated, the first two cleavage planes are always at right angles to the plane of the cut.

Subsequent development demonstrates that even more radical changes have been effected. When the gastrulation of the fragments occurs, it invariably takes place at the centre of the cut surface, and at right angles to it. The polarity of the developing egg-fragment and the axis of the resultant larva is therefore determined in relation to the cut, and not in relation to the original polarity of the whole egg.[1]

It is to be supposed that the operation intensely stimulates the cut surface, and that the resultant increase of protoplasmic activity grades away across the fragment. The activity-gradient thus produced must be able to override the original gradient within the fragment.

This is also stated to occur in the California species of *Paracentrotus*. However, in the European *Paracentrotus lividus*, meridional halves of the egg produced by isolation of the 1/2 blastomeres appear to retain the original polarity.[2] It is to be noted that in this case no raising of activity by cutting has occurred; the separation also took place at a later stage. Thus the observations on the two forms are not necessarily contradictory.

In *P. lividus* also, marked deformation as a result of centrifuging, however, is incapable of altering the original polarity. The point at which gastrulation is initiated is always at the original vegetative pole, as indicated by its relation to the subequatorial pigment-band. Thus gastrulae are produced which may be extremely elongated, flattened, or obliquely deformed in the animal-vegetative direction[3] (see also Chap. IV, p. 69).

[1] Taylor, Tennent, and Whitaker, 1925.
[2] Hörstadius, 1928.
[3] Harvey, 1933.

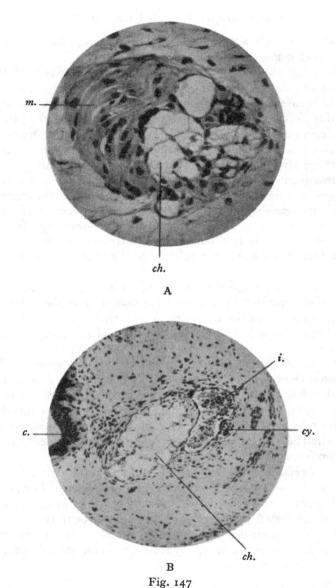

Fig. 147

Atypical (A–C) and typical (D) differentiation of portions of *Triton* early gastrulae, grafted (interplanted) into the orbit of larvae. A, Notochordal tissue from presumptive epidermis.. B, Notochordal tissue from presumptive endoderm. C, Cartilage from presumptive neural plate. D, Epithelial vesicle from pre-

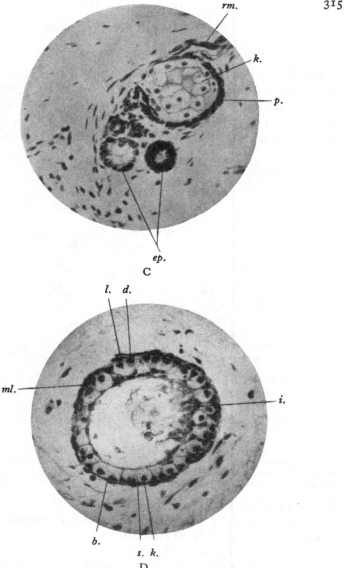

C

D

sumptive epidermis. *b*. basal membrane; *c*, cornea; *ch*. notochord; *cy*. epithelial
vesicle; *d*. covering layer; *ep*. epithelial vesicle; *i*. contents of vesicle; *k*. cartilage;
l. Leydig's cells; *m*. muscle; *ml*. pigment; *p*. pigment; *rm*. muscle; *s*. granules in
Leydig's cells. (From Kusche, *Arch. Entwmech.* cxx, 1929, figs. 9, 13, 20, 22.)

§ 2. *The dominant region in ontogeny*

(iii) Independence of the dominant region

Instances of this are difficult to obtain in ontogeny. The egg cannot regenerate new tissue like a Planarian worm, and we can therefore only compare the morphogenetic processes occurring in it to those occurring by morphallaxis in the regeneration of, for example, a piece of *Tubularia* stem. However, the gradient-system of the egg

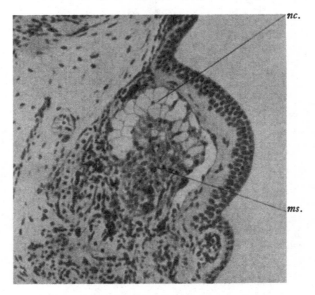

Fig. 148

Differentiation of notochord (*nc.*) and mesoderm (*ms.* muscle) from animal pole material (presumptive epidermis and/or brain) of *Triton* interplanted in the eye-socket of a larva of *Triton taeniatus*. (From Bautzmann, *Naturwiss.* XVII, 1929.)

is almost always more specialised than that of a hydroid stem, being in many cases partly or wholly determined as regards different levels along its main axis. It is also more limited in size, and therefore its gradient is presumably steeper.

However, when both animal and vegetative portions of the egg can reorganise themselves to form perfect wholes, as in some Coelenterates (p. 97), equatorial portions of the egg, originally in the centre of the gradient, and constituting a subordinate region, must in the vegetative half have turned into a dominant region and come to control the new complete gradient-field of the fragment. The curious and apparently anomalous production of notochord and mesoderm by various isolated regions of the blastula of the newt, although these regions may possess the most diverse prospective fates (p. 139, footnote), may perhaps be explained on these lines when it is remembered that isolation of a piece of tissue removes it from the control of the dominant region to which it has been subjected. As noted on p. 285, experiments on Planarians and Hydroids have shown that the tendency in such cases is for a small isolated piece to develop by self-differentiation into an isolated dominant region. The dominant and only self-differentiating region in the late blastula of the newt is the organiser, and the tissue into which it differentiates is notochord and mesoderm: other regions develop in subordination to it. On this assumption, therefore, a piece from any other region, when isolated, should, if environmental circumstances permit, come to be the site of a new dominant region, and differentiate accordingly.[1] However, the occasional differentiation of such pieces into tissues which represent neither the presumptive fate of the piece nor that of the dominant region (organiser) presents a difficulty. We should however recall that whereas in the Invertebrates only a simple field is involved, in Amphibia there are two interacting gradient-fields (pp. 310, 318).

(iv) The modifying influence exerted by the dominant
region on other parts

This is obvious in the example just given of the formation of miniature wholes from animal and vegetative portions of Coelenterate eggs. Regions originally containing but half the length of the main gradient become reorganised to contain whole gradients.

In most of the well-analysed types of ontogeny, however, conditions are more complex than in the regeneration of Hydroids or Planarians, for the main organising activity proceeds from the high

---

[1] In connexion with the environmental circumstances, it is a curious fact that pieces of presumptive neural tube tissue (which has been the tissue most frequently used in these experiments) show a much greater tendency to differentiate into notochord when interplanted into the coelomic cavity of an older larva than when explanted in an inorganic medium (Holtfreter, 1931 A). See also Huxley, 1930.

point of a secondary gradient, established after fertilisation, and the precise morphogenetic effects are due to the interaction of this with the original animal-vegetative field established in the oocyte.

The relation between the gradient-field set up by a dominant region and an amount of tissue representing a reduced range, is well shown in the experiments on newt embryos (described on p. 239) in which the early gastrula is constricted into dorsal and ventral halves. The dorso-ventral gradient is then of half the normal length, and the dorsal half-gastrulae possess neural folds of proportionately reduced size. Another example is provided by the experiments on sea-urchin larvae to be described below (p. 323), in which four micromeres are added to a single ring of mesomeres (disc *an 1*), and a properly proportioned pluteus larva is formed. The main gradient is here represented by one quarter of its original length, and in this case the amount of the dominant region has had to be reduced in order to produce a harmonic result.

§ 3. *The interaction of primary and secondary gradients*

In early amphibian development, for instance, there appears clearly to be two gradients of qualitatively different nature. One is the gradient along the primary egg-axis from animal to vegetative pole; the other, a gradient whose high point or dominant region is the organiser. The first appears to be established during the development of the oocyte in the ovary. It must in the first instance be quantitative and concerned only with some general activity of the cytoplasm: but by the time that the egg is ripe, it has in addition produced a structural effect, in the shape of the graded increase in the proportion of yolk found when passing down the egg-axis towards the vegetative pole. The existence of this gradient has been shown by susceptibility experiments. (See p. 332, and figs. 154, 155, 156.)

Per contra, although the other gradient, which is normally established as a result of fertilisation, has a sharply qualitative aspect in that the dorsal lip region alone is capable of exerting organiser capacities, yet it is also quantitative in other aspects. For instance, it is found that, as determined by cell-size in the late blastula and early gastrula, the rate of cleavage in the future dorsal side of the animal hemisphere is greater than in the ventral side.

Susceptibility experiments also demonstrate the existence of a dorso-ventral gradient in general activity, from the region of the grey crescent ventralwards over the egg. In respect of its position at the high end of a gradient, the organiser of the amphibian egg shows a further resemblance to the dominant regions of a Coelenterate, Planarian, or Annelid. (See p. 68 and fig. 28.)

As already pointed out in Chap. vi, the action of an apical region such as a Planarian head is extremely similar to that of an organiser in ontogeny. Not only does it exert a morphogenetic effect during regeneration, but also when grafted into an intact worm. But the morphogenetic action of the amphibian organiser is normally exerted in a way somewhat different from that of a regenerating head, for its definitive influence is exerted on those parts which it actually comes to underlie as a result of gastrulation, and appears to be a chemical effect, demanding contact for its realisation. In this respect the dorso-ventral gradient of the amphibian egg reveals itself as a gradient-system of secondary effect, thus differing importantly from the apico-basal gradient system (see Chap. viii, p. 310). However, the labile determination effected before the onset of gastrulation can only be the result of action at a distance, as with the effects of a regenerated Planarian head.

In Amphibia the end-result, in the shape of the main morphological organisation of the embryo, is dependent on the interaction of the organising capacity of the dorsal lip with the primary apico-basal gradient-system. The most important action of the dorso-ventral gradient, from the point of view of developmental physiology, is the production of a specific inducing substance localised in the dorsal lip region, which acts as a trigger or releasing stimulus for the differentiation of its own and other tissues. On the other hand, the most important action of the apico-basal gradient is the production of an individuation-field, which sees to it that the development released by the non-specific action of the organiser is in the first place different in different parts of the field, and in the second place is correlated into an organised whole.

Either system also appears to have minor effects of the opposite type to its main effect. The organiser, as just mentioned, appears to exert an action at a distance prior to gastrulation, and this may be comparable with that of the individuation-field set up by a Planarian head; on the other hand it may be that the individuative component of this action is really due to the apico-basal individuation-field, and that the organiser region here again only exerts a releasing action, whether by the diffusion of chemical substances, or by neuroid transmission, or by other means.

It is further probable that the dorso-ventral gradient contributes to the total individuation-field of the embryo, e.g. by introducing a dorso-ventral polarity (see p. 357 for the dorso-ventral polarity of limb-areas).

Per contra, the primary individuation-field also exerts indirect effects owing to the graded accumulation of yolk, cytoplasm, fat and other substances along its axis (p. 311). This has secondary effects upon the rate of cleavage and relative cell-size in different parts, which are of importance in the mechanics of gastrulation; and also upon the amount of raw materials available in different parts of the body. The apical region of the primary field will always attempt to form a brain: but it can only form a brain of normal type if it contains less than a certain proportion of yolk, and less than a certain proportion of fat. Thus the indirect effects of the primary gradient are adjusted to co-operate with the direct effects.

The position of the grey crescent itself is a prior example of such interaction. The point of sperm-entry decides the meridian of the grey crescent, and therefore the meridian on which the high point of the secondary gradient will lie. However, the precise latitudinal position of this high point is not sharply predetermined at a fixed level, but depends upon conditions in the primary gradient and can be experimentally modified by modifying these. For instance, exposure of the frog's egg to depressant agencies (e.g. $N/10$ LiCl) during early segmentation leads to the dorsal lip being formed nearer the animal pole than usual; in some cases even above the equator (fig. 149).[1] Temperature gradients (p. 339) applied during segmentation also influence the position of the dorsal lip.[2]

The same sort of interaction of two gradient-systems occurs in the Echinoderms, only the high point of the secondary gradient is here directly vegetative instead of dorsal. In all probability, similar processes are at work in Annelids and Arthropods (see p. 309).

[1] Bellamy, 1919.
[2] Dean, Shaw, and Tazelaar, 1928.

Obvious examples of the dependent differentiation of a sub-ordinate region under the influence of the dominant (organiser) region interacting with the primary gradient-field are seen in the Amphibia in the formation of secondary embryos after the grafting of an organiser; or the development of engrafted fragments of organs other than the organiser, when the donor has not reached mid-gastrulation, in accordance with their new position instead of their original presumptive fate. The most remarkable of all such cases are the modification of pieces of Anuran presumptive epi-dermis, grafted into the future mouth-region of a Urodele egg, to

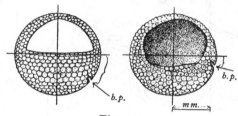

Fig. 149

Modification of the site of dorsal lip formation in the frog. Left, control egg; right, egg exposed to $m/10,000$ KCN for 24 hours from the 2-cell stage. The dorsal lip (*b.p.*) is much closer to the equator in the treated egg. (After Bellamy, *Biol. Bull.* XXXVII, 1919; modified.)

form a part of the head and jaws which is perfectly organised with the rest of the larva, but which differentiates Anuran structures (e.g. suckers and apparently teeth) never found in Urodeles[1] (p. 142).

(v) The influence of more apical (but not completely apical or dominant) regions on less apical regions

This is excellently illustrated by the experiments on the 32-cell stage in sea-urchin eggs, described in Chaps. v and vi (pp. 103, 168), in which it was shown that not only would the basalmost disc of cells (micromeres) induce gastrulation, but so would the sub-basal disc after removal of the micromeres.

It may also be recalled that the tendency of the animal disc, *an. 1*, is to produce a larva in the middle of which the cilia of the apical organ occupy much too much space and in which no gastrulation takes place, while the tendency of vegetative material is to produce

[1] Spemann, 1932, 1933; Spemann and Schotté, 1932.

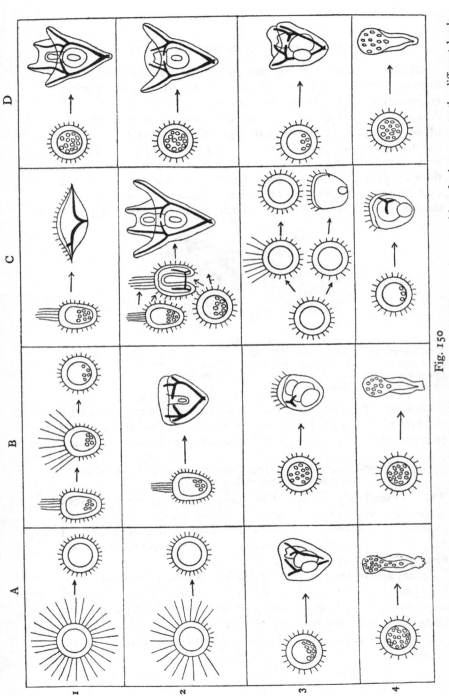

Fig. 150

The potencies of different levels of the sea-urchin egg, and the effects of adding varying quantities of micromeres to the different levels.
[...] of the whole egg; (A 1) whole; (A 2) disc. *eg.* 1.1 disc. *eg.* 2.1 disc. *eg.* 3.1 disc. *eg.* 4.1 disc. *eg.* 2. Column B, the results of

an exogastrula, without apical organ, cilia, or mouth. The situation here is complicated by the fact that the gradient-field (vegetative-animal) concerned with inducing gastrulation interacts with the previously established animal-vegetative gradient-system of the egg. As a result, not only are organising capacities graded with distance from the vegetative pole, but so are the capacities for being organised.

It has been found[1] that the production of a properly proportioned pluteus larva is dependent in the first place on the presence of some of the vegetative pole material. This material acts as an organiser, and is normally to be found in the micromeres: micromeres grafted into abnormal situations will induce gastrulation and the formation of a secondary set of main organs where they are grafted; they will organise the neighbouring tissues so as to make them conform to the normal morphology of a larva, and to the new polarity set up by the graft. But the organising capacities are not restricted to the micromeres, for, if they are removed, it is found that the next most vegetative region, disc *veg. 2*, is capable of forming a pluteus with proportions approximating to those of the normal. It is well known that lithium salts produce exogastrulation in Echinoderm larvae, i.e. a reinforcement of the vegetative potencies.[2] It is therefore interesting to find that lithium salts induce gastrulation and the awakening of organiser properties in isolated animal halves[3] (see also p. 337).

In the second place, the production of a perfect pluteus larva is dependent on a balance between animal and vegetative material, and it has been possible to study this balance quantitatively. In order to obtain a well-proportioned pluteus it is necessary to add one micromere to disc *veg. 1*, two micromeres to disc *an. 2*, and four micromeres to disc *an. 1*. Excess of material from the animal pole leads to imperfect gastrulation and abnormal enlargement of the apical organ, excess of material from the vegetative pole leads to exogastrulation and reduction in the extent of the ciliated area.

Similarly, it is possible to observe gradual approximation to normal proportions when macromeres are added to a complete animal hemisphere. Isolated, the animal hemisphere gives a

---

[1] Hörstadius, 1931.   [2] Herbst, 1895.
[3] von Ubisch, 1929.

blastula, three-quarters of the surface of which is covered by the cilia of the apical organ. Addition of half a macromere gives a larva in which the apical organ is reduced almost to normal proportions; a ciliated band and a stomodaeum are formed, but no gut. Addition of a whole macromere gives a little pluteus in which the gut is, however, too small. Addition of two macromeres gives a perfect pluteus. The addition of four micromeres produces roughly the same effect as that of one macromere. Thus, in proportion to total bulk, the organising capacity of the micromeres is far higher than that of the macromeres, since their size is only about one-thirtieth of that of the macromeres (fig. 150).

If in place of a whole animal half, an isolated disc *an. 1* had been used, the addition of four macromeres would have resulted in the formation of a perfect pluteus. This again shows that the morphogenetic effects of the organiser material are dependent on the level (within the main gradient) of the tissues which they are organising. A further proof of this is given by the following fact. An isolated *veg. 1* disc will invaginate a little gut; but the addition of an animal hemisphere to *veg. 1* prevents the latter from gastrulating at all.

If corresponding amounts are removed from both ends of the gradient, the remaining tissue is still able to form a pluteus. Thus discs *an. 2*, *veg. 1* and *veg. 2*, together, are able to form a properly proportioned larva. But the zones which have been removed, *an. 1* and the micromeres, are together also able to give rise to a proper pluteus. It is therefore possible to obtain two perfect larvae after section at right angles to the egg-axis, provided only that the balance between animal and vegetative potencies is preserved.

The importance of these facts needs no emphasising. They show that the morphogenetic properties of the organiser in the Echinoderm larva are located at one end of a gradient; that these capacities are not localised in any given tissue, but diminish gradually with increasing distance from the vegetative pole, along the gradient, and that the degree of organisation produced is quantitatively dependent, first upon the difference of level (along the main gradient) between organising material and material to be organised, and secondly upon the relative amounts of the two kinds of material.

§ 4. *Inhibition, physiological isolation, and multiple potentiality of fields in ontogeny*

(vi) Inhibition exerted by a dominant region on other parts of the system

Perhaps the most striking example of this in early ontogeny is found in sea-urchins. Here, the presence of the organiser (gastrulating) region inhibits the formation of long cilia on the late blastula and gastrula, except for a small tuft at the apical pole. In the absence of the organiser, these cilia spread over all or most of the surface of the blastula (see p. 103). The inhibition is here exerted by the dominant region of the secondary or vegetative-animal gradient (p. 320); but the principle is the same as in the example given in the preceding chapter.

No cases of resorption of a subordinate by a dominant region are known in early embryology. The resorption of parts occurring at metamorphosis (Amphibia, Echinodermata), and the partial resorption of one member of a pair of double monsters by the other are clearly of rather a different nature. However, an alteration in relative size of parts can often be obtained as the result of differential inhibition. This is so in the experiments on *Chaetopterus* larvae and Echinoid plutei, described on p. 332: it indicates that there is a competition for available food-material between the different parts of the embryo, and that the degree of success in that competition is, in part at least, regulated by the relative activity of the dominant region and other parts of the organism.

(vii) Physiological isolation and the multiple potentiality of gradient-field systems[1]

The fact that in many forms the early stages of development can be made, by appropriate fragmentation, to produce more than one normal larva was one of the earliest discoveries of the science of experimental embryology. It attracted a great deal of attention, and led Driesch to formulate his conception of "harmonic equipotential systems" (p. 353).

Numerous examples of this have been given. We need only recall

[1] And see corollary, Chap. VIII, p. 294.

that multiple development can be obtained by cutting the unfertilised egg and inseminating the fragments (p. 120); by isolating
1/2 or 1/4, and in some cases even 1/8 blastomeres (p. 97); or by
cutting and breaking the blastula into fragments (pp. 81, 89).
The most significant example of the multiplication of potencies in
the early egg is perhaps the production of double monsters from
inverted frog's eggs (p. 94). In this case there is no spatial isola-

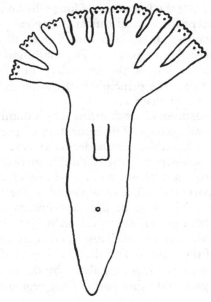

Fig. 151

Multiple potentiality in head-field in the Planarian *Dendrocoelum lacteum*. The
anterior end was partially slit by a number of cuts; the organism has produced
ten heads. (Redrawn from Korschelt, *Regeneration und Transplantation*, 1927,
fig. 269, p. 444; after Lus.)

tion of fragments; a physiological isolation between two active
regions is brought about by the intercalation of a mass of inert yolk.
The coalescence of two eggs to produce a single unitary embryo is
a converse result of the same principles. Further, just as two-
headed Planarians or bifurcated regenerated limbs can be produced
by operations, so can two-headed newt embryos be produced by
partial constriction in the 2-cell stage (pp. 75, 350).

Partial or regional fields can also give rise to more than one structure. The amphibian organiser region itself can be divided and engrafted to produce several embryos (p. 151). The limb-field can be made to produce a number of limbs. This can be done not only by grafting portions of it into new situations (p. 223), but simply by making deep cuts in the early limb-buds:[1] the result is a number of limbs growing out from the limb-area. Other regional fields also show this multiple potentiality, e.g. heart, balancer, etc. One of the most striking examples is provided by the anterior end of a Planarian, which, by making deep cuts, can be led to give rise to as many as ten heads[2] (fig. 151).

One point which may here be mentioned is the existence in all large-yolked vertebrate embryos and in all mammals of considerable areas of tissue produced by the fertilised egg but not organised into the body of the embryo. Examples of such tissues are the extra-embryonic blastoderm of selachians, reptiles and birds, and the trophoblast of mammals. These do not appear to be organised in relation to the organising centre of the embryo, and in some cases (chorion or trophoblast of amniotes) are cast away at hatching or birth, and thus never become incorporated in the field-gradient system of the organism. In other cases (yolk-sac) they do ultimately become incorporated by resorption within the body, and are then organised to produce a portion of the gut.

Such extra-embryonic structures may perhaps be looked on as composed of tissue which has grown so rapidly as to escape the organising action of the organiser, and thus to remain beyond the boundaries of the embryo. It is of interest that exposure of fowl eggs to low temperature will produce a large proportion of "anidian" blastoderms, in which no embryo is formed, but the blastoderm shows considerable powers of growth.[3]

With regard to points (viii) and (x) of our previous chapter, these only apply to cases of regeneration. They are thus not relevant to normal ontogeny.

These points lead on to a consideration of the problem of twinning. The term *twinning* in the broad sense is applied to any process by which more than one individual is produced during early ontogeny from a single zygote. We may, however, profitably

[1] Tornier, 1906.   [2] Lus, 1924.   [3] Needham, 1933.

distinguish cases in which the separation of the future individuals occurs during cleavage from those in which the process concerns later stages. In the former cases, the separate individuals are isolated by the process of cleavage itself, whereas in the latter, processes of dichotomous growth and fission are involved.

In the former category, we first have certain cases in which repeated and irregular division, leading to separation, occurs at an early stage of cleavage. This phenomenon, usually called *polyembryony*, is found in certain Hymenoptera and Polyzoa. Here, this process leads to the production of numerous separate individuals from one egg by the separation of its blastomeres or groups of blastomeres. These cases are really natural experiments of blastomere isolation, and it may be noted that axes of polarity and symmetry relations play little part in the process. As to why it is in these cases that the blastomeres separate and produce wholes on their own instead of parts, little can be said except to point out that in Hymenoptera and Polyzoa the fertilised egg undergoes cleavage within a mass of living matter, consisting in the case of the former of the tissues of a parasitised caterpillar preyed upon, and in the case of the latter, of the nutritive cells of the ovicell or brood pouch.[1] In these cases it is interesting to note that a fertilised frog's egg grafted into the body cavity of a fully developed frog undergoes modified cleavage and these products become separated and develop as far as they are able on their own.[2]

Another group of cases comprises those where the early cleavage stages are artificially interfered with in one way or another. This phenomenon leads to the formation of double (or multiple) monsters, each partner being derived from a blastomere or group of blastomeres which has been to a certain extent isolated, physiologically or physically, from the others. Here we must place the double monsters obtained in *Amphioxus* as a result of shaking and disarranging the blastomeres (pp. 79, 123): in *Tubifex* and in *Chaetopterus* as a result of inducing equal divisions of blastomere *D* containing the essential ingredients for the formation of somatoblasts (twinning in *Clepsine* is probably of this type (p. 113)): in the starfish *Patiria* as a result of spontaneous parthenogenetic development resulting in semi-independent development of both of the blasto-

[1] Harmer, 1930.   [2] Belogolowy, 1918.

meres of the 2-cell stage.[1] Here also may be included those cases in which for reasons at present unknown the heavily yolked egg of fish and of birds may exceptionally possess two blastoderms, and perhaps cases of double monsters in scorpions.[2]

Twinning in the restricted sense, however, is the result of a dichotomy setting in, not during the earliest stages of cleavage, but during later stages of development, and resulting in definite

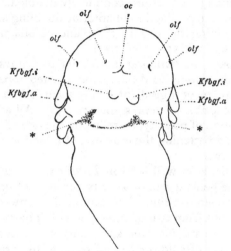

Fig. 152

Incipient twinning mechanically produced. Ventral view of *Triton* embryo from an egg slightly constricted in pre-gastrulation stage, showing slight anterior doubling. *Kfbgf.a*, outer gill-filaments; *Kfbgf.i*, inner gill-filaments; *oc*, inner eyes; *olf*, olfactory pits; * pigment overlying heart-rudiments. (From Spemann, *Arch. Entwmech.* xvi, 1903.)

fission of one embryo into two or more. Here belong the cases of twinning as found regularly in the armadillo (here resulting in the formation of four or eight embryos),[3] occasionally in other mammals including man (leading to the production of so-called identical twins), or in birds or earthworms leading to the production of double monsters: experimentally leading to the production of double monsters in the frog after reversal of the egg, in *Fundulus* and trout after subjecting the egg to cold or oxygen-deficiency

[1] Newman, 1923.   [2] Brauer, 1917.   [3] Newman, 1917, 1923.

(fig. 153), in *Patiria* after fertilisation by sperm of another species, or as a result of overcrowding.

In all cases in which twinning has been experimentally produced, it is clear that the critical stage at which dichotomy occurs is that of early gastrulation. In the reversed frog's egg the invaginated gut becomes mechanically split into two in a manner described above (p. 95) and since the gut-roof is the organiser, the resulting embryo is accordingly more or less completely doubled. Similar cases are operative in the production of anterior doubling as a result of a ligature constricting the egg in the plane of bilateral symmetry (p. 156, and figs. 32, 152, 169, 170).

In other cases, the twinning is due not to a physical but to a physiological dichotomy, and the region affected appears always to be the apical point of a gradient. This point is known to be differentially susceptible to depressants (p. 332). All agencies which make for abolition of polarity, by reducing the rate of activity of the apical point and flattening the gradient, also tend to encourage the production of twinning.

This is particularly well seen in *Patiria* where as a result of a lowering of the general rate of activity consequent upon abnormal fertilisation or overcrowding, invagination of an enteron takes place not from one, but from two or three points.[1] The same phenomenon occurs in teleosts (*Fundulus* and trout), where as a result of the depressant effects of cold, or lack of oxygen, the originally single axis of polarity is replaced by two.[2]

In the armadillo, there is, relatively to other mammals, a delay in the formation of a placenta, and consequently in the establishment of a source of supply of oxygen and nutriment for the embryo, and this occurs at a stage corresponding to the early gastrula, just before the appearance of the primitive streak. In those occasional cases in which two embryos are formed on a single blastoderm in a bird's egg, it is probable that the cold experienced by the egg after laying and before incubation is responsible for an arrest of development at a stage which corresponds to the early gastrula, shortly before the appearance of the primitive streak.

The twinned worms occasionally to be found in the cocoons of Oligochaetes are presumably to be accounted for by a delay in

[1] Newman, 1923.   [2] Stockard, 1921.

development caused by lack of oxygen within the cocoon, which is occasioned by the high mortality of the eggs and consequent foulness.

A formal explanation of twinning and the replacement of a single axis of polarity by two axes, more or less independent, is to be found in the principle of axial gradients. The maximal susceptibility of the apical point of the gradient, when acted on by

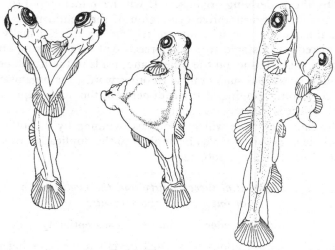

Fig. 153

Partial twinning in trout brought about by reduced oxygen supply during pre-gastrulation stages. Left, unequal components, anterior duplication. Centre, anterior duplication, unequal components: component on the left has a very small head and is cyclopean. Right, a subnormal individual, with only one eye, no mouth, gills or tail fin, and much reduced trunk, is attached to the surface of the yolk-sac opposite to the larger normal individual. (Redrawn after Stockard, *Amer. Journ. Anat.* XXVIII, 1921.)

depressant agencies, brings about the depression of its level of activity below that of the immediately neighbouring regions. These, in all cases where the original embryonic area is a flat plate or blastoderm, as in fish, birds or mammals, will be symmetrically situated right and left of the original apical point.

Interesting confirmation of the truth of this interpretation is provided by the cases of twinning presented by the Oligochaetes. As mentioned above, these worms are characterised by the presence of two gradients, with apical point to the front and hind ends respectively. Here, twinning occurs most frequently at each end of the worm, and very rarely in the middle region.[1]

The opposite to twinning is the merging together in the middle line of organs which are typically paired, a good example of which is provided by cyclopia and monorhiny (p. 348). Attention may here be called to the part played in the development of the amphibian eyes by the underlying organiser. It will be remembered (p. 245) that a piece of presumptive eye-region of the neural plate, taken from the middle line without underlying organiser, usually differentiated into a single eye. The presence of neighbouring underlying organiser tissue, on the other hand, leads to the development of paired eyes from such grafts. In other words, the organiser has brought about twinning of the rudiments in the field: explanation of this effect is, however, obscure.

The same principle which underlies twinning by dichotomy in the whole organism can also be applied to the duplication of single organs (see above, pp. 296, 327).

§ 5. *Differential susceptibility and the modification*
of ontogeny in invertebrates

(viii) The effects of differential susceptibility

Remarkable modifications of normal development have been obtained by applying the principles of differential activity to the eggs of Annelid worms.[2] In *Chaetopterus*, susceptibility experiments show that the animal pole is at first the most active region. This condition persists until the young larva begins to show elongation of the trunk, when the posterior region becomes the most active.

By immersing the developing eggs in inhibiting agents (e.g. $m/100,000$ KCN) from fertilisation onwards, microcephalic forms are produced. These forms also have their extreme posterior regions inhibited, as the treatment is continued during the period when these show high susceptibility. If the treatment is discontinued after 11 hours from fertilisation, the posterior region is better developed, while the microcephaly persists.

If, on the other hand, the treatment is not begun until the

---

[1] Hyman, 1921.   [2] Child, 1917, 1925 A.

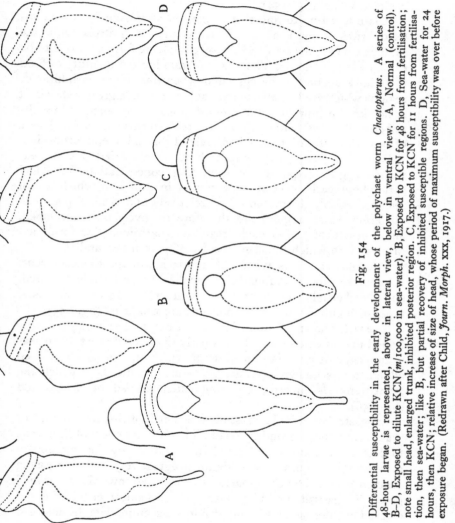

Fig. 154

Differential susceptibility in the early development of the polychaet worm *Chaetopterus*. A series of 48-hour larvae is represented, above in lateral view, below in ventral view. A, Normal (control). B–D, Exposed to dilute KCN (m/100,000 in sea-water). B, Exposed to KCN for 48 hours from fertilisation; note small head, enlarged trunk, inhibited posterior region. C, Exposed to KCN for 11 hours from fertilisation, then sea-water; like B, but partial recovery of inhibited susceptible regions. D, Sea-water for 24 hours, then KCN; relative increase of size of head, whose period of maximum susceptibility was over before exposure began. (Redrawn after Child, *Journ. Morph.* xxx, 1917.)

24-hour stage, the susceptibility conditions are reversed and mega-cephaly results (fig. 154).

It is interesting to find that in the microcephalic forms the anterior trunk region is absolutely larger than in controls, while the same is true for the heads of the megacephalic forms. This is to be explained very simply. There is a definite limited quantity of food material available in the egg; and when one region is inhibited, regions which are less affected are able to obtain a greater share. It is not known how long the modifications of proportion thus obtained will persist, although, in *Arenicola*, forms with some degree of posterior inhibition have been reared through metamorphosis.

Other interesting experiments have been carried out on sea-urchin eggs.[1] Exposure to inhibiting concentrations of KCN throughout early development results in plutei in which apical regions, notably the oral lobe, are relatively under-developed. The posterior (basal) regions are therefore relatively over-developed, and consequently after such treatments narrow-angled forms are produced, in which the arm spicules may even be parallel.

In weaker solutions, where differential acclimatisation can occur, the reverse process is found. The oral lobe is relatively enlarged, and the skeletal arms diverge at a wide angle. In extreme cases, types of highly abnormal proportions are produced (fig. 155).

The well-known "lithium-larvae" of Echinoderms may be mentioned in this connexion. It was early discovered[2] that when sea-urchin eggs are reared in a medium to which lithium salts have been added, forms known as exogastrulae are produced, in which the archenteron is present, but evaginated instead of invaginated (fig. 156).

Exogastrulation is the most obvious effect of this treatment, but it appears to be a secondary result. The primary effect of lithium is to decrease the amount of ectoderm produced by the egg, and to increase the amount of endoderm, progressively with increasing concentration. The skeletogenous cells move towards the animal pole. Exogastrulation is probably a mechanical effect, the decreased ectodermal area being unable to accommodate the enlarged archenteron in its interior. If the lithium treatment is continued until the middle blastula stage, "holoentoblastulae" may be

---

[1] Child, 1916 B.   [2] Herbst, 1895.

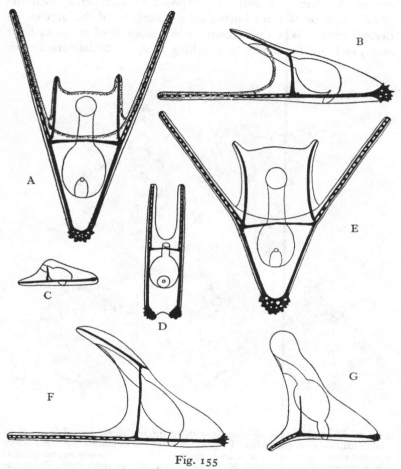

Fig. 155

Differential susceptibility in the early development of sea-urchins (*Arbacia*).
A, B, Normal pluteus. C, D, Differential inhibition by dilute KCN; C, applied
throughout development; D, applied for a short period. Inhibition of apical
region (oral lobe) with (in D) correlative increase in basal regions and consequent
parallel-armed condition. E-G, Differential acclimatisation in very dilute
solutions. The apical regions become relatively very large, with consequent
wide-angled condition of the arms. (From Child, *Physiological Foundations of
Behavior*, New York, 1924.)

produced, which are entirely composed of endoderm, with the usual exception of a tiny button at the apical end of the large endodermal vesicle. When the treatment is discontinued at the 24-hour stage, only moderate effects, resulting in exogastrulae, are found.

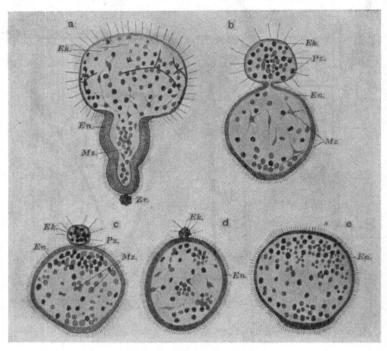

Fig. 156

Differential inhibition of the ectoderm in sea-urchin larvae reared in water to which lithium salts are added. Progressive stages of inhibition with increasing concentrations of lithium. *a*, is an exogastrula; *e*, is completely endodermised. *Ek.* ectoderm; *En.* endoderm; *Mz.* mesenchyme; *Pz.* pigment cells; *Zr.* clump of cells at base of exogastrulated gut. (After Herbst, from Schleip, *Determination der Primitiventwicklung*, 1929, fig. 323, p. 505.)

When the treatment is not begun until the late blastula or gastrula stage, death soon ensues, but without any modification of the proportions of the germ-layers, indicating that this is determined by the mid-blastula stage.

It may be suggested that this result is in part due to differential inhibition causing a flattening of the primary (animal-vegetative) gradient of the egg, the apical portions being more susceptible to lithium. However, it cannot be due entirely to this, since differential inhibition brought about by KCN does not result in a relative increase of endoderm. There must be some more specific effect of the lithium, though this again is not purely specific, since similar exogastrulae can be obtained by treatment with the salts of other alkali metals such as potassium, and such substances as carbon monoxide. Examination of sea-urchin eggs under dark ground illumination has revealed the presence of a yellow-coloured ring, the extent of which appears to coincide with the presumptive endoderm. Treatment with lithium raises the upper border of this ring towards the animal pole and thus provides a visible index of the degree of "endodermisation". The effect of lithium appears to be exerted on the colloid structure of the cytoplasm, which it coarsens; and since in normal development the ectoderm cells present a finer microstructure than the endoderm cells, it is probable that this coarsening renders differentiation along ectodermal lines impossible.[1] (See also Appendix, p. 496.)

A remarkable contrast to the "vegetativised" larvae produced by lithium are the "animalised" larvae which result from a treatment of the unfertilised eggs with sodium thiocyanide (NaSCN). Such larvae show an expansion of the ectodermal region at the expense of the endodermal: the cilia of the apical organ occupy more than the normal area; the gut is smaller or even absent; and the number of skeletogenous mesenchyme cells is reduced, even altogether to zero.[2]

In such larvae which are completely "ectodermalised", a very interesting feature is the appearance of a second apical organ at the vegetative pole: in other words, the original animal-vegetative gradient has been steepened, and the secondary vegetative-animal gradient obliterated: its place has been taken by an additional gradient of the animal-vegetative type, but with its apical point on the site of the vegetative pole. The polarity of the vegetative half of the egg has been reversed, and the larva is comparable to a biaxial head-regeneration in *Planaria* (p. 285).

If now such an "animalised" larva is subjected to lithium treat-

[1] Runnström, 1928.   [2] Lindahl, 1933 c.

ment, skeletogenous mesenchyme cells and endoderm are produced from the equator of the blastula, and two guts are formed, one in relation to each pole.

It is of further interest to note that the effects of lithium are seen on the ventral side sooner than on other meridians, thus indicating

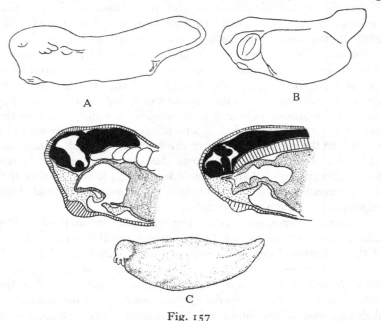

Fig. 157

A, Megacephalic, and B, microcephalic, larvae produced by exposing frogs' eggs to 10 hours' adjuvant and antagonistic temperature-gradients respectively. Above, external views; below, sections of head in region of maximum brain depth. Note difference in size of brain. C, Extremely microcephalic tadpole produced by exposure to an antagonistic temperature-gradient for 32 hours from fertilisation, then kept in water for 7 days. (Redrawn, A and B after Huxley, *Arch. Entwmech.* CXII, 1927; C, after Tazelaar, Huxley and de Beer, *Anat. Rec.* XLVII, 1930.)

the existence and polarity of the dorso-ventral axis (see Chap. IV, p. 68).

All the evidence therefore goes to show that the main gradient-systems are concerned with a number of separate physiological processes which may be variously affected by different agencies. This is an important extension of Child's views.

§ 6. *The effects of temperature-gradients*

In Amphibia, too, the primary gradient can be experimentally modified in various ways. One is by superimposing a temperature-gradient upon it during early development. This has been effected by several different methods.[1] The gradient may be applied in various directions, e.g. from side to side across the main axis (lateral) or along it (polar). In the latter case, the temperature-gradient may be *adjuvant* to the egg's original gradient, or else

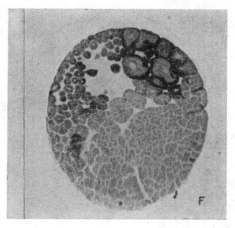

Fig. 158

Effect of a lateral temperature-gradient, applied for $5\frac{1}{2}$ hours from the 2-cell stage, on cleavage in the frog. The animal cells are larger on the right (cooled) side, small on the left (heated) side. Note the sharp demarcation between large and small cells. (From Dean, Shaw and Tazelaar, *Brit. Journ. Exp. Biol.* v, 1928.)

antagonistic. In the former case, the difference in size between blastomeres of the animal and vegetative hemispheres is accentuated at the close of cleavage, whereas in the latter it is reduced, often to the extent of leaving the animal blastomeres scarcely smaller than those at the vegetative pole. Various minor anomalies of gastrulation are produced, but the net result of adjuvant temperature-gradients is the production of embryos and young larvae with somewhat oversized heads, whereas, with antagonistic gradients, the head region is subnormal. This shows the plasticity of the

[1] Huxley, 1927; Castelnuovo, 1932.

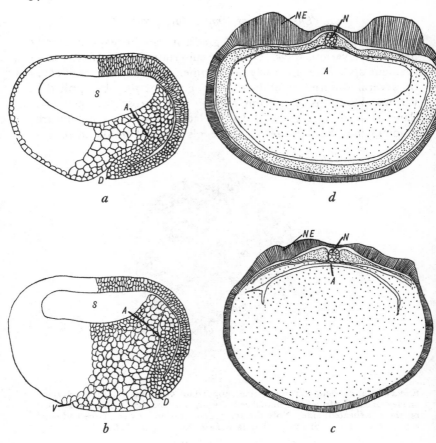

Fig. 159

Effects of temperature-gradients applied to frogs' eggs from soon after fertilisation. *a, b*, Preserved during gastrulation. *a*, After adjuvant gradient for 12 hours:
note marked overgrowth by the dorsal lip and absence of the ventral lip. *b*, After
antagonistic gradient for 16 hours. Note larger animal and smaller yolk-cells,
presence of ventral lip, and slight overgrowth by dorsal lip. *c* and *d*, Preserved as
early neurulae. *c* (below), After adjuvant gradient for 12 hours. Note small
neural folds (see text), no ventral differentiation of mesoderm. *d* (above), After
antagonistic gradient for 16 hours. Note large neural folds (see text), well-
formed ventral mesoderm but poor differentiation of notochord and myotomal
mesoderm. (From Dean, Shaw and Tazelaar, *Brit. Journ. Exp. Biol.* v, 1928.)

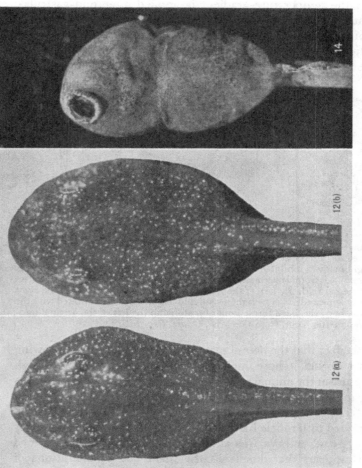

Fig. 160

Permanent effects of a lateral temperature-gradient applied during one day in the early tail-bud stage. Left, a tadpole which had been heated on the right side. Centre, the same some weeks later, showing some diminution of the asymmetry. Right, ventral view of another specimen which had been heated on the left side, showing the marked asymmetrical position of the mouth, and a bending of the body-axis in the anterior region. (From Dean, Shaw and Tazelaar, *Brit. Journ. Exp. Biol. v*, 1928.)

primary field-system: in the first case presumptive trunk regions actually become head, and *vice versa* in the second case (figs. 158, 160).

When the temperature-gradient is applied after mid-gastrulation, antagonistic gradients often produce neural folds which are much bulkier than normal, while the opposite effect is produced with adjuvant gradients.[1] This is also true of the mesoderm. As Gilchrist suggests, this apparently paradoxical effect is presumably

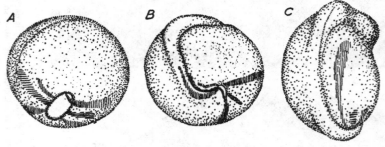

Fig. 161

Three stages in the development of an *Amblystoma* embryo treated from the 4-cell stage for 3 days after being symmetrically marked with vital stains, the whole right half was inhibited by being subjected to abnormally low temperature. A, On removal from treatment, yolk-plug stage; the normal side shows an incipient neural fold. B, Later; the left neural fold is well developed, the right has still not appeared. C, The right neural fold has arisen and has united with the left; it has, however, been formed out of material to the left of the original mid-dorsal line. (After Vogt, from Gilchrist, *Quart. Rev. Biol.* IV, 1929.)

due to the fact that the neural plate is determined by the ingrowing organiser region, whose high point is vegetative, so that high temperature at the animal pole is really antagonistic to the processes leading to neural plate formation (fig. 159).

In a series of experiments in which lateral temperature-gradients were applied to Urodele blastulae,[2] the plane of bilateral symmetry of the egg and embryo was actually shifted towards the warmed side. It appears that this is in the main due to alterations of growth of the invaginated organiser; however, since no experiments seem to have been performed in which the application of

[1] Gilchrist, 1929; Dean, Shaw and Tazelaar, 1928, text-figs. 6 and 7.
[2] Gilchrist, 1928; Vogt, 1928 B, 1932.

the temperature-gradient was concluded prior to the onset of gastrulation, we do not know whether the primary gradient-system may not also be directly deformed (fig. 161). See Appendix, p. 494.

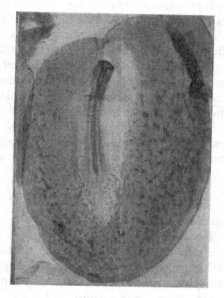

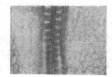

Fig. 162

Effects of a lateral temperature-gradient (3 days from beginning incubation) on the development of the chick. Above, the whole blastoderm on the heated (right) side, the area vasculosa much larger and more differentiated, the optic vesicle moderately larger. The somites on the heated side have been "stepped up" so as to alternate with those on the other side: this is shown below on a larger scale. (From Tazelaar, *Quart. Journ. Micr. Sci.* LXXII, 1928.)

Somewhat similar results were obtained by applying temperature-gradients to chick embryos. In some specimens treated with lateral gradients, the mesoblastic somites on the heated side were slightly shifted anteriorly, so as to alternate with those on the cooled side: the precise meaning of this is not clear[1] (fig. 162).

A curious effect upon cleavage has been noted in some of these temperature-gradient experiments. It was not infrequently found that in two sets of eggs from the same batch, one exposed to an adjuvant and the other to an antagonistic gradient, the yolk-cells were no more divided in the latter than in the former case, although, of course, they had been exposed to a much higher temperature. The cells of the animal hemisphere, on the other hand, were very much smaller in the adjuvant series. In other words, the development of the adjuvant series was more advanced, although its mean temperature had been the same. This can only be explained by postulating some effect of the rapid division of the heated animal cells which stimulates division in other parts of the egg.[2]

§ 7. *Differential susceptibility in the ontogeny of vertebrates*

Experimental modification of the primary gradient of the vertebrate egg has also been achieved by the method of differential acceleration. Certain treatments produce an acceleration of development in all parts, but the acceleration is disproportionately high in the more apical regions. For instance, by exposing the eggs of the fish *Macropodus* to atropin sulphate for an hour and three quarters during cleavage, the size of the head is increased relatively and absolutely and it also has altered proportions, for the relative width of the extreme anterior portion of the animal between the eyes is much increased.[3] Similar results have been obtained in experiments on the frog,[4] notably with weak acids, and by means of differential acclimatisation to very weak poisons (figs. 163, 164).

Equally interesting, and in some ways more instructive, results have been obtained by the use of depressants on early stages, causing differential inhibition. The depressant first used was magnesium chloride,[5] acting upon the fish *Fundulus*, and it was originally thought that the effects were the specific result of that particular substance; but later work has shown that essentially similar effects

---

[1] Tazelaar, 1928. [2] Huxley, 1927; Castelnuovo, 1932.
[3] Gowanloch, in Child, 1924, pp. 85–6.
[4] Bellamy, 1919, 1922. [5] Stockard, 1910.

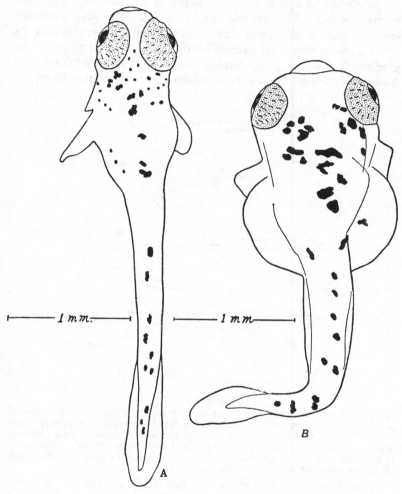

Fig. 163

Differential acceleration in the development of the teleost fish *Macropodus*. A, Control. B, Exposed to dilute atropin sulphate for 1¾ hours during early cleavage. Note large head, relatively shorter posterior trunk region. (From Child, *Physiological Foundations of Behavior*, New York, 1924, after Gowanloch.)

are produced by a wide variety of depressant substances. This fact is characteristic of the experimental modification of gradients: any specific effect of the agent employed is usually overridden by its general effects which are exerted on the shape and the slope of the gradient (but see p. 337). Similar results have been obtained with Anura.[1] In toads, remarkable malformations of the mouth region are to be noted (figs. 165, 168).

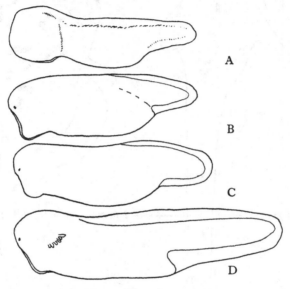

A

B

C

D

Fig. 164

Differential susceptibility in the early development of the frog. A, Differential acclimatisation of frog embryo exposed for 4 days from fertilisation to very dilute KCN. Note very large head. B and D, Differential acceleration. Frog embryos after 4 and 6 days respectively in $N/5000$ HCl from the 2-cell stage. Note relatively large head and accelerated development, as against control at 6 days (C). (Redrawn after Bellamy, *Amer. Journ. Anat.* xxx, 1922.)

In moderate concentrations, the result of exposure of *Fundulus* eggs to depressant substances is the production of a head of reduced size, the reduction being disproportionately great in the inter-ocular region—in other words, the exact converse of the experiments with stimulants. But when more marked effects are produced, they consist in the complete non-formation of a greater or

[1] Bellamy, 1919; Cotronei, 1921.

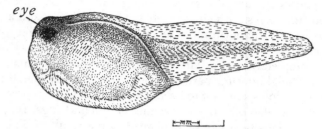

Fig. 165

Cyclopic frog tadpole produced by treatment with $M/7$ lithium chloride for 3 hours in the early gastrula stage. The single median eye is beneath the surface. The mouth is rudimentary. (From Child, *Physiological Foundations of Behavior*, New York, 1924, after Bellamy.)

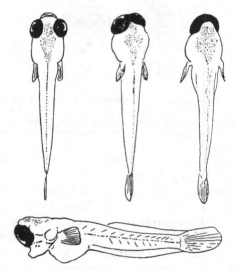

Fig. 166

Cyclopia induced by depressant agencies in *Fundulus*. Above: left, normal young fish; centre, partial cyclopia, and right, complete cyclopia, induced by treatment with magnesium chloride in stages prior to eye-determination. Below, side view of the completely cyclopic specimen, showing malformed and ventrally situated mouth. The treatment leads to the non-formation of the most anterior and median regions. (After Stockard, from Wells, Huxley and Wells, *The Science of Life*, London, 1929.)

less extent of the apical regions, resulting in animals with eyes in contact, fused eyes, or a single median eye (cyclopia), and a single median nostril (monorhiny). The mouth undergoes corresponding modifications. In Amphibia, the effects may go so far as to give rise to completely eyeless larvae, often with markedly malformed mouths. Neighbouring parts are only very slightly affected, and the trunk region seems not to be affected at all, or to a degree which would be revealed only by precise measurements[1] (figs. 166, 167).

These curious facts can be explained as the result of differential susceptibility of the different regions of the gradient. A certain level of activity is needed for the formation of apical (anterior) structures, a slightly lower level for those next posterior, and so on.

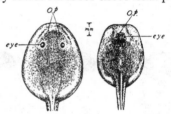

Fig. 167

Effect of lithium chloride on apical structures in anuran development. Left, control frog tadpole. Right, tadpole from an egg exposed for 3 hours to $M/7$ LiCl during early gastrulation; the anterior head region is inhibited, the external nostrils (*o.p.*) are fused, and the eyes close together. (Redrawn after Bellamy, *Biol. Bull.* xxxvii, 1919.)

While chemo-differentiation proceeds apicalwards under the influence of the organiser, the posterior levels of the body can all be determined. But the extreme apical end, being the most susceptible to depressant agents on account of its high rate of activity,[2] is now in a state which will not permit of the formation of high-level organs. The material of the apical (animal) region is, however, not destroyed, and is used up in the construction of subapical structures. This will explain why certain definite structures are absent from an embryo which has been exposed to depressants in the early stages of cleavage, i.e. long before the structures in question have become determined, let alone differentiated. Another way of putting this interpretation is to say that the whole gradient has been flattened out in such a way that its apical end no longer reaches the threshold potential value needed for the production of extreme

[1] Cotronei, 1921.   [2] See Child, 1915 A; Bellamy and Child, 1924.

apical structures. This has been confirmed by first of all finding the most susceptible region of Anuran gastrulae, then, in another experiment, staining this region *intra vitam*, and subsequently producing cyclopia with LiCl and finding that the stained region gives rise to the prechordal part of the brain.[1] A similar explanation will apply to cyclopia in regenerating Planarian heads (p. 301).

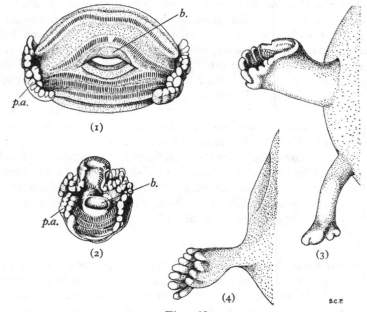

Fig. 168

Effect of lithium chloride, applied for about 24 hours during late gastrulation, on the mouth region of *Bufo vulgaris*. (1) Mouth of normal larva, showing horny beak (*b.*), rows of horny teeth, and lateral papillae (*p.a.*). (2–4) Mouth of lithium larvae; (2) showing fusion of the two parts of the beak across the aperture, and great lateral compression; (3) and (4), mouth reduced to two or one projections, in some cases without horny teeth; no beak. (Redrawn after Cotronei, *Riv. Biol.* III, 1921.)

In such poikilothermal systems it is clear that the action of the gradient-fields cannot be concerned solely with the absolute values of some fundamental process such as oxidation, but with something more complex, involving primarily the relative values at different points—in other words, with the form of the gradient rather than the absolute intensity of the processes constituting it.

[1] Guareschi, 1932.

We may here mention some other experimental results which may be interpreted on similar lines. It has been seen (p. 156) that a constriction of the blastula of the newt in the plane of symmetry will lead to the formation of two miniature but complete embryos if the constriction is complete, or of a double monster in which there are two perfectly formed heads joined on to a single posterior region of the body, if the constriction is incomplete. Sometimes, however, the plane of the constriction is not exactly coincident with the plane of bilateral symmetry, and one half comes to contain more of the region of the animal pole (i.e. the top of the gradient) than the other. In such cases, while one of the heads of such a monster is normal, the other is cyclopic[1] (fig. 169).

The explanation is based on the same considerations as those already used above. Since by the constriction, one half has been deprived of the region of the extreme animal pole, that half has a gradient of which the top is not relatively high enough to form a perfect head, complete with extreme apical structures; the other half, with the complete gradient, is capable of doing this (see fig. 170).

Certain lines of evidence indicate that it is the high point of the organiser gradient which is affected by lithium, not the high point of the eggs' primary gradient.[2]

§ 8

From what has already been said in regard to power of regulation, either in isolated blastomeres (p. 102) or in particular organ-fields, it should now be clear that regulation in early ontogeny can only occur while the system in question is in the form of a gradient-field: it cannot occur when the system is split up into a mosaic of independent chemo-differentiated regions. A system, be it egg, blastomere, or field, can only make good the loss of material in so far as that which was lost only formed part of a field, and was not a definitely localised determination forming part of an established mosaic. In regeneration, the new dominant region may override and remodel what remains of the original organisation.

From this point of view, power of regulation ceases to be a mysterious force striving for a return to the normal: systems that can regulate are merely in the same case as the egg, viz. gradient-

---

[1] Spemann, 1904.   [2] F. E. Lehmann, 1933, *Rev. Suisse Zool.* XL, 251.

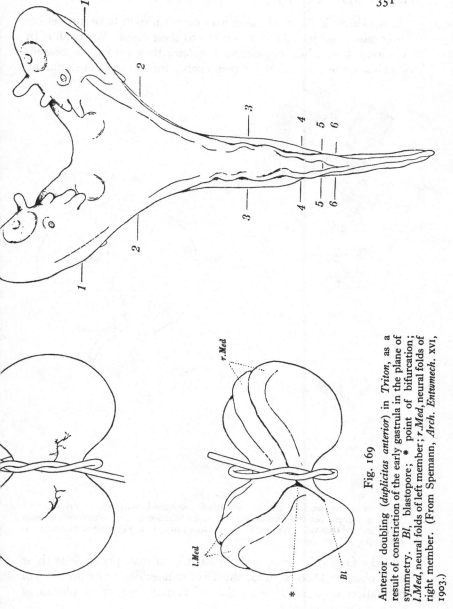

351

Fig. 169

Anterior doubling (*duplicitas anterior*) in *Triton*, as a
result of constriction of the early gastrula in the plane of
symmetry. *Bl*, blastopore; * point of bifurcation;
l.Med, neural folds of left member; *r.Med*, neural folds of
right member. (From Spemann, *Arch. Entwmech.* XVI,
1903.)

fields, which, if the expression may be permitted, have not yet cut their coats, but will do so according to their cloth. It will thus be apparent that when organisms regulate, they do so for reasons which are the same as those responsible for normal development,

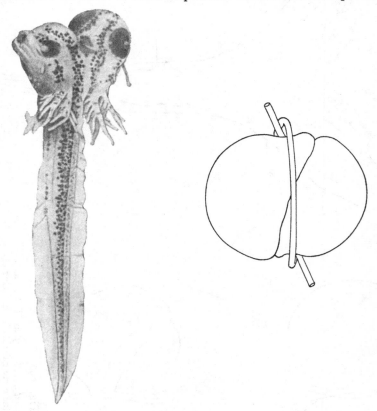

Fig. 170

Cyclopia in one member of a pair of anterior-doubled monsters in *Triton*. The result of oblique constriction of the egg, and exclusion of the animal-pole region from the half that will give rise to the cyclopic member. (From Spemann, *Zool. Jahrb. Suppl.* VII, 1904.)

and the facts call for no transcendent regulative principle such as is invoked by Driesch in his theory of entelechies. The problem of regulation is identical with that of certain important phases of normal development.

From the experiments in which isolated sea-urchin blastomeres develop into perfect larvae, Driesch was led to formulate his principle of "harmonic equipotential systems": equipotential since parts can give rise to wholes and must therefore possess equal and complete potencies: harmonic since the product is of normal proportions and affords evidence of a definite relation-equilibrium within the system.

Driesch asserted that such systems afforded proof of vitalism. We may however point out that the requirements of harmonic equipotential systems are met by the theory of gradient-fields: relative *quantitative* differences in activity-rate leading to *qualitative* differentiation: localisation being due to relative position along the total length of a gradient. But it may be doubted whether true harmonic equipotential systems have any existence in fact. The 1/4 blastomere of the sea-urchin regulates because it possesses the whole extent of the gradient: divide it transversely, or, an even more demonstrative case, divide the egg transversely (equatorially), and no perfect larva will be formed (p. 101). The parts are not all equipotential, although it may be possible, as in the case of blastomeres at the 4-cell stage, to effect subdivisions of a system without segregating regions of different potencies.

The limb-disc has been claimed to be a "harmonic equipotential partial system", but it does not appear that this connotation serves any more useful purpose, or even carries the analysis as far as the simpler concept of gradient-field, since, as already mentioned (p. 223), limb-forming potency is unequally distributed round a sub-central high point.

The results of this chapter may be briefly summed up by saying that in ontogeny the developing egg, prior to the stage of primary chemo-differentiation, possesses an organisation in the shape of a field-gradient system. The unitary and plastic nature of such a system may be partly obscured by the unequal deposition of raw materials, or by some degree of determination (though not an irreversible chemo-differentiation) having taken place before cleavage begins. Further, matters are often complicated, notably in vertebrates, by the existence of a second gradient-system connected with the organiser.

Chapter X

GRADIENT-FIELDS IN POST-EMBRYONIC LIFE

§ 1

Chap. VIII was concerned with phenomena which could only be explained by postulating the existence in adult Hydroids, Planarians and Annelids of gradient-fields concerned with morphogenesis and reproduction. In higher animals, such as Arthropods and Vertebrates, in which asexual reproduction does not occur and in which total regeneration is no longer possible, the existence of gradient-fields in adult life is not easy to detect. In such forms, the presence of total axial gradient-fields is especially noticeable during the earliest stage of development when they constitute the only or at least the major organisation of the developing embryo. Similarly the presence of partial (regional) fields is especially noticeable during the immediately succeeding phase, when the organism consists essentially of a patchwork of chemo-differentiated regions, each with its own field but as yet not differentiated into organs.

It might be reasonably supposed that these gradient-systems were only operative during the stage when they are most noticeable, and that the organisation of one stage does not persist, but is wholly supplanted by that of the next stage. This, however, does not in point of fact appear to be the case, and there is considerable evidence for the persistence of the gradient-fields of the embryo throughout life, even in the highest animals.

There is the natural presumption that the gradient-field in Hydroids and worms is directly derived from the primary gradient-field of the egg which has persisted into the adult phase. But even if this be so, in less plastic and more complex types the gradient-fields might be imagined to fade out at a certain stage of development. In what follows, various lines of evidence to the contrary will be presented.

Examples of the persistence of the main axial gradient of the organism, as evidenced by its influence upon the polarity of the later developed regional fields, are to be found in the differentiation of the lateral-line, the limbs, ear, gills, and heart, in Amphibia. The lateral-line arises from an epidermal rudiment or placode situated close behind the ear at the early tail-bud stage. It extends down the side of the body by free growth. This can be observed in experiments where an anterior half of the body of an embryo of the dark-coloured *Rana sylvatica* is grafted on to the posterior half of the body of an embryo of the light-coloured *Rana palustris*. The lateral-line then grows back as a dark structure on a light background.[1]

The determination of the path along which the lateral-line will grow is of special interest in connexion with the concept of field-gradient systems. If the tail of a frog embryo is cut off and replaced in an inverted position so that its ventral side is a continuation of the dorsal side of the trunk, one of two things may happen. If the tail heals on to the trunk perfectly, the lateral-line will grow back on to the tail and remain at the same level at which it was on the trunk. This means that it grows along a line on the side of the tail, along which it would not have grown if the tail had not been inverted. But if the healing of the tail on to the trunk is imperfect, and the continuity between them is obstructed by scar-tissue, the lateral-line, as it grows back on to the tail, changes its level for the one proper to the tail-region before its inversion (fig. 171).

It is clear that the track along which the lateral-line grows is not rigidly predetermined, for it can follow a line along which it would normally not have grown. At the same time, the growth of the lateral-line is controlled in relation to the field-system of the body so that it grows along the antero-posterior axis of the organism at a certain definite level on the dorso-ventral axis. If the inverted tail heals on perfectly, it appears that it comes under the control of the main gradient-system of the embryo, so as to form part of a single unitary field. This will allow the lateral-line to grow back in a straight line without changing its level. If, on the other hand, scar-tissue intervenes between the trunk and the inverted tail, the latter remains in some important way isolated from the main gradient-system of the former, and preserves its old field-organisation to which the lateral-line conforms when it comes into its sphere of influence.

Confirmation of this view is obtained by the experiment of

[1] Harrison, 1904.

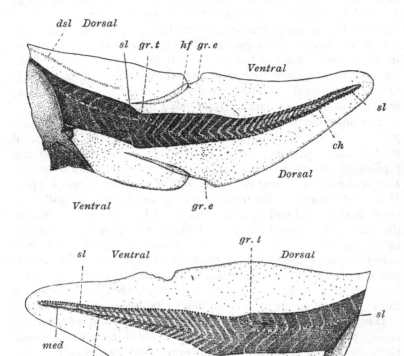

Fig. 171

Modification of gradient-fields in grafted tail-fragments of frog tadpoles. The tip of the tail was removed in the embryo and regrafted upside down. When smooth healing occurred (as in lower figure) the lateral-line growing down from the trunk assumed a position normal for the intact organism on the dorsally directed (originally morphologically ventral) side of the muscles of the grafted piece. When, however, much scar-tissue was formed (as in upper figure) the lateral-line grew along the morphologically dorsal side of the muscles of the graft, i.e. in relation to the field-system of the graft, not of the organism as a whole. *ch*, notochord; *med*, neural tube; *gr.e*, line of fusion of epidermis; *gr.t*, line of fusion of myotomes; *hf*, fold in fin; *sl*, lateral line; *dsl*, dorsal branch of lateral line. (From Harrison, *Arch. Mikr. Anat.* LXIII, 1904.)

grafting the anterior half of an embryo of *Rana sylvatica* into the back of an embryo of *Rana palustris* from which the rudiment of the lateral-line has been extirpated. The *sylvatica* embryo has its antero-posterior axis at right angles to that of the *palustris* embryo. The lateral-line of the *sylvatica* head grows back normally under the influence of its own gradient-system, until it reaches the tissues

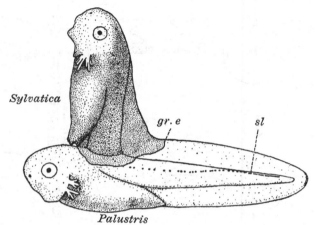

Sylvatica

gr. e   *sl*

Palustris

Fig. 172

Effect of the main gradient-field on the direction of growth of the lateral-line. An anterior half-embryo of *Rana sylvatica* (dark) was grafted on to the back of an embryo of *Rana palustris* (light). The *sylvatica* lateral-line (*sl*), on growing back to reach the *palustris* component, bent back to assume the position normal for a lateral-line in the posterior region of the body. (From Harrison, *Arch. Mikr. Anat.* LXIII, 1904.)

of the *palustris* embryo. Here it bends round when it has reached the appropriate level, and continues growing back under the influence of the gradient-field of the *palustris* embryo[1] (fig. 172).

§ 2

The fore-limb rudiment of *Amblystoma* at the early tail-bud stage is in the form of a disc of mesodermal tissue at the side of the body (see Chap. VII, p. 222). To each disc there can be ascribed two invisible axes—the antero-posterior axis, and the dorso-ventral axis, defined relatively to the axes of the whole embryo. If a left limb-

[1] Harrison, 1904.

disc is grafted on to the right side of the body, the proper way up
and the proper way out ("heteropleural, antero-posterior, dorso-
dorsal"), only the antero-posterior axis has been interfered with
and reversed. In such case, the disc develops into a limb with a
left-hand asymmetry on the right side of the body, with elbow

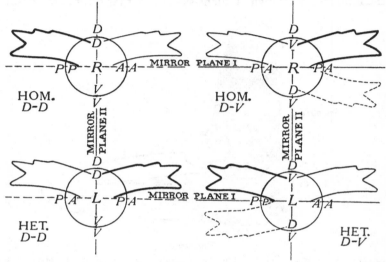

Fig. 173

Diagram illustrating experiments on the symmetry of limbs. The circles
represent the limb-buds as grafted on to the *right* side of the body. The letters *R*
and *L* in the centre of the circles indicate the side of origin of the bud (right or
left). The letters *A, P, D, V inside* the circle indicate the antero-posterior and
dorso-ventral axes of the grafted bud, these letters *outside* the circle refer to the
same axes of the body of the organism. The limb which develops is shown with a
thick outline. The position of a reduplicated limb (should one develop; see
footnote, p. 224) is indicated by the fine outline; the dotted line refers to the
form which the limb would have taken if the dorso-ventral axis of the bud had
been fixed at the time of grafting. Only medio-medial combinations are shown.
(After Harrison, from de Beer, *Biol. Rev.* ii, 1927.)

pointing forwards and hand pointing backwards: it has preserved
its prospective antero-posterior polarity, in spite of the reversal of
this relative to the body as a whole. But if a left limb-disc is grafted
on to the right side of the body, the proper way out but upside
down ("heteropleural, antero-anterior, dorso-ventral"), the antero-
posterior axis has been respected and only the dorso-ventral axis

has been reversed. Such a disc, however, develops into a limb with right-hand asymmetry on the right side of the body, although it originally came from the left side, and it is the proper way up: the palmar surface of the hand is turned down.[1] It has failed to preserve its prospective dorso-ventral polarity, and has acquired a new one in conformity with its new surroundings (fig. 173).

These experiments show that the antero-posterior axis of the limb-disc was irreversibly fixed before the time of the operation. The polarity thus imposed on the limb-disc determines where a preaxial border (that marked by the radius and first digit) will be. But the dorso-ventral axis is not yet fixed, and the determination as to which side will be the palm and which the back of the hand depends on the orientation of the disc with regard to its host. In the antero-posterior axis of the limb-disc it is easy to recognise the primary axis of polarity of the embryo. The main axial gradient of the egg persists, and permeates the limb-disc. The dorso-ventral gradient of the embryo, however, appears to be less powerful or to become active only at a much later stage.

As regards the medio-lateral axis, it is found that a limb-disc will always develop outwards, away from the body, whether it was grafted the proper or the wrong way out ("medio-medial", or "medio-lateral"), and this shows that the medio-lateral polarity is not fixed in the limb-disc stage.[2]

[1] Harrison, 1921 A; Ruud, 1926.
[2] A further point of interest in connexion with the grafts of limb-discs is that, at these early stages, it is not "right-handedness" or "left-handedness" that is determined at all. This is made quite clear from the fact that a left limb-disc can be made to differentiate into a right-handed limb on the *left* side of the body by reversing the antero-posterior axis. (Either, "homopleural, antero-posterior, dorso-ventral, medio-medial"; or "homopleural, antero-posterior, dorso-dorsal, medio-lateral".) The geometrical configuration of right- or left-handedness is the result of the determination of three axes. One of these, the antero-posterior, is already determined at the stage operated upon. The second axis, the dorso-ventral, is determined later, so that grafts of limb-buds of a more advanced stage of development show a determination not only of the preaxial border, but also of the palmar surface (Brandt, 1924). The third axis, the medio-lateral, seems throughout life to be dependent on the orientation of the limb-rudiment relative to the whole organism, and never to be irrevocably determined. In an adult newt, the left leg may be cut off, and planted into the dorsal side of the animal in such a way that the end which was originally proximal now points outwards. Regeneration takes place from this end, and a bud is formed which proceeds to differentiate into a right leg. The preaxial border and the palmar surface being determined as a result of the original antero-posterior and dorso-ventral axes, a

A state of affairs which presents many similarities with the determination of the axes of the limb is found in connexion with the development of the ear (see p. 233). The auditory vesicle in Amphibia arises as a sac formed from the epidermis on each side of the neural tube, behind the eye. When it is first determined in normal development, the ear-rudiment is a field the constituent parts of which are not yet fixed[1] and it is therefore capable of regulation. But the field is polarised with reference to the main axis of the embryo, so that if the ear-rudiment is grafted in such a way that the antero-posterior axis is reversed, it develops with reversed asymmetry.[2] Comparable grafting experiments have shown that in the rudiments of the external gills and of the heart[3] of Amphibia,[4] and of the operculum in Anura,[5] the antero-posterior axis is already determined at a stage when the rudiment is still in the condition of a field, capable of regulation after losing a portion of itself.

§ 3

Another fact of morphogenesis which appears to depend upon the persistence of the main gradient-system of the organism is the phenomenon of serial heteromorphosis. As is well known, after amputation of an appendage, certain Arthropods may regenerate one of another type, e.g. an antenna in place of an eye in *Palaemon*, or a leg in place of an antenna in various Orthopteran Insects. Natural examples of this have also been found in various groups. The abnormally located appendage is, in almost all cases, one belonging properly to a more posterior region of the body.[6] This could be accounted for if it is assumed that the original morphogenetic gradient persists throughout life, but becomes flattened during later development, so that anterior structures now come to correspond to a lower level of the gradient than they formerly did during early development. Since it is known that cold and depressant chemical agents will flatten a physiological gradient (p. 337), it is by no means unreasonable to assume that increasing age will have the same effect.

reversal of the medio-lateral axis in the regeneration-bud results in a reversal of the asymmetry of the limb regenerated (Milojević and Grbić, 1925). See also Harrison, 1925 A.

[1] Kaan, 1926.   [2] Tokura, 1925.
[3] Stöhr, 1925; Copenhaver, 1926.   [4] Harrison, 1921 B.
[5] Ekman, 1913.   [6] Przibram, 1931 B.

Experimental evidence in support of the view that age is concerned is provided by the fact that if the antennae of *Dixippus* are amputated in the first instar they regenerate as antennae, but if they are amputated in later instars, they regenerate as leg-like organs. Meanwhile *Sphodromantis* provides evidence supporting the view that the effect is correlated with general metabolic activity. In this form, an amputated antenna will regenerate as an antenna if the animal is

Fig. 174

Diagram to illustrate serial heteromorphosis. In *Palaemon* (above) removal of the eye without removal of its ganglion (*a*, distal cut) leads to regeneration of an eye (*b*); with removal of the ganglion (*a*, proximal cut), to that of an antenna-like organ (*c*). In *Mantids* (below) amputation of the antenna in the region of the flagellum (*d*, *I*) leads to regeneration of a fresh flagellum (*e*); in a basal joint (*d*, *III*), to that of a leg (*f*). (From Przibram, *Handb. norm. u. path. Physiol.* xiv (1), (i).)

kept at 25° C., but at lower temperatures a leg-like organ is formed. These heteromorphoses are thus presumably produced when the main gradient of the animal is flattened. The flattening would be primarily due to age, but can be accentuated by external conditions. The function of these heteromorphoses is of great interest. (Lissmann and Wolsky, 1933.)

Another main gradient of the early vertebrate embryo is the asymmetry-gradient (Chap. iv, p. 77), which is responsible for the asymmetrical disposal of the heart and viscera. Further evidence for the existence and persistence of this gradient is to be found in the asymmetry of the reproductive system. It will be remembered that the asymmetry-gradient gives rise to a general preponderance of the left side. This is apparent also in the gonads. When only one gonad becomes functional, as in female birds and monotremes, it is the left. Further, in normal development, e.g. of frogs, the left gonad in both sexes is usually the larger.[1] The left testis is larger than the right in many species of birds.[2] With this may be associated the fact that in intersexual mammals the left gonad tends to be more female, the right gonad more male.[3]

We may also mention the interesting fact that in genetic polydactyly in birds, when, as sometimes occurs, the extra digit is formed only on one leg, this leg is usually the left.[4]

§ 4

The persistence of regional fields to later stages has been demonstrated in adult Vertebrates capable of regeneration, by experiments in which nerves are deflected from their normal course and left to end in various regions close under the skin. In the newt, for instance, if a brachial nerve is diverted from its normal course and led away so as to end freely within a certain area surrounding the arm, the growth of a supernumerary arm is initiated: if it is led into an area close to the dorsal fin (or crest), a supernumerary piece of crest is induced. Similarly, a sciatic nerve deflected into the region of the arm or of the tail causes an extra arm or an extra tail to arise. In the lizard the area at the base of the tail can be stimulated to form a supernumerary tail by the sciatic nerve.[5]

Thus round the arm, in the newt, there exists an area which retains the potency of arm-production even in adult life. This area has been appropriately called the arm-field.[6] Similar fields exist for the tail, leg, dorsal crest, etc. Other evidence, confirming this, is provided by the experiments recorded in Chap. VIII (p. 271), in which undetermined regeneration-buds of newts grafted into abnormal situations produced organs characteristic of their new situations, and not the type of organ by which they had been budded

---

[1] Cheng, 1932.   [2] Friedmann, 1927.
[3] Baker, 1926.   [4] Bond, 1920, 1926.
[5] Guyénot, 1928.   [6] Guyénot and Ponse, 1930.

out. Each field occupies only a certain definite zone surrounding
the structure to which it gave rise during development. If nerves
are deflected to "frontier" regions between the fields, mixed
structures or chimaeras are produced, partaking of the nature of
both fields.[1]

From various lines of evidence, it appears that the action of the
deflected nerves in these experiments is in no way specific, but
merely trophic. What the nerve does is to stimulate proliferation:
the type of structure proliferated is a function of the specific field.

A   B   C

Fig. 175

Effects of deflected nerves ending freely in fields. A, In the limb-field, leading to
the formation of a limb. B, In the dorsal-crest field, producing dorsal crest.
C, In the tail-field, giving rise to extra tail. (From Guyénot, *Rev. Suisse de Zool.*
XXXIV, 1927.)

This view is confirmed by other work, carried out on non-breeding
newts, in which a fine silk ligature was tied tightly round the body,
passing over the amputated stumps of the hind-limbs. This was
done in order to produce a mechanical division of the limb
regeneration-buds. In addition to succeeding in this object it
caused an unexpected effect in promoting a local proliferation on
the mid-dorsal line, which developed into a typically crest-like
structure[2]. This occurred whether the ligature was superficial, or
was passed through below the surface in the dorsal region. (See
also Chap. XIII, p. 430.)

[1] Locatelli, 1925; Bovet, 1930.   [2] Milojević, Grbić and Vlatković, 1926.

A remarkable effect sometimes occurs after implantation of a foreign limb-bud in Urodele larvae. The grafted limb may degenerate, but its presence may stimulate the host-tissue to proliferate and replace the grafted tissue. This was proved by grafting haploid limb-buds on to diploid larvae. After a time all the haploid tissue had been replaced by diploid: the formation of a supernumerary limb by the host had been induced.[1] In three cases it appears that a grafted fore-limb which degenerated after transplantation into the hind-limb field was replaced by host-tissue which then differentiated into a hind-limb. This recalls the movements of cells induced by grafts in *Hydra*, etc. (p. 301).

Normally, however, the regenerate is formed definitely from the remainder of the organ; this is shown in cases where a haploid arm has been grafted on to a diploid body in *Triton*, and then the graft is cut through: the regenerate is entirely haploid.[1] Similar results are found with the regeneration of *Triton* limbs grafted heteroplastically on to Salamanders.

It is important to note that the morphogenetic properties of the regional field itself, once they have been determined, are not influenced by position relative to the whole organism. For instance, in Salamander larvae, fore-limbs grafted into the hind-limb field, and then cut through, produce fore-limb structures in regeneration, and *vice versa* for hind-limbs grafted into the fore-limb field and made to regenerate. A portion of the determined field has here been transplanted, and continues to produce structures of its proper type irrespective of its position.[2]

Only an extremely small portion of a determined field is needed to determine the character of the structures regenerated. If a limb regeneration-bud, in the stage in which it is still undetermined, together with a small portion of stump, be grafted into a foreign field, it will regenerate in accordance with the character of the stump, not in accordance with the character of the new field as would have happened if it had been grafted alone[3] (see p. 271).[4]

---

[1] G. Hertwig, 1927.   [2] Weiss, 1924 B.   [3] Milojević, 1924.
[4] The existence of sharply delimited fields differing in their histological and physiological properties is also known from studies on Anuran metamorphosis (see p. 427) and from work on bio-electric phenomena in the regions of the mammalian brain (Kornmüller, 1933).

§ 5

The loss of power of regeneration has been studied in connexion with the tail and limbs in Amphibia. As is well known, the Urodela will regenerate tails and limbs even in the adult, but in the Anura the adult has lost this power, which is present only in the young tadpole. It is further to be noticed that the Anuran tadpole loses the power to regenerate its limbs before it loses the power of

Fig. 176

Absence of regeneration after total extirpation of the field. *Triton* from which the entire tail-field has been removed. No regeneration at all. (From Guyénot, *Rev. Suisse de Zool.* XXXIV, 1927.)

regenerating its tail. This may be compared with the fact that adult lizards can regenerate a tail, but not a limb.

In analysing the problem as to why the power of regeneration in the Anura is limited, it is possible straightway to discard the view that the degree of histological differentiation of the tissues is the deciding factor. The differentiation of the leg of the adult newt, with its bony skeleton, functional muscles, and fibrous connective tissue, is much greater than that of the leg of the tadpole which has already lost its regeneratory power, in which the leg consists simply of a cartilaginous skeleton, muscles in process of differentiation, and mesenchymatous connective tissue. It is necessary therefore to look for another explanation.

Grafting experiments have shown that tails and limbs of tadpoles, transplanted on to adult frogs and then amputated, can regenerate in their new position provided that the tadpole from which they were taken had not already lost its regeneratory power.[1] The internal environment of the adult Anuran, therefore, does not provide any factor specifically inimical to regeneration. Nor, on the other hand, does the internal environment of the adult Urodele provide any factor specifically helpful to regeneration, for a limb of a tadpole of the toad (*Bufo*) taken after the power of regeneration is lost, grafted on to an adult *Salamandra* and amputated there, fails to regenerate.[2]

The conclusion, is, therefore that loss of power to regenerate, however it may originate, comes to operate regionally within the fields themselves. It is not without interest to find that, in the Urodeles, power to regenerate is effectively stopped if the whole field is extirpated.[3] This has been proved in respect of the snout, the tail, and the limbs.

§ 6

Further evidence for the persistence of a total field is derived from a study of *growth-gradients*.[4]

In the first place the relative growth of parts, including the phenomena seen in their regeneration, is regulated with reference to a "growth-equilibrium" which concerns the organism as a whole. The precise size of any part at any time depends on a partition-coefficient of material as between the part and the rest of the body (i.e. all the other parts). The value of this growth-coefficient differs for different parts of the body, and depends primarily on factors inherent in the tissues of the organ. If the growth-coefficient is above unity, the part will increase in relative size (positive heterogony); if below unity, it will decrease (negative

---

[1] Naville, 1927.   [2] Guyénot, 1927.
[3] Schotté, 1926 A; Guyénot and Valette, 1925; Bischler, 1926.
[4] Huxley, 1932.

heterogony); if equal to unity, it will stay constant (isogony). The external conditions, such as temperature[1] and, notably, nutritive level, will modify the partition of material between various parts of the body; but in every case a total equilibrium is concerned in the process.

Such an equilibrium does not constitute a field-system. However, it is further found that the growth-potencies of various regions are frequently graded in a quantitative way, so that the body appears to be permeated by a field-system of interconnected *growth-gradients*.

The most clear-cut examples of such growth-gradients are derived from the study of the large chela of Crustacea. When, as in the males of many species and both sexes of others, these show marked positive heterogony, they always exhibit a growth-gradient with subterminal high point. When they are not disproportionate in their growth (approximately isogonic), all their joints are growing at approximately the same rate—i.e. their growth-gradient is almost flat. The same is true of the abdomen of female Brachyura, which shows marked positive heterogony, and has a well-defined growth-gradient with subterminal or terminal high point, whereas the male abdomen is almost isogonic and has a very slight growth-gradient, with central rise[2] (fig. 177).

In limbs which show negative heterogony, the sign of the growth-gradients is reversed. For instance, the limbs of sheep decrease in relative size after birth; here the girdle is the high point of the growth-gradient, the foot the low point[3] (see fig. 198, p. 414). In other cases, growth within an organ is regulated in a more complex way, though still in a graded pattern. A good example of this is seen in the antennae of Copepods. (For further details, see Huxley, 1932.)

These gradients may not only act within an appendage, or a region of the body, but may permeate the body as a whole. Examples are seen in the relative growth of the appendages along the axis of the body in hermit-crabs, or in the growth-profiles of male and female stag-beetles. It is probable that the growth of the different regions of the body in Planarians also occurs in relation to a simple gradient (with posterior high point); but the available data only concern themselves with the proportions of the head and of the

[1] Przibram, 1917, 1925.   [2] Huxley, 1932, p. 83.   [3] Huxley, 1932, p. 88.

rest of the body. It is of interest to note that during the reduction in size that accompanies starvation in these animals, the trend in change of proportions is reversed, the trunk becoming relatively more reduced in size than the head, so that the proportions of an animal of a given size are the same whether it has been growing larger or becoming smaller.[1]

Special cases of great interest resulting from graded growth are those of the shells of Molluscs and Brachiopods. In these cases

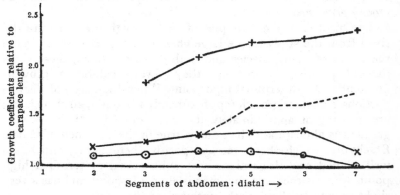

Fig. 177

Growth-gradients in the abdomen of crabs. The abscissae represent growth-coefficients (differential growth-ratios) of linear dimensions of abdominal segments relative to carapace length. The ordinates refer to the abdominal segments; 7, telson. Solid line, breadth of segments: ⊙ *Telmessus*, ♂; × the same, ♀; + *Pinnotheres* (pea-crab), ♀. The dotted line refers to segment-length in *Pinnotheres*, ♀. (From Huxley, *Problems of Relative Growth*, London, 1932.)

growth takes place at a definite growing edge, and the new material laid down solidifies and takes no further part in growth. A similar type of growth is found in other hard structures such as the horns of mammals, teeth, etc.

D'Arcy Thompson[2] first pointed out that the form and size of the horns of two-horned rhinoceroses could only be understood on the assumption of a growth-gradient, decreasing posteriorly, in the head region, affecting the proliferation of epidermal structures. This is of some general interest, as it can only manifest itself where

---

[1] Abeloos, 1928.   [2] *Growth and Form*, 1917, p. 612.

the centres for horn-formation (which are doubtless specifically chemo-differentiated regions) are present. In many other mammals, presumably, similar gradients are present, but we are ignorant of their existence, as no horn-centres exist by which they can manifest themselves.

In all these cases, if growth-potency is evenly graded along the growing zone, the resultant hard structure assumes the form known

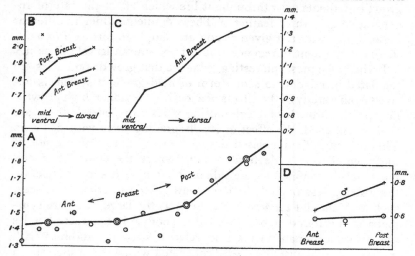

Fig. 178

Persistent gradient-fields affecting feather-growth in adult birds. In all, the ordinates represent growth of regenerating feathers per day. The abscissae represent distances within the breast-region, in A and C antero-posteriorly, in B and D ventro-dorsally. A, In a capon (⊙ single feathers, ⦾ means). B, In a cock, in two regions of the breast. C, In a capon. D, In a cock and a hen. (Based on data of Juhn, Faulkner and Gustavsen, from Huxley, *Problems of Relative Growth*, London, 1932.)

mathematically as the logarithmic spiral. Slight departures from a straight-line growth-gradient give rise to departures from strict logarithmic-spiral form.

The most important of such departures is seen in Molluscs. The growing edge of the mantle here makes a more or less circular aperture. If growth is equally graded on the two sides of this aperture between high and low point of growth-activity, the shell produced is a plane spiral, as in Ammonites or Scaphopods. If, however, the growth-gradient on the one side of the mantle aperture is concave upwards, on the other side concave downwards, the result is what is known as a turbinate spiral—i.e. a form such as that of a whelk- or snail-shell, characteristic of most Gastropods. The corkscrew horns of sheep, goats, etc. are due to similar asymmetrical growth-fields.

Experiments on fowls have shown that here (fig. 178), too, growth-gradients exist throughout life which affect the rate of regenerative growth of feathers.[1] These gradients differ in different regions, but within a given region are simple in form. Similarly, there is a gradient in regeneration-rate of anuran larval tail skin.[2]

Perhaps the most interesting evidence that growth-potencies are regulated in relation to some form of field-gradient system is derived from a study of the effects of a localised region of high growth-rate on the growth of neighbouring parts. In general, these are slightly enlarged, the effect gradually grading away with distance. This is seen in the increased size of the walking legs on the side of the large claw in male fiddler-crabs, where the enormous male-type chela is confined to one side of the body.[3] A similar effect on the walking legs behind the large claw is seen in other Crustacea, such as *Maia* and *Palaemon*, but here, as the large claws are symmetrical, it is found on both sides of the body.[3] In male stag-beetles the disproportionate increase in relative size of mandibles with increase in total absolute size is correlated with a slight increase in relative size of antennae, and of first as against third legs[3].

As regards Crustacean limbs, this effect of a localised region of intensive growth appears only to be exerted posteriorly, while anteriorly the result is partly or wholly reversed. In some cases the induced increase of growth is less in limbs immediately anterior to the region of intense growth-rate than in those immediately posterior; in other cases, their growth is even slightly inhibited. Examples of this positional effect are seen in the second and third maxillipeds of male spider crabs (*Maia* and *Inachus*), and in the first pereiopod of male prawns (*Palaemon*) in which the second pereiopod is enlarged as a large claw.[2] Such a differential action

---

[1] Juhn, Faulkner and Gustavsen, 1931; Lillie and Juhn, 1932. In this case, an additional point of great interest is the correlation found between regeneration-rate and susceptibility to hormones.

[3] Clausen, 1932.   [3] Huxley, 1932.

anteriorly and posteriorly to a region of high growth-intensity can only be explained by postulating some polarised agency connected with growth-regulation, which extends through the body as a whole (fig. 179).

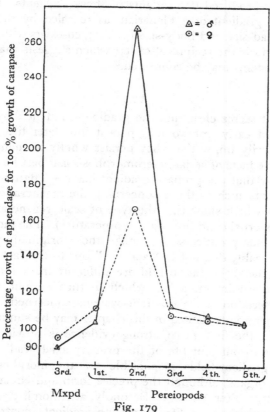

Fig. 179

Polarised effect of the presence of a region of high growth-rate upon the general growth-gradient, in the prawn *Palaemon carcinus*. Abscissae, growth-rate (percentage increase for a hundred per cent. increase in carapace length) of linear dimensions of the organs represented along the ordinates: third maxilliped, and first to fifth pereiopods. The large claw is here the second pereiopod (not the first as in crabs and lobsters), and is much more enlarged in males (solid line) than in females (dotted line). Correlated with this, in males the appendages posterior to the large claw show an increased growth-rate (the increase diminishing posteriorly), those anterior to it a decreased growth-rate. (From Huxley, *Problems of Relative Growth*, 1932, after Tazelaar.)

These growth-fields continue to operate so long as growth continues. The processes underlying them are clearly of a different nature from those concerned with the gradient-systems of the early embryo, and in higher animals it is uncertain whether they are even the directly-produced descendants of those gradients. However, the growth-gradient of a Planarian, as revealed by the relation between head-size and body-size (p. 287), co-exists with the axial gradient, and is the reciprocal of it;[1] which suggests that the two gradient-systems may be connected.[2]

§ 7

Although it seems clear that the gradient- and field-systems of the egg and early embryo may persist into later life, this does not necessarily imply that they persist wholly unchanged. For instance, the facts of serial heteromorphosis can best be explained on the view that the primary gradient has been flattened. Then we have facts such as those concerning the regeneration of skin in lizards,[3] which show that the type of scale regenerated varies with the external conditions (e.g. temperature). The fact that regenerated tails produce scales unlike those originally present is thus presumably due not to "atavism" but to the fact that conditions in the regeneration-bud are different from those in the original tail-rudiment, a fact which in turn may be correlated with an alteration of the gradient-systems concerned.

The chief points elicited in this chapter may be summed up as follows. In the first place, strong evidence is provided for the persistence throughout life of the primary axial gradient and of focalised gradient-fields responsible for the morphogenesis of particular organs, although the precise form and effects of these gradients may alter with age. Secondly, attention is drawn to the persistence throughout life of growth-gradients controlling the relative growth of parts of the body. Here again, both total growth-gradients and local growth-gradients appear to exist. It is possible, though not certain, that these growth-gradients stand in some close relation to the morphogenetic gradients previously described.

[1] Abeloos, 1928.
[2] For a more detailed discussion see Huxley, 1932, Chap. VI.
[3] Noble and Bradley, 1933.

Chapter XI

THE FURTHER DIFFERENTIATION OF THE
AMPHIBIAN NERVOUS SYSTEM

§ 1

The differentiation of the amphibian nervous system presents a number of special problems of great interest for the physiology of development. A large number of experiments have been made on this subject, and they illustrate so many of the principles which operate to bring about differentiation, that a chapter may be profitably devoted to it.

At the blastula stage, as already mentioned, the presumptive neural fold material occupies a zone in the form of a transverse band, at right angles to the plane of bilateral symmetry, and passing close to the animal pole of the egg. This presumptive neural fold region appears to have received a partial and labile determination *in situ* before gastrulation, and this is more marked in the region of the brain than in that of the spinal cord. Then, during gastrulation, a streaming movement of the cells of the animal hemisphere takes place, which results in a shifting of the presumptive neural fold material, so that it comes to occupy the position of a band running down the dorsal side of the embryo. At the same time, the organiser has become invaginated, and having become the notochord, gut-roof and mesoderm, it underlies the neural fold region and determines it irrevocably to develop by self-differentiation. As already mentioned (p. 28) the definitive neural tube arises from the anterior 4/5th of the neural folds, while the hindmost fifth becomes caudal mesoderm.

The definitive determination of the neural fold field as a whole does not prevent the possibility of a considerable degree of regulation taking place within it. This implies, as explained above (Chap. VII, p. 239), that its various constituent structures have not yet been individually localised, delimited, and determined. Such further determination soon follows, however; the region of the cerebral hemispheres is now determined to evaginate to form vesicles,[1] and the eye-cup, with its stalk, retina, and tapetum, becomes qualitatively and quantitatively determined. In the remainder of the neural tube, centres of differentiation of neurons from the neuro-epithelial cells, and of their greater or lesser degree of proliferation, are determined at certain definite places.[2] The main lines of the regional determination of the nervous system are thus completed when the neural folds have fused with one another to give rise to the neural tube, and the optic cups have been formed.

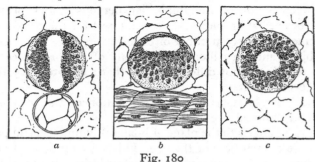

a   b   c

Fig. 180

Diagram showing the effect on the differentiation of the neural tube of a, proximity of a notochord without myotomes; b, proximity of myotomes without notochord; c, absence of notochord and myotomes (mesenchymal environment); as seen in transverse section. (From Holtfreter, *Arch. Entwmech.* cxxvii, 1933.)

At the same time, certain features of the differentiation of the neural tube are not independent of the presence of other structures. For instance, the notochord is responsible for the formation of the ventral sulcus of the central canal, i.e. it determines the formation of a thin floor on the side of the neural tube immediately overlying it.[3] On the other hand, the myotomes which flank the neural tube are responsible for the formation of the thick lateral walls, and for the radial arrangement of the cells in them[4]. These facts emerge from experiments in which embryos were obtained possessing a notochord but no myotomes, or with myotomes but no notochord. It will be noticed that the action on the differentiation of the neural tube of both notochord and myotomes tends to the same result. If

[1] Nicholas, 1930.   [2] Coghill, 1929; Detwiler, several works.
[3] Mangold and Seidel, 1927; Bautzmann, 1928; Bytinski-Salz, 1929.
[4] Lehmann, 1926.

the notochord is absent and the myotomes join one another in the middle line beneath the neural tube, the latter has a very thick floor and thin roof, and the central canal extends horizontally instead of vertically (fig. 180; see also p. 220).

It is clear, therefore, that the normal bilateral symmetry of the neural tube is dependent on the presence and normal relative positions of notochord and myotomes. Not only does the notochord induce the formation of the neural tube (p. 135), but it plays a part in determining its subsequent differentiation.

If a portion of neural tube is made to develop in a region of mesenchyme (i.e. deprived of the proximity of notochord and myotomes) it differentiates with radial symmetry: the walls are of equal moderate thickness, and the central canal is of circular cross-section.[1]

It follows that an environment of myotomes is unlikely to be conducive to the formation of the vertebrate brain and its numerous outgrowths and vesicles, and, in point of fact, myotomes are absent from the neighbourhood of the fore-brain, where the somites are destined to become the extrinsic eye-muscles. Conversely, in *Amphioxus*, where myotomes flank the neural tube right up to its anterior end, there is a minimum of cerebral differentiation.

§ 2

The conditions of the histological differentiation of the neural tube must now be considered. This consists of the formation in definite regions of accumulations of the cell-bodies of the neurons or nerve-cells forming the grey matter, and of the development from these cells of axons or fibres in definite directions or tracts forming the white matter. The subsequent morphological differentiation of the brain is really only the result of the histological differentiation of neurons in particular places, and the directed growth of their axons. The prefacial, postfacial, and hemispheric centres in the brain, and the anterior region of the spinal cord, appear to be places at which a certain definite number of neurons are determined at the early neural tube stage to develop by self-differentiation.[2] The proof of this for the centre in the anterior region of the spinal cord is given

---

[1] Holtfreter, 1933 B.　　　　[2] Detwiler, 1925 B; Coghill, 1929.

by the following experiment. If a region of the spinal cord corresponding to trunk-segments 1–3 is removed and grafted into the spinal cord of another embryo in place of the region of segments 4–6, the amount of neuron proliferation shown by it remains roughly the same as it would have exhibited in its normal position.[1]

Other regions, however, show dependent differentiation in the proliferation of neurons. The spinal cord of a newt tapers from front to back, which means that the tube in the region of the more posterior segments of the body contains fewer neurons and axons

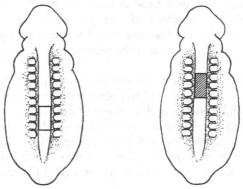

Fig. 181

Diagram to show the operation of exchanging a region of the spinal cord of *Amblystoma* corresponding to segments 3–5 for a region corresponding to segments 7–9. (From Detwiler, *Naturwiss.* xv, 1927.)

than do the anterior regions. If a region of the spinal cord corresponding to trunk-segments 3, 4 and 5 is cut out, rotated about its long axis, and planted back again in the order 5, 4 and 3,[2] the result is the normal differentiation of the spinal cord, with the proper taper and the proper number of neurons and axons in the various regions. The same is true if a region of the cord corresponding to trunk-segments 7, 8 and 9 is grafted in place of the region of segments 3, 4 and 5[3] (fig. 181). It is clear, therefore, that the factors which govern the proliferation of neurons in these regions of the spinal cord reside elsewhere. In other words, while the spinal cord is qualitatively self-differentiating, it is still dependent-

[1] Detwiler, 1925 B, 1928 A.   [2] Detwiler, 1923 B.
[2] Detwiler, 1923 A.

Fig. 182

Tissue-culture of nerve-cells of *Rana palustris*, showing free outgrowth of the axon fibres from the cells, and the amount of growth obtained after different lengths of time, given in hours. (After Harrison, *Journ. Exp. Zool.* IX, 1910; from de Beer, *An Introduction to Experimental Embryology*, Oxford, 1926.)

differentiating as regards certain quantitative features.[1] But before dealing with these factors, attention must be paid to the conditions under which tracts of axons are formed and the direction of their growth controlled, for they play an all-important part in the problem (fig. 182).

From the pioneer experiments on tissue-culture,[2] and those of grafting limb-buds (" aneurogenic ") from embryos whose spinal cords had previously been extirpated,[3] it is known that axons grow out as free projections from the cell-bodies of the neurons. It is also important to notice that experiments in which neurons have been made to produce axons in tissue-cultures through which an electric current is passing, show that the direction of outgrowth of the axon is controlled by the direction of the current. Further, if a conductor carrying an electric current is passed through the culture, the axons grow out from their cell-bodies in a direction at right angles to the axis of the conductor.[4] The strengths of current used (about 2 billionths of an amp.) correspond in range with those found in living embryos.

Now, experiments on the differential susceptibility of the parts of young embryos of *Amblystoma* at the early tail-bud stage show that two axial gradients are present. One of these appears in the ectodermal tissue of the dorsal side of the body and has its high end at the head, decreasing posteriorly. This gradient is clearly a derivative of the original gradient of the primary egg-axis of polarity. The other gradient is situated in the tissues of the notochord and mesoderm underlying the neural tube, and has its high end at the high end of the embryo, decreasing anteriorly. This has been proved by susceptibility experiments.[5] Since the hind end of the embryo, where this second gradient has its high end, corresponds to the point of closure of the blastopore, this gradient must represent that of which the organiser was the top during earlier stages of development (fig. 183).

In the embryo of *Amblystoma* at the early tail-bud stage there are therefore two gradients, working in opposite directions. Now the ventral part of the neural tube is in intimate contact with, and is

[1] Yamane, 1930.   [2] Harrison, 1907 B, 1910.
[3] Harrison, 1907 A.   [4] Ingvar, 1920.
[5] Coghill, 1929.

even firmly adherent to, the underlying notochord and mesoderm, and is under the influence of the second gradient with the high end posteriorly. The dorsal part of the neural tube is under the influence of the first gradient with the high end anteriorly. An illustration of the action of these two gradients can be obtained from a simple study of the development of the vertebral column in trout larvae. The basidorsal cartilages can be seen to develop in cranio-caudal succession, while the basiventral cartilages appear in caudo-cranial succession.[1] The order of development of the cartilages is presumably another expression of the gradients.

Several experiments have shown that one of the manifestations of axial gradients is a difference of potential when the high and low ends of a gradient are connected with a galvanometer.[2] Further,

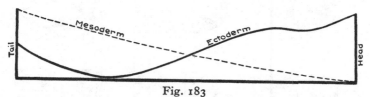

Fig. 183

Graphs showing the gradients in ectoderm and mesoderm of *Amblystoma* embryos, revealed by susceptibility experiments (KCN). (From Coghill, *Anatomy and the Problem of Behaviour*, Cambridge, 1929.)

it is known that an electric current can induce an axis of polarity and a consequent gradient in tissue exposed to it (p. 63).[3] Since, again, an electric current is known to be able to direct the outgrowth of axons, it seems very probable that the gradients in the body determine the direction of growth of the tracts of axons which constitute the white matter running up and down the neural tube.[4]

Careful observation of the initial stages of neuron-differentiation in *Amblystoma* have shown that the axons and dendrites arise from the neurons as processes which creep along the inner surface of the membrane lining the neural tube, and this creeping always takes place along the long axis of the tube, i.e. either in an anterior or a posterior direction. It is therefore very probable that the direction of outgrowth of these processes from the neurons is governed by

[1] de Beer, unpublished. [2] Hyman and Bellamy, 1922.
[3] Lund, 1923 A, 1924. [4] Kappers, 1917, 1921.

the gradients[1] and made to coincide with their axes. The processes grow up and down the gradients (fig. 184).

But there is a further point to notice. The axons in the dorsal half of the neural tube conduct impulses forwards towards the anterior end and the brain, and form part of the afferent or sensory system. The axons in the ventral half of the neural tube conduct impulses backwards, away from the brain, and form part of the

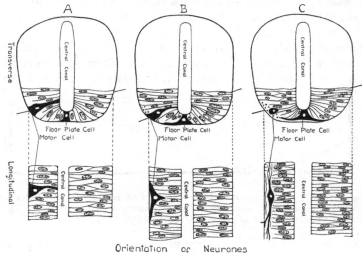

Orientation of Neurones

Fig. 184

Sections showing three stages (A, B, C) in the differentiation of neuro-epithelial cells into neurons. The axon and dendrite processes of the neuron creep along the inner surface of the limiting membrane of the spinal cord, along its long axis. The floor-plate cells, in the mid-ventral line, form processes which grow laterally and then backwards. (From Coghill, *Anatomy and the Problem of Behaviour*, Cambridge, 1929.)

efferent or motor system. Reversal end-for-end of a section of the spinal cord at the tail-bud stage does not alter this plan, and it must therefore be concluded that the polarisation (as well as the direction of growth) of the neuron processes is determined by the axial gradient under whose influence the neuron is situated, in such a way that a process of a neuron which grows up the gradient from the low to the high end becomes an axon, and later on conducts in

[1] Coghill, 1929.

this direction, while the processes which grow down the gradient become dendrites. It can then be easily understood why the dorsal half of the spinal cord (subjected to the ectodermal gradient from front to back) should contain afferent axons conducting forwards, and the ventral half (subjected to the mesodermal gradient from back to front) efferent axons conducting backwards.[1]

§ 3

A further application of the principles stated above gives a formal explanation of the main architecture of the peripheral nervous system, characterised by the formation of paired nerves growing out in each segment of the body, at right angles to the spinal cord. By this time, tracts of axons are present running along the spinal cord, and one of the results of the passage of an impulse through these tracts is the setting up of an electric disturbance, analogous to the passage of an electric current. A neuron under the influence of such a current will produce an axon which will grow out at right angles to the direction of the current, as in the tissue-culture through which a conductor carrying an electric current is passed. In the chick it has been observed that this outgrowth of neurons at right angles to the spinal cord normally occurs as the axons of the "activating bundle" reach their level.[2] In *Amblystoma*, it has been found that isolation of a portion of spinal cord from the medulla (by grafting it into the side of the body), with consequent reduction in the number of descending fibres, leads to quantitative reduction in the development of the ventral nerve-roots.[3]

On each side of the neural tube, the mesoderm becomes segmented into myotomes, or muscle-segments, and within each of these there is evidence of a gradient: the high point being in the centre and the activity-rate grading off forwards and backwards from this central point. The existence of these gradients is expressed by the distribution of the pigment, since, in the development of Amphibia generally, pigment is formed most abundantly in regions of high activity-rate. In the developing muscle-segments, pigment is usually accumulated near their centres. The septa between the segments are therefore regions of low activity.

[1] Coghill, 1929.   [2] Bok, 1915.   [3] Yamane, 1930.

These conditions have a bearing on the direction of growth of the peripheral nerve-fibres when they have emerged from the spinal cord. Those fibres which emerge in the ventral region of the cord, continuing to grow along a gradient (and eventually becoming differentiated into axons, since they are growing up the gradient), will accordingly grow to the centre of each muscle-segment, and innervate it. On the other hand, the regions of the septa, between the muscle-segments, will attract the dendrites of the sensory neurons,[1] which will then grow to the ends of the muscle-segments (thus providing their proprioceptive innervation), and continue in the septum between the muscle-segments to the skin.

Both in the sensory and motor systems, therefore, the distribution of the peripheral nerves can be interpreted in terms of gradients: axons growing towards a region of higher rate, and dendrites towards a region of lower rate. Within the central nervous system itself, the same principle can be applied. Experiments of differential susceptibility on the spinal cord indicate that a strip of tissue occupying the ventral mid-line, and forming the so-called keel, has, during late embryonic life, the highest activity-rate at any given level of the cord: this is also proved by the fact that the keel is the site of the most rapid differentiation of neurons in the spinal cord. It is most interesting to find that during this period any axon outgrowths formed in the transverse plane are directed towards the keel.

In the brain, other centres of differentiation of neurons are the postfacial and prefacial centres, already mentioned, and, further forward, the dimesencephalic, the postoptic and the hemispheric. Up to the early swimming stage, the postfacial centre is the most active, as evidenced by the relative rate at which neurons are differentiated there, compared with the rate in other centres. Correlated with this fact, it is found that the first neurons to become polarised in the dimesencephalic centre send out axons towards and into the prefacial centre. In a similar way, all over the brain, neurons which are differentiated in the neighbourhood of a centre grow axons to-

[1] The sensory neurons considered here form part of the transient sensory system of Rohon-Beard. They differ, of course, from the sensory neurons of the definitive system in that they are situated in the neural tube instead of the dorsal-root ganglia. Eventually, the Rohon-Beard neurons are superseded by the latter.

wards that centre, and a number of tracts and commissures are formed, including the olfactory paths, the posterior and postoptic commissures, and paths between the thalamus and hypothalamus.

An interesting but as yet unexplained point is that the relative rates of activity of the various centres, measured by the rate of neuron-differentiation, do not remain constant. At one period, the hemispheric centre is more active than the olfactory, but later on the olfactory is more active than the hemispheric. This state of affairs allows of the formation of the reciprocal paths which are so characteristic of various parts of the brain. It is clear, therefore, that a simple application and extension of the principles of axial gradients go a long way towards explaining the problems connected with the laying down of the main lines of the systems of tracts in the central nervous system, and in the peripheral nervous system, although the determination of the time-relations still remains obscure.

§ 4

It is now time to revert to the question of the factors which control the proliferation of neurons in the spinal cord, in regions other than those in which their proliferation at certain definite centres is the result of a previous determination, followed by self-differentiation. It has been found that the sensory load, as given by the number of receptor-organs, is the governing factor controlling the number of sensory neurons, but that the motor load, as given by the number of muscle-fibres to be innervated, has no effect upon the number of motor neurons. It is the number of axons which end in any given place that determines the proliferation of neurons at that place, but the endings of dendrites have no such effect. Thus, planting an extra limb in the side of the body increases the amount of muscular and epidermal tissue present; it has no effect on the number of motor neurons in the ventral region of the spinal cord, but it results in an increase in the number of sensory neurons in the dorsal-root ganglia[1] at the level of the graft.

Removal of the skin from one side of the body (effected by grafting together side by side two embryos each of which has had the skin removed from one side) does not affect the number of motor

[1] Detwiler, 1920 A; Carpenter, 1932, 1933; Carpenter and Carpenter, 1932.

384

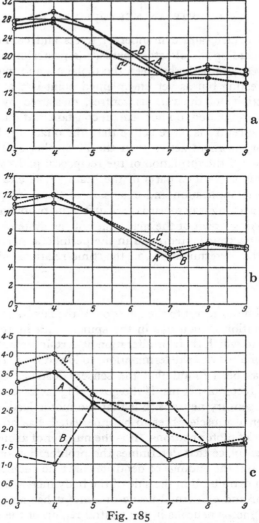

Fig. 185

Graphs illustrating the degree of proliferation in the spinal cord and in the dorsal-root ganglia of the trunk region of axolotl larvae. Abscissae, trunk-segments 3 to 9. Ordinates, weight as estimated by wax model reconstructions of the various regions. Graph *a*, Weight of entire spinal cord in segments indicated. Graph *b*, Weight of grey matter of spinal cord in segments indicated. Graph *c*, Weight of dorsal-root ganglia in segments indicated. Curve *A* (full line), normal larva; curve *B* (broken line), larva in which the fore-limb was grafted farther back and innervated from segments 5, 6 and 7; Curve *C* (dotted line), larva in which spinal cord segments 7, 8 and 9 were substituted for segments 3, 4 and 5. Note that the motor cell-area (ventral region of grey matter of spinal cord) is little or not affected by interchange of segments or transposition of limb, but that the sensory cell-area (dorsal-root ganglia) is markedly affected by transposition of limb. (From Mangold, *Ergebn. der Biol.* III, 1928; after Detwiler.)

cells in the ventral region of the spinal cord, but it results in a 60 per cent. decrease in the number of sensory neurons in the dorsal-root ganglia of that side.[1] Unilateral removal of the muscles (which of course contain proprioceptive receptor organs), without injuring the skin overlying them, results in a 40 per cent. decrease in the number of sensory neurons in the dorsal-root ganglia of that side, again without affecting the number of motor neurons in the ventral part of the spinal cord.[2] Incidentally, it may be observed that these results give interesting information as to the proportion in which exteroceptive and proprioceptive

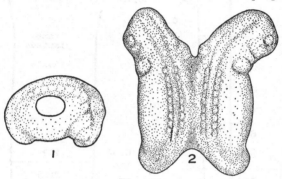

Fig. 186

1, *Amblystoma* embryo showing area of epidermis removed. 2, Two embryos grafted together and lacking epidermis, and therefore sensory load, on their inner sides. (From Detwiler, *Journ. Exp. Zool.* XLV, 1926.)

neurons occur in the dorsal-root ganglia, since the skin contains exteroceptors and the muscles proprioceptors.

All these experiments have an effect only on the number of neurons in the dorsal-root ganglia, i.e. of sensory neurons, and the effect has in all cases been due to an increase or decrease in the sensory region from which the sensory cell-area (the ganglia) in question receives impulses. On the other hand, the number of motor neurons in the ventral region of the spinal cord, i.e. the multiplication of cells in the motor cell-area, is controlled by the number of endings of axons of the descending tracts of the cord (tractus bulbo-spinalis). Thus if the region of the first five segments of the spinal cord is removed, and in its place an extra medulla oblongata

[1] Detwiler, 1926 A.   [2] Detwiler, 1927 A.

and first two spinal segments from another embryo are grafted (fig. 187), there will be two medullas, and an increased number of fibres in the descending tracts as compared with the normal. The result is an increase in the number of motor neurons in the ventral

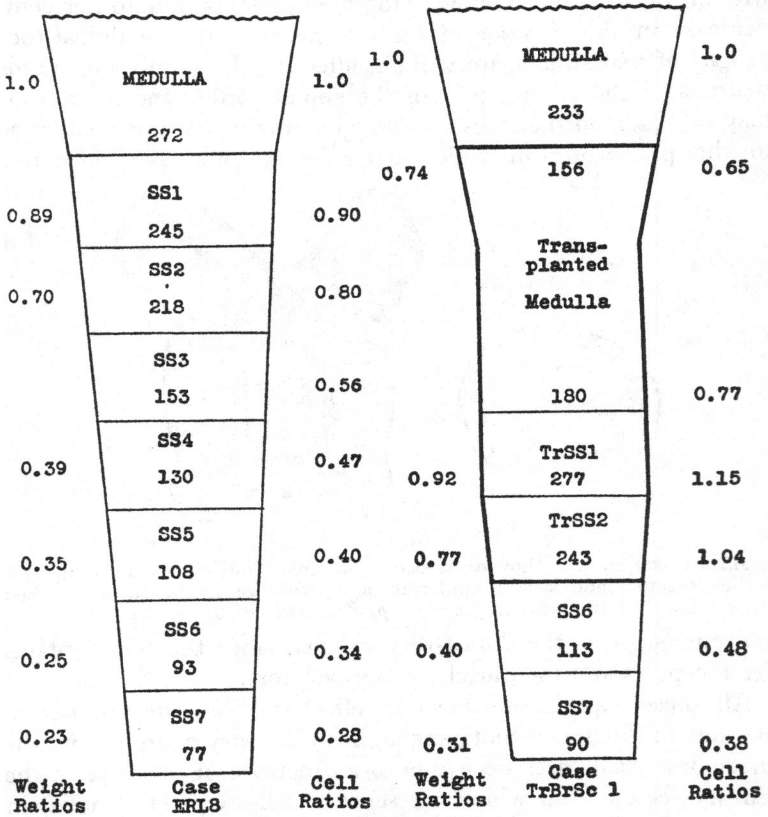

Fig. 187

Diagram showing the number of cells in various regions of the medulla and spinal cord of *Amblystoma* in a normal embryo (case ERL8) and in an embryo from which spinal segments 1–5 were removed and replaced by an extra medulla and spinal segments 1 and 2. SS 1, SS 2, etc., spinal segments 1, 2, etc. TrSS 1, 2, grafted spinal segments 1, 2. The figures within each segment represent the average number of cells seen in a transverse section. The weight ratios are estimated from the weights of reconstructed wax models of the various regions. (From Detwiler, *Quart. Rev. Biol.* 1, 1926.)

region of the grafted first and second segments;[1] this increase is quantitatively proportional to the size of the extra region implanted.[2] That the medulla is the region from which the tracts responsible for this proliferative effect originate, follows from the fact that interference with higher levels of the brain (e.g. removal of the mid-brain) produces no effect on the normal proliferation of neurons in the spinal cord,[3] while removal of the medulla is attended by a reduction in the number of neurons in the ventral region of the spinal cord. At the same time, the nature of the action exerted by the medulla is complex, for if a medulla is grafted in place of segments 4, 5 and 6, of the spinal cord of another embryo, the graft exerts no proliferative effect on posterior regions, and itself undergoes no more proliferation than is typical for its position.[4]

The centre of independent high rate of proliferation at the anterior end of the spinal cord (see p. 375) appears to exert an influence on the rate of multiplication of neurons in the dorsal or sensory cell-area anterior to it; for when the first three segments of the cord are grafted into a more posterior position, so as to occupy the position of the third, fourth and sixth, the intact segments anterior to the graft show a higher rate of proliferation than normal, and in some respects come to resemble a medulla oblongata. It is therefore possible that normally the medulla may be dependent for its rate of neuron-proliferation on influences emanating from the anterior end of the spinal cord, possibly by way of the neurons of the spino-bulbar tract.[5]

The proliferation of neurons is, we see, under the control of factors which are situated "upstream" relatively to the direction in which nervous impulses will eventually be conducted by the axons to the cell-area in question. But it is not the passage of ordinary nervous impulses that is responsible for this effect, for embryos can develop normally in a solution containing narcotics which prevent the passage of impulses.[6] The proliferative effect must therefore be due to some other activity of the axon-endings. It is very possibly identical with the trophic effect of adult nerves, which, also, is not identical with the ordinary conducting function (see p. 431).

[1] Detwiler, 1926 B.   [2] Nicholas, 1931.   [3] Nicholas, 1930.
[4] Detwiler, 1927 B.   [5] Detwiler, 1928 A, 1929 B, 1930 A.
[6] Matthews and Detwiler, 1926.

§ 5

The discovery of the factors controlling neuron-proliferation is of theoretical interest from two further points of view. In the first place, it is clear why the brain differs in form from the spinal cord. The anterior end of the body is occupied by the organs of special sense, from which an enormous number of fibres enters the neural tube. These fibres induce the multiplication of neurons where they

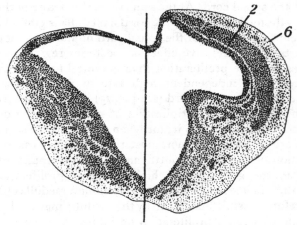

Fig. 188

The effect of eye-extirpation on the development of the mid-brain. Transverse section through the mid-brain of a larva of *Rana fusca* from which at an early stage an eye was removed. The roof of the mid-brain on the operated side is markedly under-developed (left half of figure), as may be seen by comparing it with that of a normal control (right side of figure); the outer molecular layer (2) and the stratum medullare superficiale (6) are absent on the operated side. (From Dürken, *Biol. Gen.* VI, 1930.)

end, and the large numbers of neurons so formed find expression in the bulges and prominences familiar as the optic lobes, restiform bodies, and olfactory lobes, which differentiate the brain morphologically from the spinal cord. It has been shown that the grafting of an extra eye or an extra nasal pit on the head results in an ingrowth of fibres from the graft to the brain, and an increase in the number of neurons in the brain at that point.[1] Conversely, the

[1] May and Detwiler, 1925; May, 1927.

extirpation of sensory organs may cause a reduction in size of the brain-centres to which their fibres normally run[1] (fig. 188).

Secondly, these results help to supply a partial explanation of some of the phenomena of neurobiotaxis. It has been observed in comparative studies that corresponding centres of neurons in different animals may occupy different positions in the brain, or, in other words, that certain nerve-centres have shifted their position during the course of evolution. The centre of origin of the motor fibres of the facial nerve is situated near the centre of the medulla oblongata in the selachian, but it lies on the floor of the medulla in mammals.[2] In each case, the nerve-centre lies close to the endings of the axons from which it habitually receives axons. Actually, this displacement of the nerve-centre in phylogeny (the "march to the sound of the firing", as it has fancifully been called) is only the result of a phylogenetic change in the positions of the axon endings. The cause of such change is another question, still obscure, but its effect has been the proliferation of neurons and the formation of nerve-centres in the changed positions, in each successive ontogeny. The nerve-centres are localised and differentiated afresh in each generation, and this may take place in new positions if the axons (from which the centre habitually receives impulses in the passage of reflex arcs) end in new positions.

§ 6

With regard to the peripheral nervous system, interesting results have been obtained bearing on the question as to how the nerve-fibres become connected up with their end-organs. Two different kinds of factors appear to be at work. In the first place, the out-growth of the nerve-fibre in the direction of the end-organ is controlled by non-specific factors; while its intimate functional connexion with the end-organ is controlled by factors specific to the organ.

As an example of the general directive effect which is exerted by the presence of an organ, we may take that of the limb-rudiment. If in an embryo of *Amblystoma* the limb-rudiment is moved some distance forwards or back from its normal position, the nerves

[1] Dürken, 1912.　　　　[2] Kappers, 1930.

which normally supply the limb do not grow to the place where it ought to be, but to the place where it is, provided that this is not more distant than two or three segments away from its normal position.[1] If this distance from the normal position is exceeded, the limb becomes innervated by other nerves, corresponding to the level of its position, which would not normally have supplied a limb at all. In *Amblystoma*, the normal supply to the fore-limb is composed of fibres from spinal nerves 3, 4 and 5, forming the

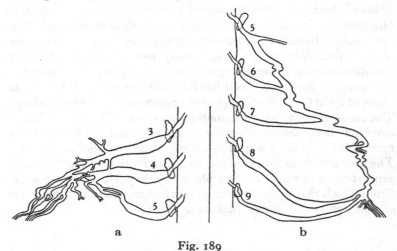

Fig. 189

The attraction of outgrowing nerve-fibres towards an abnormally situated limb. *a*, The constitution of the normal brachial plexus of an axolotl formed from spinal nerves 3–5. *b*, The brachial plexus of an embryo in which the limb-bud was moved five segments further back; the plexus is formed by spinal nerves 5–9. (From Mangold, *Ergebn. der Biol.* III, 1928, after Detwiler.)

brachial plexus. But the plexus can be formed from spinal nerves 2, 3 and 4, or 5, 6 and 7 (sometimes with the co-operation of additional nerves) (fig. 189).

The attraction which the limb exerts on the outgrowing nerve-fibres is shown still more clearly by experiments in which the whole of one half (the right) of the rudiment of the spinal cord of the frog is removed at the neurula stage. No nerves at all grow out from the right side towards the hind-limb, but fibres from the sciatic plexus

[1] Detwiler, 1920 B, 1922.

of the left side turn back across the mid-line, and innervate the right hind-leg. In general, it appears that the pattern of the plexus formed is largely independent of the amount, origin, and direction of ingrowth of the immigrant nerve-fibres, and is determined by factors intrinsic to the limb.[1]

In yet other experiments, on newt larvae in which a fore-limb is grafted into the flank of the body close to an intact hind-limb, and the nerve to the hind-limb is severed, the nerve regenerates and innervates both autochthonous hind-limb and grafted fore-limb. The actual details of innervation vary in each experiment. It is possible for the grafted fore-limb to be completely innervated by branches of the third lumbar nerve, which normally supplies only the adductors of the femur and the flexors of the knee. This shows that nerves may be attracted towards and innervate muscles different from those which they normally supply.[2] The same conclusion emerges from experiments on *Amblystoma* in which a limb-rudiment is partially removed at the early tail-bud stage, and grafted back into the same embryo at a distance of four segments posterior to the normal position. From the remainder of the rudiment in the normal position a limb is also formed, so that the embryo has two fore-limbs on the same side, and the nerves of the brachial plexus may be supplied to both.[3] In these cases, an additional point of interest is the fact that both limbs show simultaneous movement of homologous muscles, although the actual nerve-fibres which innervate them may be quite different, and their distribution varies in each individual case.[4]

The attraction which is exerted by a limb on a growing nerve is even less specific than would appear from the experiments just mentioned, for it is also exerted by an eye or a nasal pit, grafted on to the side of the body of a larva (in *Amblystoma*), after removal of the limb-rudiment.[5] In these cases the nerve-fibres which would normally have innervated the limb grow towards the eye or the nasal pit as the case may be, and end in the tissue immediately surrounding it.

[1] Hamburger, 1927, 1929.   [2] Weiss, 1924 A.   [3] Detwiler, 1925 A.
[4] Experiments of this type have led to the so-called resonance theory of nerve action. See Weiss, 1924 A, 1928; Versluys, 1927, 1928; Detwiler, 1926 C, 1930 B, C; Detwiler and Carpenter, 1929; Detwiler and McKennon, 1930.
[5] Detwiler, 1927 C.

The same non-specific attraction has been shown in the case of grafts of rudiments of chick embryos on to the chorio-allantois. If the rudiments include those of the mid-brain, muscle-segments, cartilage, and mesonephros, it is found that nerve-fibres grow out from the mid-brain towards them. Normally, the neurons of the mid-brain produce axons which do not emerge from the central nervous system, but form visual association neurons. Under the conditions of the experiment, however, they are attracted towards the various structures which happen to be differentiating in proximity to them.[1] It may also be noted that in these experiments the mid-brain is not enclosed in a connective tissue capsule, so that there is no mechanical obstacle to the outgrowth of axons.[2]

It would appear that this non-specific attraction is a result of a high degree of physiological activity on the part of the structure exerting the attraction; and in a general way the growth of a nerve-fibre towards such a structure may be compared with its growth up and down the gradients within the neural tube. It should also be noted that the deflection of nerves to an abnormally situated graft is greater if the graft is a limb than if it is an eye.

A structure or organ which is already innervated appears to exert no attractive effect on a growing nerve; it is, as it were, saturated. This fact emerges clearly from experiments on *Amblystoma* in which the limb-rudiment is removed and a tail-rudiment is grafted on to the side of the body, some distance behind the normal limb position. Contrary to what happens when a limb, an eye, or a nasal pit is grafted, no nerves grow out towards the tail. This is presumably because the tail contains its own little piece of neural tube, the nerves from which provide for its own innervation.[3] It must be for this reason that in those cases where a limb is transplanted to an abnormal position, the brachial nerve (which is attracted by the

[1] Hoadley, 1925.
[2] See also Detwiler, 1928 A. A similar alteration of morphological process in the absence of a retaining capsule is seen in the lens. When lens-rudiments are grafted into blastulae, they develop as regular spheroids if their limiting membrane remains intact. If, however, it is locally damaged, a large irregular protrusion of fibre-elements occurs (Krüger, 1930). In a somewhat similar way, the normal absence of capsule round the thyroid of teleost fish permits a pseudo-malignant growth of the organ if it is induced to hypertrophy, while this is impossible with the encapsulated thyroid of higher forms (Marine and Lenhart, 1911).
[3] Detwiler, 1928 B.

muscles of the limb) is not attracted by the muscle-fibres of its segmental myotomes, for the latter are already innervated whereas the muscles of the limb are not.

It is to be noted that when a limb is grafted to an abnormal position, nerve-fibres are not only attracted to it, but they form intimate functional contact with its muscles. An eye or a nasal pit, on the other hand, can attract the nerve-fibres to their vicinity, but no more; no intimate functional contact is established. These facts have led to the view that the establishment of functional contact and innervation is controlled by factors of a specific kind for each type of structure, possibly chemical in nature.[1] If this hypothesis should turn out to be justified, then, in the outgrowth of a nerve-fibre and its functional innervation of an end-organ, both non-specific and specific factors would be involved.

§ 7

We may now turn to the differentiation of the cells of the neural crest. Many of these, of course, give rise to the neurons of the dorsal-root ganglia, but it appears that the metamery and differentiation of the ganglia is dependent on the presence of the segmented myotomes. If at the tail-bud stage of *Pleurodeles* the myotomes are removed from one side of the trunk without damaging the neural crest, the resulting embryo lacks spinal ganglia on the operated side.[2] Similarly, the spinal ganglia fail to develop normally in experiments in which portions of spinal cord are grafted without myotomes into the flank of other embryos.[3] Conversely, the interpolation of an extra myotome as a result of grafting leads to the formation of an extra spinal ganglion.[4]

In addition to giving rise to neurons, some of the cells of the neural crest have been experimentally shown to produce the sheath cells, which enclose the peripheral nerves. If in *Amblystoma* the neural crest is removed in the region of the trunk, no dorsal nerve-roots or ganglia are developed: the ventral nerve-roots develop normally, but have no sheaths. On the other hand, if the ventral

---

[1] Cajal, 1906; Tello, 1923.
[2] Lehmann, 1927.
[3] Yamane, 1930.
[4] Detwiler, 1932, 1933 B.

half of the spinal cord is removed, the dorsal nerve-roots are unaffected, and their nerves possess sheaths in the normal way.[1]

Removal of the neural crest in the region of the head leads to results which are in many ways remarkable, and difficult to interpret.
It is found that embryos of *Amblystoma punctatum* from which the
neural crest of the head has been extirpated on one side show deficient chondrification of the anterior part of the trabecula cranii
and of the cartilages of the visceral arches, including the jaws and

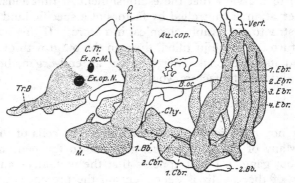

Fig. 190

Left side view of the chondrocranium of a larva of *Amblystoma* showing (shaded
by dots) the regions which fail to develop after extirpation of the neural crest.
Au.cap. auditory capsule; *B.oc.* basal plate; *Cbr.* ceratobranchial; *Chy.* cerato-
hyal; *C.Tr.* orbital cartilage; *Ex.oc.M.* oculomotor nerve foramen; *Ex.op.N.*
optic nerve foramen; *M.* Meckel's cartilage; *Q.* quadrate; *Tr.B.* trabecula;
Vert. first vertebra; 1 *Bb.*, 2 *Bb.* first, second basibranchial; 1–4 *Ebr.* first to
fourth epibranchial. (From Mangold, *Ergebn. der Biol.* III, 1928, after Stone.)

branchial arch skeleton.[2] These results have been confirmed on
Amblystoma mexicanum[3] and *Rana*.[4] It is known that derivatives
of the cells of the neural crest extend ventrally at early stages into
the region of the visceral arches, and it would seem from these experiments that these cells became directly converted into cartilage
cells. Conclusive proof would be obtained if *intra vitam* stains in
the neural crest at the neurula stage could be found in cartilage cells
at subsequent stages: some authors, indeed, working with descriptive methods only, have professed to see special histological

[1] Harrison, 1924 B.   [2] Stone, 1926.
[3] Raven, 1931 B.   [4] Stone, 1929.

characteristics in the cells of visceral arch cartilage, and to have traced them back to the neural crest cells[1] (figs. 190, 191).

Experiments in which the neural crest cells were stained *intra vitam* have not yet demonstrated the presence of the stain actually

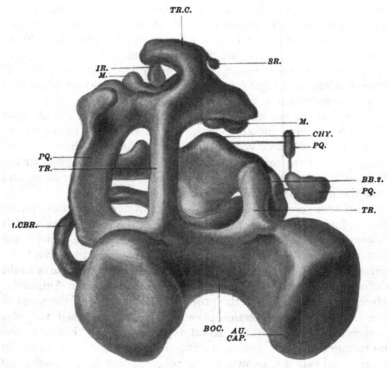

Fig. 191

Chondrocranium of a larva of *Rana palustris* from which the neural crest was removed on the right-hand side; note reduction of trabecula and visceral arches. Letters as in fig. 190. Also: *IR.* infra-rostral; *PQ.* pterygo-quadrate; *SR.* supra-rostral. (From Stone, *Arch. Entwmech.* CXVIII, 1929.)

in the cartilage cells, although the colour can be seen in the correct position in the living state.[2] Presumably, by the time the cartilage is differentiated, the stain has been dissipated. However, definite proof of this potentiality of neural crest cells has recently been

[1] For the morphological bearing of these facts, see de Beer, 1930.
[2] Stone, 1932.

provided by grafting experiments[1] in which portions of neural fold from the head-region of the early neurula were grafted into the ventral epidermis of other neurulae, and there produced cartilage as well as nerve-cord and ganglia. Grafts of the corresponding presumptive region of the late gastrula only produced nerve cord and ganglia: it would appear that the capacity to produce cartilage is determined later than that to produce neural elements. From other experiments, it appears that neural crest tissue has the power of determining other tissue (e.g. presumptive epidermis) to differentiate into cartilage[2] (see p. 193), and this might be taken as a case of homoiogenetic induction.

Experiments on heteroplastic grafts of axolotl tissues into *Triton* hosts have shown that the neural crest cells in the trunk-region also may have various prospective fates. While some of them give rise to the trunk spinal ganglia, others migrate in the form of mesenchyme to the outer side of the myotomes, and into the dorsal and ventral fins.[3]

Further differentiations of the nervous system may occur under the influence of hormones. Strictly speaking, such cases fall beyond the scope of this book. But we may mention the well-known fact that human cerebral development is incomplete without the presence of a sufficiency of thyroid hormone. Another case of brain differentiation under the influence of thyroid is seen in Amphibia. Here a marked change in the proportions and shape of the parts of the brain occur at metamorphosis. Thyroidectomised tadpoles preserve in the main the larval type of brain.[4] Further, the morphogenetic changes occurring in the amphibian brain at metamorphosis are known to be accompanied by psychological changes. Salamander larvae can be tamed and trained to take food out of the human hand; but this habit vanishes completely from the day of metamorphosis.[5] This 'forgetting is clearly due, not to a psychological process of suppression' (as suggested by W. H. Rivers in his *Instinct and the Unconscious*, 1920), but to morphological changes in the nervous system.

[1] Raven, 1933 A.   [2] Holtfreter, 1933 B.   [3] Raven, 1931 B.
[4] B. M. Allen, 1924.   [5] Flower, 1927.

Chapter XII

THE HEREDITARY FACTORS AND DIFFERENTIATION

§ 1

One of the most important results obtained from the experimental study of development is the fact that all the evidence points to the equality of nuclear division as being the general rule. Also, many of the results of regeneration would be unintelligible except on this idea. Now, genetic research has revealed the existence of unit hereditary factors or genes, whose only visible effect is upon some local characteristic of the organism. For instance, in *Drosophila*, there exist genes whose primary effect is to modify the colour of the eye, while other genes are more particularly concerned with the shape of the wing. But since the factors which control the formation of an eye are present not only in the cells of the eye but also in the cells of the wing and everywhere else in the body, the question immediately arises as to why the genetic effects are localised in particular regions. It is useless to appeal to other hereditary factors in order to account for this phenomenon, for such factors, on the same evidence, must be present in all cells, and therefore will be unable by themselves to establish a differential anywhere.

The answer to this question has already been provided. It is that primary differentiation is not an effect of the hereditary factors, but of external factors. Their first effect is to establish a system of gradients, as a result of which the various regions of the developing egg come to exhibit differences of a quantitative nature, both in respect of the activity of their processes, and of the proportion of materials such as yolk which they contain. There are several gradient-systems in the pre-mosaic stage of development of a newt's egg —the primary apico-basal (animal-vegetative, or future antero-posterior) with high point at the animal pole; the dorso-ventral gradient with high point at the grey crescent; the exterior-interior gradient, presumably with low point at the centre of the egg; and, apparently, the asymmetry gradient with high point to the left. These interact to form a complex compound system, no two points in which will be in entirely identical conditions.

It is these quantitative differences between regions of the embryo which are responsible for initiating the processes of differentiation. Of themselves, the hereditary factors are insufficient to account for differentiation, and their action must be considered in relation to the external factors and to the new internal factors which are constantly arising as a result of antecedent processes of development: internal factors which as such were not present in the undifferentiated oocyte.

A clear-cut example of the direct influence of the cytoplasmic environment upon the chromosomes is furnished by the development of *Ascaris*. Here, a process takes place known as the diminution of the chromatin, which occurs in all the blastomeres except that one which will give rise to the reproductive organs. The fertilised egg has normal chromosomes which divide at the first cleavage, but in one of the resulting two blastomeres the ends of the chromosomes are thrown off into the cytoplasm and their middle portion breaks up into fragments. In the other blastomere the chromosomes remain entire. In the subsequent divisions of the blastomere with diminished chromosomes, all the chromosomes appear in the diminished form. On the other hand, in the division of the blastomere with entire chromosomes, one blastomere retains the entire chromosomes, while those in the other blastomere undergo diminution. A similar process occurs in the subsequent divisions of the blastomere (always a single one) in which the chromosomes are entire, until it gives rise to the gonads (fig. 192).

It has been shown by experiment that the presence in any blastomere of the cytoplasm of the vegetative pole of the egg (containing the so-called "brown granules") prevents the diminution of the chromosomes. Normally, since the first cleavage division in *Ascaris* is in the equatorial plane of the egg, the division spindle being vertical in the plane of the egg-axis, only one blastomere of the 2-cell stage contains the vegetative-pole cytoplasm, and therefore only one blastomere preserves the entire chromosomes. If a ripe egg is placed in a centrifuge apparatus and rotated at 3800 revolutions per minute for several hours, the egg, being free to revolve, orientates itself with its axis along a radius of the centrifuge, and the stratified distribution of its contents is accentuated. Further, the egg becomes flattened, and the cleavage-spindle, adapting itself to the longest axis of available cytoplasm, lies horizontally instead of in the vertical position. The result is a cleavage division in the vertical

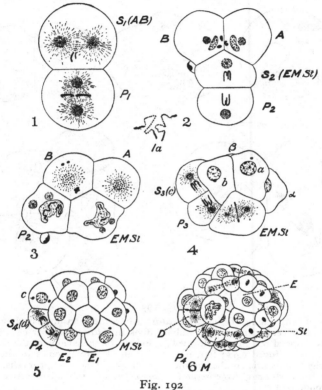

Fig. 192

Cleavage and chromatin-diminution in the normal egg of *Ascaris*. 1, 2-cell stage. The first cleavage is latitudinal; chromatin-diminution is taking place in the animal cell. S_1 (*AB*), first somatoblast rudiment of the primary ectoderm. 1 *a*, Enlarged view of the diminution process. 2, 3, 4-cell stage; 2, T-shaped phase. 3, Lozenge-shaped phase. Note extra-nuclear chromatin resulting from diminution in *A* and *B*. S_2 (*EMSt*), second somatoblast (endo-meso-stomodaeal rudiment). 4, At the next cleavage, chromatin-diminution occurs in the second somatoblast. 5, 6, Later stages. S_4, secondary and tertiary ectoderm rudiments. P_4, germ-cell with undiminished chromatin. (After Boveri, from Jenkinson, *Experimental Embryology*, Oxford, 1909.)

plane, or the plane of the egg-axis, and both the resulting two blastomeres contain a portion of the original vegetative-pole cytoplasm; it is further found that both retain the entire chromosomes.[1] Each of these two blastomeres then behaves like the single blastomere of the 2-cell stage which contains the vegetative-pole cyto-

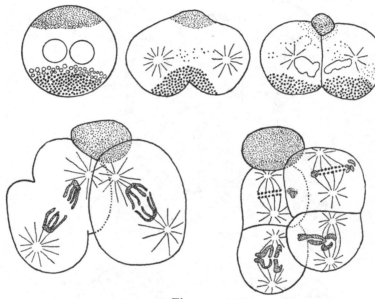

Fig. 193

Results of centrifuging the egg of *Ascaris*. Above: left, an uncleaved egg after centrifuging; centre and right, resultant division into two similar cells (plus a small centripetal mass containing yolk). Below: the behaviour of the chromosomes in centrifuged eggs; left, no diminution of chromosomes in the 2-cell stage; right, diminution of the chromosomes in both of the two upper cells. (After Hogue, from Morgan, *Experimental Embryology*, Columbia University Press, 1927; modified.)

plasm in normal development, and the embryos resulting from such treatment are double monsters (fig. 193; and see p. 101).

The conditions controlling the retention of entire chromosomes in the blastomeres of *Ascaris*, therefore, reside not in the nuclei but in the cytoplasm. The cytoplasm produces a situation to which the

---

[1] Boveri and Hogue, 1909.

chromosomes of the nuclei react, by undergoing or not undergoing diminution.

A different but equally interesting method of chromosome elimination is found in the fungus-fly *Sciara coprophila*. In this species the cells of the male germ-line possess five pairs of chromosomes. In the somatic tissues of the male, only seven of these ten chromosomes are found, one pair of large chromosomes and one single member of another pair being eliminated. In the female the somatic and probably the germ-cells contain eight chromosomes. It is probable that here too elimination occurs, but extends to germ-cells as well as to soma, and is confined to the pair of large chromosomes which is also eliminated in the male. There would then exist not one but two types of chromosome elimination. It appears that the decision as to which shall occur is predetermined in the zygote by the genes in one particular chromosome of the mother. Undoubtedly the reduced chromosome-complexes must differ from each other and from the unreduced complex in their morphogenetic and physiological effects, and the elimination process is thus here a true link in the chain of differentiation. However, it seems certain that this constitutes a highly exceptional method, but it is of interest as showing that qualitative changes in the total gene-complex may arise during early development in different parts of the embryo.[1]

§ 2

The effect of external environmental factors, in co-operating with the hereditary factors (and other internal factors) in producing development, is shown by experiments in which embryos are made to undergo development in abnormal environments. A simple and striking case is that of sea-urchin eggs made to develop in seawater which is deficient in calcium. The blastomeres resulting from the cleavage of such eggs do not remain in contact with one another, but become separated as isolated and independent cells, so that normal development of the original embryo is of course out of the question (although each of the blastomeres of the 4-cell stage if replaced in normal sea-water can produce a normally proportioned but diminutive larva)[2] (figs. 44, 194).

---

[1] Metz, 1931.   [2] Herbst, 1897, 1900.

Another example is provided by the exposure of the eggs of the frog or of certain fish to the action of weak toxic substances. In such cases (already noted in Chap. IX, p. 348) the animals develop with one median cyclopic eye instead of the normal pair.[1] Since it

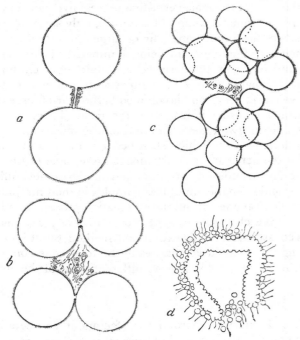

Fig. 194

Absence of cohesion in the blastomeres of sea-urchin eggs in calcium-free sea-water. *a–c*, Successive stages in one egg. *a*, 2-cell stage. *b*, 4-cell stage. *c*, 16-cell stage. The cell-membrane has become radially striated, and the cells fail to remain united. *d*, Disintegration of a blastula into its component cells when placed in the same medium. (After Herbst, from Jenkinson, *Experimental Embryology*, Oxford, 1909.)

it known from palaeontological evidence that fish have possessed paired eyes since the Silurian epoch, these experimental results are an illuminating example of the fact that hereditary factors, however long the time during which they have been transmitted to successive generations, can only produce their normal effects by interacting

[1] Stockard, 1910; Bellamy, 1919.

with a specific normal environment.[1] An equally good case is that of the adult characters of the axolotl. As is well known, the adult characters (the genes controlling which have been inherited for countless generations) normally fail to appear, as the animal is almost invariably neotenous and does not undergo metamorphosis. But spontaneous metamorphosis does occasionally occur under certain conditions of the external and internal environment; in particular, the administration of thyroid hormone. In the absence of these environmental conditions, the genes are powerless to produce the adult characters.

§ 3

While the genes are by themselves incapable of initiating the processes of development and differentiation, it is obvious that they play an active part in the control of these processes, once development has been started, and their presence is essential. A good illustration of this is provided by sea-urchin eggs when fertilised by two sperms. Each sperm brings with it an aster which divides, with the result that there are four, and a quadripolar spindle may be formed in the egg. Such an egg contains three nuclei, and since each is haploid, there will be three n chromosomes spread at random over the four spindles. Each chromosome divides, thus producing six n chromosomes in all, to be distributed between the four blastomeres into which the egg divides at once. On the average, therefore, there will be $6n/4$, or $1\cdot5n$, chromosomes to each blastomere.

It is known from experiments on parthenogenesis that the haploid number of chromosomes, or n, is sufficient to enable development to occur, and therefore, if all the chromosomes were equivalent, any blastomere which received at least n chromosomes might be expected to develop. But such is not the case. If, on the other hand, it is assumed that each chromosome of each genome is functionally different, so that when a particular chromosome is absent its place cannot be taken by any other chromosome of the same genome, but it can be supplanted by the corresponding chromosome of one of the other genomes, then it is possible to calculate the chances in favour of any one blastomere receiving at least one

[1] Goodrich, 1924, p. 56.

complete set of all the chromosomes. As has already been seen, the blastomeres of the sea-urchin can be separated, and the hypothesis can be tested by seeing how many of such blastomeres of dispermic eggs are capable of development. As a matter of fact, the observations are in accordance with the calculated probabilities. Further, in some dispermic eggs, there is formed not a quadripolar but a tripolar spindle, and the egg cleaves into three. Here, the probabilities of any blastomere receiving a complete set of chromosomes are different, but again, observation accords with calculation. Thus the chromosomes of any haploid set (genome) are functionally different, and the presence of all of them is essential.[1]

The problem has also been attacked from another angle by means of experiments on frogs' eggs which have been subjected to X-rays or mechanical injuries to the nucleus, and which are fertilised by sperms subjected likewise to X-rays, ultra-violet rays, or trypaflavine. The effect of such treatment on the sperm is to incapacitate the nucleus from playing any further part in development, without destroying the activating power of the sperm. In no case can normal development ensue if both the egg and the sperm nuclei have been affected, but it has been possible to determine the stages at which the normality of the developmental processes breaks down. In the first place, it has been found that the presence of a certain amount of chromatic material on the spindle is necessary if cleavage is to take place at all.[2] Next, it appears that as a result of slight irradiation of the egg (the sperm having been treated with trypaflavine), a normal though retarded cleavage may take place, but gastrulation is seriously affected. Either the blastopore closes very slowly and nothing more happens, or the blastopore lip is merely ephemeral, or it does not even appear at all. In all these cases it is clear that the damaged nuclear apparatus is responsible for the failure to develop.[3]

Further evidence is supplied by experiments with larval hybrids, i.e. larvae resulting from the fertilisation of eggs of one species by sperm of another. This is well shown in some sea-urchins, where the larval skeleton may show considerable specific differences. The pluteus of *Echinus microtuberculatus* is of an elongated

[1] Boveri, 1904, 1907. [2] Dalcq and Simon, 1932.
[3] Dalcq and Simon, 1931.

pyramidal form, the arms being supported by simple rods. The pluteus of *Sphaerechinus granularis* is of a more rounded form, with two of its four arms longer than the others: the skeleton is in the form of a rough framework made up of several rods interconnected. The hybrid obtained by fertilising eggs of *Sphaerechinus* with sperm of *Echinus* is intermediate in shape between the parental types, and its structures show some of the characteristics of both parents.[1] Analogous results have been obtained from a study of hybrids between fish species.[2] It is clear that those characters in which a hybrid resembles its father are due to paternally inherited genes.

In heteroplastic experiments in which a piece of tissue from an embryo of one species is grafted into an embryo of another species, artificial embryonic or larval chimaeras are produced. When the two species are closely related, as are for instance *Triton cristatus* and *Triton taeniatus*, the result is the production of fairly normal embryos.[3] Chimaeras may also be formed by mixing regeneration-buds of the black and the white varieties of the axolotl[4] (fig. 195).

In all such cases, when the operation is performed before irreversible determination of the tissues has taken place, the general pattern of differentiation is imposed by the field-system of the organism or region, acting as a unit. But the detailed peculiarities of the differentiated tissues are determined by the hereditary constitution of the species to which the tissue originally belonged. This

[1] A related yet separate problem is the question as to the relative importance of the parts played by nucleus and cytoplasm in controlling the development of the larval hybrid. The method used to investigate this matter has been to fertilise enucleated eggs with foreign sperm. Experiments of this kind have been performed on Amphibia (Baltzer, 1920), where, however, the embryos do not live long enough to enable definite conclusions to be drawn, and on Echinoderms, where until recently the technical difficulties involved have introduced uncertainties, particularly as to whether the nucleus really is eliminated from the egg. These difficulties have now been overcome, and it appears that the cytoplasm of an enucleated egg can exert some effect on the characters of the larva, although the nucleus seems to be more powerful (Hörstadius, 1932). The presence of hereditary factors in the cytoplasm of the oocyte has been revealed in experiments on sex-determination in moths and on the inheritance of dextrality in snails, and in each case there is reason to believe that these factors are the persistent results of genes situated in the chromosomes at a previous stage. The same may be true in the case of the Echinoderm hybrids just mentioned. See also Boveri, 1903.

[2] Newman, 1914.　　　　　　　　　[3] Spemann, 1921.
[4] Schaxel, 1922 A.

may concern not only such characteristics as pigmentation, but also cell-size, specific growth-intensity, specific structures (see Chap. VI, p. 142 and Chap. VII, p. 236), or the time-relations of development.

An example of this last type is provided by experiments in which a portion of presumptive neural tube material of *Triton taeniatus* is grafted into the side of an embryo of *Triton cristatus*. It may there undergo differentiation into gills, but such gills preserve a feature of their specific origin, although the tissues from which they have arisen would normally never have given rise to gills. In *Triton taeniatus* the gills develop relatively earlier than in *cristatus*, and in the experiment just described the gills which are formed from the graft of *taeniatus* tissue show a greater precocity of differentiation than the host *cristatus* gills of the other side.[1] The *taeniatus* tissue, in its differentiation into a structure which it would normally never have formed, is still controlled by certain of its hereditary factors. Still more demonstrative results have been obtained by xenoplastic grafting between Anura and Urodela (Chap. VI, p. 142). Here, then, is additional evidence of the fact that the hereditary equipment of all the cells of the organism is the same (see Chap. V, p. 85).

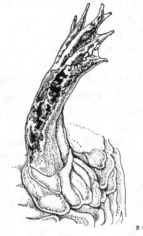

Fig. 195

Sectorial limb-chimaera in an axolotl, produced by combining the dorsal half of a hind-limb regeneration-bud from a black specimen with the ventral half of a hind-limb regeneration-bud left *in situ* on a white specimen; one year after operation. (Redrawn after Schaxel, *Arch. Entwmech.* L, 1922.)

It is possible to make up a compound embryo by grafting together an anterior half-embryo of *Rana virescens* and a posterior half-embryo of *Rana palustris*, or *vice versa*. The compound organism behaves as a unit in regard to its general physiology and can undergo metamorphosis and develop into a full-grown frog. But the two components retain some of their specific characters, not only as regards pigmentation, but also as regards structural

[1] Spemann, 1921.

features such as details of head-shape[1] (fig. 196). In an analogous experiment in which lateral halves of gastrulae of *Triton taeniatus* and *Triton cristatus* are grafted together, it can be shown that although the compound organism is, here again, a functional physiological unit which can develop into a full-grown newt, the tissues

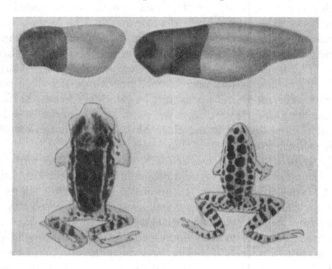

Fig. 196

Compound organisms produced by grafting together half-embryos of two species of frogs in the early tail-bud stage. Above, anterior component *Rana sylvatica*, posterior component *Rana palustris*. Left, shortly after operation. Right, later embryonic stage (note the differential growth of the anterior component). Below: left, a compound frog produced in the same way, but with *Rana virescens* as anterior component; 4½ months after operation. Below: right, a normal *Rana virescens*, showing pigmentation and pattern of trunk and hind legs differing markedly from those of the *palustris* component of the compound organism. (After Harrison, from Wells, Huxley and Wells, *The Science of Life*, London, 1929.)

not only retain some of their specific histological characteristics, but certain specific postural characteristics are retained as well, for the manner in which the limbs are held is typical of the species.[2]

[1] Harrison, 1898.
[2] Spemann, 1921.

§ 4

Another line of work concerns the time at which the hereditary factors in the chromosomes begin to exert their action in differentiation. *Cidaris* and *Lytechinus* are two species of sea-urchins which differ considerably in the times at which corresponding processes take place during their developments. The larva of *Cidaris* gastrulates about 20 hours after fertilisation, and later, mesenchyme is formed from the inner end of the archenteron at about 23 hours. The larva of *Lytechinus* gastrulates after about 9 hours, but mesenchyme has already been formed at 8 hours after fertilisation; this mesenchyme therefore cannot be produced from the formed archenteron but is derived from the outer surface of the larva before gastrulation has begun, at the place where the archenteron will later begin to invaginate.

The hybrid obtained by fertilising eggs of *Cidaris* with sperm of *Lytechinus* begins by developing as a larva of typical maternal (i.e. *Cidaris*) character, up to the end of the blastula stage. This indicates that the paternal factors have not yet exerted any effect up to this stage. But the mesenchyme is produced just as the archenteron begins to invaginate, not from its inner end (as in *Cidaris*) but from the sides of its base, near the outer surface of the larva, thus resembling the conditions in *Lytechinus*. In this respect the hybrid is intermediate between the two parent-species, and it is clear that the paternal factors begin to make their effects observable just at the beginning of the gastrula stage.[1]

It is probable, therefore, that it is in the immediately preceding stage, that of the late blastula, that the action of the hereditary factors in the nuclei commences. In this connexion it is most interesting to note that the late blastula is precisely the stage at which the ratio of cytoplasm to nucleus in the blastomeres reverts to the value at which it stood in the oocyte, before maturation of the egg took place (see Chap. v, p. 132). It may therefore perhaps be suggested that the time of onset of the action of the hereditary factors of the nuclei depends upon the reversion of the cytoplasmic-nuclear ratio to its initial value.[2]

[1] Tennent, 1914, 1922.
[2] Boveri, 1905.

§ 5

As to the intermediate steps in the chain of processes by means of which the hereditary factors influence differentiation, little is known. It is, however, becoming clear that many genetic differences, including certain apparently qualitative effects, depend upon quantitative differences in the rate of action of the factors. The hereditary control of the rates of certain developmental processes has been studied in the insect *Lymantria*, and in the crustacean *Gammarus*.

In *Lymantria* it has been shown that sexual differentiation is conditioned by a competition between two sets of processes: those controlling the production of structures that characterise the female, and those which characterise the male. These in their turn are controlled by hereditary factors, the female-determiners which seem to be lodged mainly in the *Y*-chromosome, and the male-determiners lodged in the *X*-chromosome. In normal development, one or another of these sets of processes wins before the time at which differentiation takes place, but, by making appropriate crosses between individuals of different races, and in pure strains under extreme experimental conditions, it is possible to alter the circumstances in such a way that an animal will develop along the female line up to a certain point, but is then switched over to the male type, or *vice versa*. The sooner this switching over takes place, the more complete is the sex-reversal.[1]

In *Gammarus*, it has been demonstrated that the difference between adult black, chocolate, red-brown, and red eye-colour, is an effect of quantitative differences in the rate of deposition of melanin pigment in the facets of the eye, and that these differences are controlled by hereditary factors.[2] The interaction between genetic factors and environment to produce a given character is also here very well illustrated. At normal temperatures, the " rapid-darkening red" factor or gene produces adult chocolate eyes. But at temperatures below a certain threshold, no melanin at all is produced, and the eyes remain pure red. At intermediate temperatures, intermediate shades of adult eye-colour are produced. One and the same gene leads to different rates of melanin-formation in

[1] Goldschmidt, 1927.   [2] Ford and Huxley, 1927.

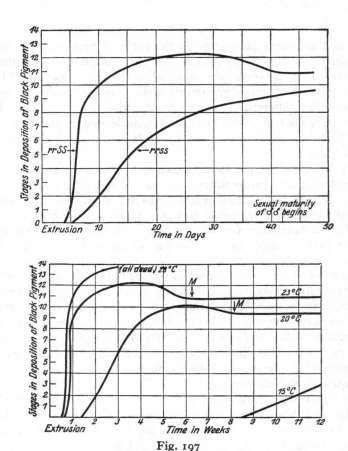

Fig. 197

The action of rate-genes in determining eye-pigmentation in *Gammarus*. Ordinates, grades of colour between pure red (o) and black (14). Abscissae, time. Above: smoothed curves for rapid-darkening (*rrSS*) and slow-darkening (*rrss*) red-eyed types (average of 1000 specimens for each curve) at standard temperature (23° C.). Below: variation of gene-expression due to temperature. All curves refer to animals of the same pure stock (inbred *rrSS*). At 10° C., no melanin is deposited, and the eye remains scarlet; at 13° C., melanin deposition only begins at 20–24 weeks. The figure shows the facts over the range from 15° C. to 28° C. *M.* sexual maturity. (From Huxley, *Problems of Relative Growth*, London, 1932; after Ford and Huxley.)

different conditions; there is a lower threshold of temperature below which no melanin is produced, and an upper threshold above which no further increase in the rate can be produced by this gene (although an allelomorphic gene causes a far more rapid rate, and, as a matter of fact, has quite a different relation to temperature).[1] Further, the precise shade of adult eye-colour produced may also depend upon a relation between the factors controlling melanin-deposition and those controlling rate of eye-growth. When melanin-formation is only moderate, the greater the area of the facets, the more dilute the colour will be (fig. 197).

The way in which genetic factors can exert their characteristic effect only in a particular cytoplasmic environment is also well shown in *Gammarus*. The so-called "albino" and "colourless" mutants have no melanin in their eyes. This is due to the fact that this pigment can only be deposited in the retinular portion of the eye, and in these types this portion of the eye is absent. The mutation has not altered the genes which produce pigment, as in true albinos, but has prevented the appearance of the only regions in which pigment-producing genes can exert their effects.[2]

§ 6

Finally, it is important to note that the cytoplasm of the egg may be modified by specific factors in the maternal hereditary constitution. One of the best examples of this is afforded by the asymmetry of the Gastropod *Limnæa peregra*.[3] As mentioned in Chap. IV (p. 71) the spiral coil of the body and shell in this species is normally right-handed (dextral), but a left-handed (sinistral) type also exists, and it has been shown that the difference between them is controlled by a pair of allelomorphic genes: a dextral-determiner and a sinistral-determiner.

A necessary result of the effect being due to genes present in the mother is the fact that the effects of these genes are delayed by a generation, so that the familiar 3 : 1 ratio is obtained, not in F_2 by individuals, but in F_3 by families. If a snail has had one dextral parent, it is found (neglecting certain special complications) that, after self-fertilisation, all its own offspring are dextral, but of these

[1] Ford, 1929.   [2] Ford, 1929; Huxley and Wolsky, 1932.
[3] Boycott, Diver, Garstang, and Turner, 1930.

offspring 75 per cent. will produce dextral and 25 per cent. will produce sinistral forms. It is clear that segregation has taken place in the snail in question, but the dextral-determiner has acted upon the cytoplasm of the oocyte before maturation in such a way that, regardless of whether the dextral-determiner or its sinistral allelomorph has been extruded with the polar body, the embryos into which those oocytes will develop when matured and fertilised will be dextral. Owing to segregation, 25 per cent. of these embryos will possess the sinistral-determiner only; their oocytes will be subjected to the action of this sinistral-determiner, and all their offspring will be sinistral.

A similar case is found in silkworms. Here, the pigmentation of the serosa membrane of the embryo is determined by the mother's genetic constitution, and not by that of the embryo. Mendelian segregation for this character occurs, but a generation later than for ordinary characters.[1]

§ 7

In other cases, precursor substances may be formed in the cytoplasm of the egg under the influence of the maternal gene-complex. An example of this is found in *Gammarus*. A mutant type known as *white body* contains no carotinoid pigments, neither red in the eyes nor green in the body: it is recessive to the pigmented type.

If a male of the white-body type is crossed with a red-eyed green-bodied female, the offspring are red-eyed and green-bodied from the start. But if the reciprocal cross is made, the young begin their career without any carotinoid pigment, and the eyes and body darken to the normal red and green shades only after some time. In this case it would appear that a gene controls the production of substances needed for the making of red and green pigment. When these substances are absent from the egg, the dominant normal gene introduced from the father takes time to produce these pigment-precursors. But if the mutant white-body gene is introduced from the father and the normal allelomorph from the mother, the precursors have been already manufactured by the mother and a store of them is present in the egg-cytoplasm.[2] It is probable that the white-body mutation renders the animal incapable of utilising carotinoids.

[1] Tanaka, 1924.   [2] Sexton and Pantin, 1927.

§ 8

The last example shows how a detailed analysis is often required to discover the mechanism by which genes exert their effect. Indeed, it is necessary to think in terms of development before it is possible to discover what is the fundamental process with which a given gene is concerned. In such an analysis, the old concept of Mendelian *characters* will disappear. The visible character is not Mendelian in any real sense: it is the resultant of the interaction of a particular gene-complex with a particular set of environmental conditions. In investigating the effect of a given gene, it is usual to study the difference in development and end-result obtained by substituting one allelomorph of the given gene for another in the gene-complex. By doing so in different environmental conditions, it is possible to obtain an idea of the fundamental process influenced by the gene in question. By paying proper attention to the development in this analysis, this fundamental process is seen to be something very different from what would have been expected if only the end-results in the adult had been studied. The resolution of the red-black series of adult eye-colours in *Gammarus* into the effects of genes controlling relative rates of melanin-deposition is a case in point; and this in all probability has a bearing upon other eye-colour series, as in *Drosophila* and in man.

Again, the fundamental process resulting in white ("albino") eyes in *Gammarus* concerns the failure of the embryonic eye to differentiate any rudiment of the retinula region: only a close study of the developmental physiology of the eye-region will be able to shed further light on the processes involved.

This is, in a certain sense, obvious. What has not been adequately recognised, however, is that the converse holds true, and that the study of developmental processes will of itself shed light upon genetics. To illustrate this point, an example may be taken from among growth-processes. The empirical study of relative growth has shown that a change in relative growth in an organ or region appears always to be brought about by a change in a growth-gradient affecting that region. For instance, in Crustacea, the differences between a small purely female type and a large male type of chela, and between the small male abdomen and the large female abdomen, are both brought about developmentally by the substitution of a steep growth-gradient with subterminal high point for a flat growth-gradient with subcentral high point. Any genes controlling chela size and shape will act first by controlling the general form of the gradient involved, and secondly by influencing its steepness. In addition, there will doubtless be other genes

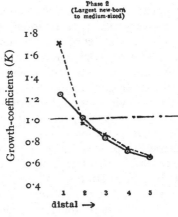

Fig. 198

Growth-gradients in the limbs of domestic sheep, from birth to half-grown specimens. Ordinates: growth-coefficients (differential growth-ratios) for weights of parts of limb relative to weight of vertebral column. The horizontal broken line represents isogony (growth-coefficient = $1 \cdot 0$); values above it signify positive heterogony, values below it negative heterogony. Abscissae: 1, limb-girdles; 2, humerus or femur; 3, radius and ulna or tibia and fibula; 4, carpals or tarsals; 5, metacarpals or metatarsals. Solid line, fore-limbs; dotted line, hind-limbs. (From Huxley, *Problems of Relative Growth*, London, 1932, based on data of Hammond.)

modifying the growth of local regions of the gradient, and influencing detailed characters such as bristles, ridges, etc.; but the main factors operative will concern the gradients as a whole.

The importance of this way of regarding the facts is well shown in sheep.[1] Here, in the first place, the limbs during postnatal development show a marked growth-gradient with terminal or subterminal low point, and high point in the limb-girdles: the growth not only of the bones but also of the muscles is affected by

[1] Hammond, 1929; Huxley, 1931.

this gradient. In the second place, one of the main differences between wild species, unimproved domesticated breeds, and improved domesticated breeds, consists in larger carcass, shoulder, and thigh size (and therefore greater proportion of meat) in relation to limb size in the improved breeds, and this on analysis is found to depend on an accentuation of the slope of the original gradient. Owing to this, the relative growth-intensity of the terminal portions of the limb is decreased, that of the central portions in the region of the limb-girdles is increased. In improving the meat qualities of the sheep, it is necessary to search for genes affecting the growth-gradients of the limbs.

In a similar way, it will undoubtedly be found that there are genes which affect the primary gradient-fields of the early embryo, and therefore the relative sizes of the chemo-differentiated fields in the next stage, and thus consequently the proportions of the developed animal.

Thus a knowledge of the nature and effects of gradient-fields will guide the geneticist in his search for Mendelian gene-differences and his analysis of the way in which they exert their effects.

§ 9

In the analysis of the genetics of qualitative characters, a knowledge of developmental processes may be of very great importance to the geneticist. In many cases, for example, the relative size of a part does not vary in linear relation with the absolute size of the body, but is proportional to the size of the body raised to a power. In such a case, to take percentage size of part as a "character" to be analysed could only lead to erroneous conclusions. To put it mathematically, if developmental study shows that the growth-formula of the part (y) relative to the body (x) is of the form $y = ax^b$, then the geneticist must search for genes modifying not only the value of the constant a, but also that of b: and if he does not know the formula, he is not likely to search for the right constants.

Again, linear dimensions would appear to be the simplest "characters" to deal with in making a genetical analysis of quantitative differences in the size and proportions of an organ. But developmental analysis appears to show that the two main variables which are here concerned are, first, the total amount of material in the organ (which itself is likely to be related to the total bulk of the organism by a non-linear formula), and, secondly, the relative intensity of growth in the different planes of space within this mass of material. The fundamental processes are concerned with the ratios of the linear dimensions, not with the linear dimensions

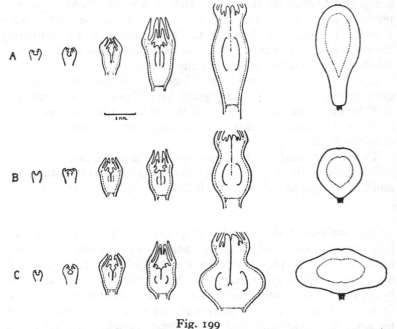

Fig. 199

Shape-genes in gourds (*Cucurbita*). Five stages in the development of ovary and fruit in A, elongate; B, spherical; and C, disc types; showing progressive change of shape of the fruit-rudiment. (From Sinnott and Durham, *Bot. Gaz.* LXXXVII, 1929.)

separately. This concept has been applied to the analysis of the size and shape of gourd fruits,[1] where it is found that a genetic analysis on the basis of linear dimensions leads to confused results, whereas an analysis on the basis of ratios between length and breadth permits of a simple interpretation of the results in terms of a few clearly defined "shape-genes" (figs. 199, 200).

[1] Sinnott and Hammond, 1930.

These examples will serve to show the relations between the sciences of genetics and of developmental physiology. Hitherto, neo-Mendelism has been concerned mainly with the manœuvres of the hereditary units, and in large part with their manœuvres

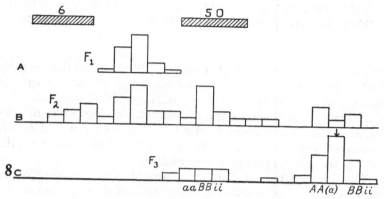

Fig. 200

Shape-genes in gourds (see also fig. 199). The abscissae give the form-indices of the fruits, expressed as breadth/length ratios, running from very elongated shape (small breadth/length ratio) on the left to very flattened (disc) shape on the right. The ordinates represent frequencies. Top line, range of form-indices of parent types: 6, a long type (elongate); 50, a rounded type (sphere). The shape-genes involved are *A*, *B* and *I*. *A* and *B* produce flattening, while *I* inhibits their action. The constituent of line 6 is *aaBBII*, of line 50 *AAbbii*. The F_1 is intermediate and unimodal. The F_2 is multi-modal: the extreme right-hand group represents a new recombination comprising the *ABii* forms, resulting in disc fruits. An F_3 from one of these (bottom line) shows a sharp 3 : 1 segregation. The parent must have been *AaBBii* and the offspring 3 *ABBii* : 1 *aaBBii*. (From E. W. Sinnott and D. Hammond, *Amer. Nat.* LXIV, 1930.)

during the two cell-generations in which the reduction of chromosomes is brought about. It is now beginning to concern itself with the mode of action of the hereditary units during the much larger number of cell-generations involved in building up the adult organism from the egg: and this task it can only accomplish satisfactorily in close contact with developmental physiology.

Chapter XIII

THE PREFUNCTIONAL AS CONTRASTED WITH THE FUNCTIONAL PERIOD OF DEVELOPMENT

§ 1

It has already been noted that some, at least, of the field-organisation, both total and partial, characterising the early stages of development, appears to persist throughout life, side by side with the organisation characteristic of later stages. However, the developmental consequences of the new processes initiated in the functional period are very striking and overshadow most of the effects dependent upon field-organisation.

These new processes fall under several main heads—growth, true functional modification, the unification of the organism by the nerves, and endocrine influences. It is impossible within the scope of this book to give any detailed treatment of development during this functional period, but a few instances may be presented which will serve to make its main characteristics clear.

The true growth-period of the embryo or larva does not begin until the organism can either feed for itself, draw upon a store of accumulated food material (as in meroblastic eggs), or be nourished by its parent. Previous growth takes place only by imbibition of water, or by slow contact absorption of yolk. Without quibbling over precise definitions of growth, however, it may be pointed out that the determination of organs may take place without any process of growth being involved, and that growth may and normally does continue long after tissue-differentiation has occurred.

It appears, however, at least in some cases, as in that of axolotl limb-buds, that degree of differentiation is correlated with absolute size of the rudiment. If the rudiment is experimentally enlarged, as by grafting one limb-bud on to another, the resulting single enlarged limb (see p. 223) shows accelerated differentiation as compared with the normal limb of the unoperated side[1] (fig. 201).

[1] Filatow, 1932. See also Guyénot and Schotté, 1923.

During the early stages of development, when the whole organism or its major organ-systems are still in the gradient-field condition, removal of a small portion of tissue will not result in the absence of any particular structure, for regulation is possible within the gradient-field. At this stage, no structures have been locally determined, and loss of tissue does not imply loss of any definite rudiment. It is only later, during the mosaic stage of development when the various rudiments are chemo-differentiated, that regulation is impossible. Later on, again, the power of regeneration

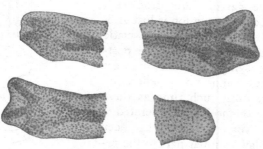

Fig. 201

Correlation of size with rate of development in fore-limb rudiments of the axolotl. The very early limb-bud of one embryo is removed and superposed on the mesodermal portion of the limb-bud of a host embryo of the same stage. The two rudiments fuse to produce a single enlarged limb, in which differentiation is more advanced than in the normal limb of the other side. Top: the host limbs; left, unoperated normal limb; right, experimentally enlarged limb with larger digit-rudiments and more advanced skeletal condensation. Below: the limbs of the donor; left, unoperated normal limb; right, small limb-rudiment regenerated from the remainder of the limb-field. (From Filatow, Zool. Jahrb. (Abt. allg. Zool. Physiol.), LI, 1932.)

appears. Regeneration, as pointed out by Przibram,[1] is intimately bound up with growth, and the onset of the capacity for regeneration after the mosaic stage of development is connected with the onset of the capacity for growth at this stage. Regulation and regeneration must therefore be carefully distinguished, since they involve developmental processes which are very different, and are operative at different periods of the life-cycle.

Regeneration also, in some cases at least, appears to be connected with the development and function of the nervous system.

[1] Przibram, 1919.

If the limb of a post-larval or adult newt is amputated it will re-generate, provided that the fibres of the autonomic (sympathetic) nervous system are intact.[1] The dorsal nerve-roots can be severed and the dorsal ganglia destroyed, or, the ventral nerve-roots can be severed close to their exit from the spinal cord, without destroying the power of regeneration of a limb. But if the sympathetic ganglia are destroyed, the power of regeneration is lost also. If the nerves of the brachial or sciatic plexus are simply severed, the post-ganglionic sympathetic fibres are thereby cut, and no regeneration takes place until such time as these fibres have themselves regenerated.[1]

Fig. 202

The morphogenetic influence of the nervous system. The anterior end of an earthworm is ampu-tated and then an incision made on the ventral surface so as to remove the ventral nerve-cord from several segments. No head is regenerated from the anterior cut surface of the trunk, but one may form in relation to the an-terior end of the nerve-cord. (Redrawn after Morgan, *Arch. Entwmech.* XIV, 1902.)

The nervous system has been found to play a similar part in the regeneration of the earthworm, for the nerve-cord must be present at the cut surface if regeneration is to take place from that surface. If the anterior end of a worm is cut off, and, in addition, the nerve-cord is ex-tirpated for a short distance behind the cut surface, an anterior end may be regenerated from the place where the nerve-cord ends, but never from the original cut surface[2] (fig. 202).

The precise rôle of the nervous system in many such cases of regeneration is unknown, but the example of the newt's limb is a warning that the relation may be difficult of analysis, and that only fibres of a particular component of the nervous system may be involved in these morphogenetic processes.

[1] Schotté, 1926 B.
[2] Morgan, 1902.

§ 2

Growth is also directly responsible for a certain type of further differentiation, namely, change of proportions. There are at least five factors involved here. One concerns the specific growth in-

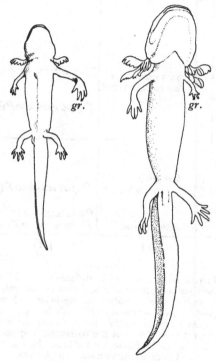

Fig. 203

Inherent growth-rates in limb-rudiments. Left, a larva of *Amblystoma punctatum* and right, one of *Amblystoma tigrinum* between which the left fore-limb rudiments were exchanged at the tail-bud stage: 50 days after operation, after maximal feeding of the larvae. The grafted limbs (*gr.*) are approximately of the same size as the corresponding unoperated limbs of the donors. (Redrawn from photograph in Twitty and Schwind, *Journ. Exp. Zool.* LIX, 1931.)

tensities of the organs, which will determine the main features of the growth-equilibrium between them and the body: this has been dealt with in Chap. x (fig. 203). The second concerns growth-gradients, which will influence the growth of parts within single organs (such as crustacean chelae), within single regions of the body (such as in the crustacean abdomen), or within the body as a whole (as in stag-beetles or Planarians).[1] The third is concerned

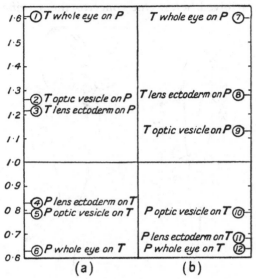

Fig. 204

The mutual influence of regions with different specific growth-intensities. Whole eyes, or their parts (optic vesicle and lens ectoderm), were grafted reciprocally between embryos of *Amblystoma tigrinum* (*T*) with high growth-intensities, and *Amblystoma punctatum* (*P*) with low growth-intensities. The ordinates represent the ratios of the diameters of the parts of the eye on the side receiving the graft to the diameters of the corresponding parts of the intact eye of the other side, (*a*) for optic vesicle, (*b*) for lens. Fast-growing whole eye on slow-growing host (1 and 7) gives high ratios. The association of a slow-growing host-lens with fast-growing grafted optic vesicle (2 and 9), or a slow-growing host optic vesicle with grafted fast-growing lens (3 and 8) reduces the ratios. Similarly, low ratios are found for slow-growing whole eyes on fast-growing hosts (6 and 12); 5 and 10, 4 and 11 show the increase of ratio when a slow-growing grafted component is associated with a fast-growing host component. (From Huxley, *Problems of Relative Growth*, London, 1932; based on data of Harrison.)

with the time-relations of development. In general, development occurs in an antero-posterior direction, so that at a given time anterior organs are further differentiated than those at a more posterior

[1] See Huxley, 1932, Chap. III.

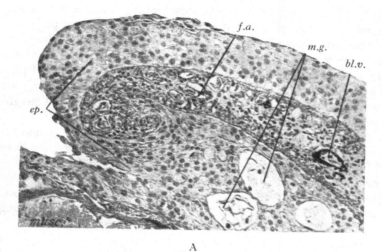

A

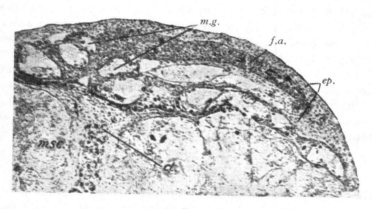

B

Fig. 205

The effect of mechanical conditions on morphogenesis. In larval axolotls kept out of water, the dorsal fin disappears. This is due to its falling over and becoming fused with the skin of the back. A, Section of early stage of fusion. The meso-dermal fin-axis (*f.a.*) is bent at the tip. B, Section of a stage showing complete fusion. The fin-axis still shows a curved tip. The limit of the fused fin is marked by a sudden thinning of the epidermis (*ep.*); *bl.v.* blood-vessel; *m.g.* mucous glands; *musc.* muscles; *c.t.* connective tissue. (From Huxley, *Proc. Roy. Soc.* B, XCVIII 1925.)

level. Within vertebrate limbs, development takes place centri-
fugally, and, as a result of this, the later-differentiating parts will
increase in proportionate size during development.[1] Fourthly,
there are growth-processes directly concerned with the functional
demands made upon an organ: these also involve change of proportion and will be dealt with later. Fifthly, the growth of one
structure may be modified by the specific growth-rate of neighbouring structures.

In illustration of this last point, it is found that the structures
composing the eyes in *Amblystoma punctatum* and *tigrinum* have
different specific growth-intensities (Chap. x, p. 366). By making
grafts of eye-cups and of lens-forming epidermis between these two
species, it is found that the presence of a fast-growing eye-cup is
correlated with an increase in the growth-rate of a slow-growing
lens associated with it, and *vice versa*[2] (fig. 204).

Mechanical modification of growth-processes is readily brought
about. It is only necessary to recall the artificial deformations of
skull, lips, waist, feet, etc., practised by various human societies.
In this connexion may be mentioned the fact that when axolotl
larvae are reared in dishes with only a little water so that their
backs protrude above the level of the water, the dorsal fin falls over
owing to its weight, and becomes completely united to the skin of
the back. But, internally, the structural and histological features
of this finless condition are quite distinct from those produced as
a result of normal metamorphosis,[3] although externally they are
more or less similar (fig. 205).

§ 3

The unification of the organism by means of the nervous system
brings the various parts into more intimate relations with each
other as regards their functional activities, and brings the organism
as a whole into a more intimate and more delicately adjusted relation with the environment. This is responsible for a greater
delicacy of functional adjustment on the part of the various organs.

[1] See Huxley, 1932, Chap. IV.
[2] Harrison, 1929; Twitty 1930; Twitty and Schwind, 1931. This is also true
of limbs (Rotmann, 1931, 1933), but does not happen with parts of the shoulder-
girdle.   [3] Huxley, 1925.

The unification of the organism by means of the circulatory system has in some ways a similar effect. It also makes possible a competition between organs and regions for available nutriment, and this may have marked effects upon development.[1] The proportions of parts of growing mammals, (a) fed maximally, (b) fed so as to permit of only slight growth, and (c) fed so as to permit only of maintenance of weight, are quite different.[2] In extreme cases, whole regions may disappear as a result of being drawn upon by the rest. For instance, if a zooid together with an attached piece of stolon of the Ascidian *Perophora* are isolated and starved in normal conditions, the stolon will be completely resorbed by the zooid; but when placed in dilute toxic solutions the zooid is more affected, and is then resorbed by the stolon (p. 294).[3] In organisms without a skeleton, starvation may produce reduction in total size, and then different parts will be reduced at different rates, as for instance in Planarians[4] and in jelly-fish[5] and hydroids.[6]

§ 4

The establishment of the circulation has a further consequence which in vertebrates at least has far-reaching effects upon development. It permits of the transport of hormones, some of which have striking morphogenetic functions. Some hormones may be liberated more or less continuously into the blood. This is apparently the case with that amount of thyroid hormone needed to produce normal development in man: when this threshold is not available, the child is a cretin, stunted in growth and subnormal in intelligence.

In other cases, the hormones may be produced cyclically, and this appears to apply to the hormone of the anterior pituitary concerned with stimulating the cyclical growth of the ovarian follicles. Or the hormones may be produced in markedly different amounts as a result of nervous impulses to the gland, which in their turn are controlled by external stimuli. In Amphibia, for instance, darkness stimulates the post-pituitary to liberate the hormone which causes expansion of melanophores: and while growth is taking place, this

[1] Roux, 1881.
[2] Jackson, 1925; Hammond, 1928; Huxley, 1932.
[3] Huxley, 1921 B.   [4] Abeloos, 1928.
[5] de Beer and Huxley, 1924.   [6] Huxley and de Beer, 1923.

also causes extra multiplication of melanophores.[1] A similar result, doubtless brought about in the same way, is seen in fish (*Lebistes*). Specimens reared on white backgrounds have contracted melanophores, few in number; specimens reared on dark background have expanded melanophores in large numbers. Functional activity increases the rate of multiplication[2] (fig. 206). Similarly in salamander larvae (*S. maculosa*), yellow backgrounds

 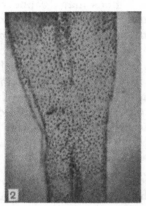

Fig. 206

Functional activity and rate of multiplication of pigment-cells. Dorsal view of the trunk region of two specimens of the teleost fish *Lebistes reticulatus*, one (1) reared for 6 months on a white background, the other (2) for the same length of time on a black background. In both cases the pigment-cells (melanophores) have been induced to assume the contracted state by adrenalin treatment. Note the much larger number of melanophores in the black-adapted specimen, in which during life they were expanded normally, while in the white-adapted specimen they were contracted. (From Sumner and Wells, *Journ. Exp. Zool.* LXIV, 1933.)

favour the increase of the yellow areas, black backgrounds that of the black areas. After metamorphosis, however, a gradual regulation towards the control type sets in, indicating that what we may call "functional multiplication" of pigment-cells is only important in certain stages.[3]

A sudden change in the activity of a gland may take place at a certain stage in development, as occurs with larval Urodela, in which the sudden onset of metamorphosis is brought about by the

[1] Smith, 1920.   [2] Sumner and Wells, 1933.
[3] Herbst, 1924.

thyroid throwing its stored secretion into the blood.[1] In this respect, the Urodele may be contrasted with the Anuran, where the thyroid becomes progressively more active during larval life, without any such extreme change in its activity.

The morphogenetic effects of hormones are varied. Some of the most marked are those concerned with amphibian metamorphosis, in which the growth or differentiation of some organs and the

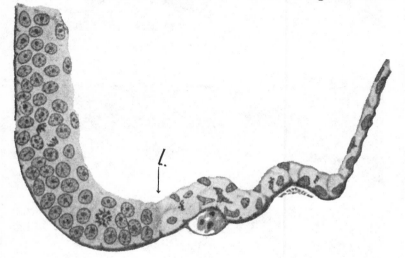

Fig. 207

Sharply delimited fields in a thyroid-treated frog tadpole. Section showing on the left the epidermis of the fore-limb bud, on the right the lining of a branchial cleft. The former has reacted to the thyroid hormone by growth (mitoses, crowded nuclei); the latter by degenerative changes (vacuolation, shrunken nuclei). The limit (*l.*) between the two zones is clear-cut, without transition. (From Champy, *Arch. Morph. Gén. Exp.* IV, 1922.)

atrophy of others will only take place under the influence of the thyroid hormone. All gradations are to be found, however, between such marked morphogenetic effects and effects of a transitory physiological nature. The morphogenetic effect of hormones may be linked with the pre-existence of qualitatively different fields. E.g. in the frog, one region of epidermis will proliferate, and another degenerate, under the influence of thyroid[2] (fig. 207).

[1] See Huxley, 1923; Uhlenhuth, 1922.   [2] Champy, 1922.

These regional differences in reactivity are established very early (Schwind, *J. Exp. Zool.* LXVI. 1933). The relation between hor-

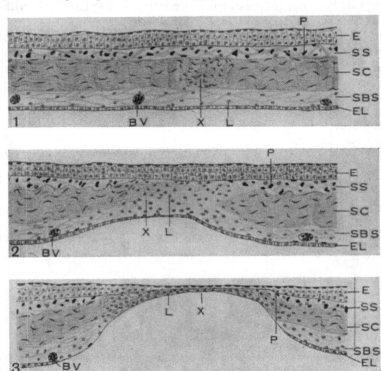

Fig. 208

The perforation of the operculum in the frog (*Rana clamitans*). Sections showing the histolysis leading to normal perforation. 1, First sign of histolysis (at *X*). 2, Histolysis well under way; the stratum compactum and stratum spongiosum have become interrupted; there has been a marked invasion of the area by lymphocytes, and the skin in this region is decreasing in thickness. 3, The skin is reduced to the epidermis, which the fore-limb then ruptures. *BV*, blood-vessel; *E*, external epidermis; *EL*, epidermal lining of branchial chamber; *P*, pigment; *L*, lymphocytes; *SC*, stratum spongiosum; *SBS*, connective tissue; *SS*, stratum spongiosum; *X*, site of histolysis. (From Helff, *Journ. Exp. Zool.* XLV, 1926.)

mones and growth-gradients is shown by studies on regeneration-rate and hormone-susceptibility in birds' feathers.[1]

[1] Lillie and Juhn, 1932.

An interesting half-way stage between chemical effects due to contact, as in the determination of a lens by the eye-cup, and those due to circulatory hormones, is seen in the perforation of the right-hand side of the operculum in Anuran tadpoles during metamorphosis. As is well known, the rudiments of the fore-limbs develop beneath the operculum, and while the left fore-limb makes its way out through the open spiracle, the right protrudes through a special perforation. After extirpation of the right fore-limb rudi-

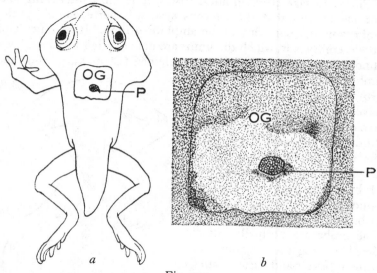

Fig. 209

Perforation of opercular skin of *Rana palustris*, grafted on to the back, over pieces of atrophying tail-muscle. The histolysis of the opercular skin leading to perforation is the same as that normally due to the atrophying gills, though slower. *a*, Larva, showing graft of opercular skin (*OG*), perforated (*P*). *b*, Enlarged view of graft showing atrophying tail-muscles seen through the perforation. (From Helff, *Journ. Exp. Zool.* XLV, 1926.)

ment, perforation of the operculum still occurs,[1] thus demonstrating that it is not due to mechanical pressure. Actually, it is a substance liberated by the gills during their atrophy that is responsible for the perforation, as is shown by experiments in which metamorphosing gills are grafted beneath the skin of the back and cause perforation here too.[2] Other atrophying organs, such as the

[1] Braus, 1906.   [2] Helff, 1926.

muscles of the tail during its resorption, will produce the same effect, but more slowly. Thus presumably some substance produced during autolysis is the agent responsible (figs. 208, 209).

§ 5

The next subject to consider is the trophic effects of the nervous system. In view of the fact that innervation (by fibres of the autonomic nervous system) is a prerequisite condition for regeneration of limbs to take place in adult newts, it is most interesting and curious to find that the nervous system is not essential for the embryonic development of the amphibian limb. It is difficult to obtain embryos in which the limbs are not supplied by some, even abnormal, nerves, for, as already explained (Chap. XI, p. 389), the limb exerts an attraction on the growing axon. But limb-rudiments have been seen to develop when free of any nerve-fibres. This condition can be realised by grafting the limb-rudiment of a frog into a lymph-space of another larva, or by extirpating the neural tube opposite the limb region on one or both sides in the neurula stage. The limbs are normally differentiated as regards all their constituent tissues and parts: cartilage, muscles, skin, blood-vessels, and the joints between the skeletal segments, all these are normally differentiated in the absence of innervation, but the limb as a whole is too small.[1] In other words, the nerves have a trophic but not a morphogenetic effect on the development of the limb (fig. 210). In

Fig. 210

The trophic effect of the nervous system on the development of the limb. Ventral view of a larva (shortly before metamorphosis) of *Rana fusca* from which at the neurula stage the rudiment of the lumbo-sacral region of the spinal cord was extirpated on the right side. Note normal form of right leg but subnormal size and development. (From Hamburger, *Arch. Entwmech.* CXIV, 1928.)

this respect the effect of the nerve is similar to that of thyroid hormone on limb-growth in larval Anura[2] (see also Chap. X, p. 363).

[1] Lebedinsky, 1924; Hamburger, 1929.   [2] Champy, 1922.

The stimulation of the multiplication of the nerve-cells in the spinal cord (Chap. XI, p. 383) in *Amblystoma* is another example of the effects of nerve-endings. There are also the cases in which the presence of a nervous connexion is necessary for the maintenance of structure in an organ. As is well known, muscles atrophy when the motor nerves to them are cut. But the best-analysed examples concern the lateral-line organs, and the taste-buds on the barbels of the catfish *Amiurus*. When the nerves to these organs are cut, the organs themselves undergo marked dedifferentiation, and redifferentiation when the regenerating nerve restores their nerve-supply.[1] The trophic stimulus has been found to pass down the nerve from the cell-body at a rate of 2 cm. per day, and the indications are that it is due to percolation of a hormone-like substance.[2] It clearly cannot be due to normal impulse-conduction (see p. 387).

Though the precise mechanism of their action is still obscure, the interest of these examples for the present purpose is clear. They demonstrate that once the nervous system becomes functional, new methods of influencing development are available in the organism. These methods concern such diverse processes as local cell-multiplication, large-scale regeneration, and the maintenance of differentiation in organs.

§ 6

Finally, there are the effects of function *per se*. This is perhaps the most pervading of all the new effects which take their origin at the onset of the functional period.

Function can influence the multiplication of cells and the size of organs, the histological appearance of cells, and the arrangement of cells and tissues within an organ. Often more than one of these processes is involved at one time. The most obvious example of purely quantitative change concerns compensatory hypertrophy. When a portion of a functioning organ complex is removed, the remainder increases in bulk in response to the increased demands made upon it. The simplest instance concerns the kidneys. When one kidney is removed, the other enlarges; the enlargement is considerable, though not to double its original bulk.[3]

[1] Olmsted, 1920.   [2] G. H. Parker, 1932 A, B.
[3] Ribbert, 1894.

Conversely, when extra demands are made upon an intact organ, it also may respond by increased growth. The excess growth of striated muscle under the influence of heavy work is the most familiar case. The heart, too, is an excellent example. In small birds, the relative heart-size is greater in specimens from high latitudes than in those of the same species from milder climates, owing to the greater demands made upon the circulation in cold conditions.[1]

In voluntary muscle, it is probable that the direction of the fibres is also influenced by function, in the first instance by the tension to which the muscle is exposed by the growth of the skeletal parts to which it is attached.[2] The directive effect of stress has been experimentally demonstrated in connective tissues. By subjecting thin tissue-cultures of fibroblasts to variations in surface tension it has been possible to show that whereas in regions free from directional stress, fibres are formed at random in all directions, in regions subjected to directional tension the medium is condensed along the lines of stress. The fibres orient themselves along these condensations, and the cells multiply more rapidly in these regions[3] (fig. 211). This case falls perfectly into line with the experiments on regenerating tendons. If the achilles tendon of an animal is cut, the space between the cut ends is filled with debris, blood, and phagocytes, and resembles a tissue-culture. Fibroblasts soon grow into it, and the fibres which they produce are at first chaotic; next they form a meshwork with diagonal interlacings; and finally form parallel bundles. The muscle, exerting a pull on one of the cut ends of the tendon, sets up lines of stress in the ground-substance, and this orientates the growth of the fibres.

But if the muscle also is cut, so as to abolish the pulling effect, no tendon is formed. If now a silk thread is drawn through the regenerating tissue, in a direction at right angles to that of the original tendon, constant gentle pulling on the silk thread will produce a bundle of fibres orientated according to the artificially-produced lines of stress. A tendon has here been formed, but at right angles to its normal direction.[4]

In respect of the orientation of the cells to the lines of stress, and

---

[1] Hesse, 1921.
[2] Carey, 1921 A.
[3] Weiss, 1929, 1933.
[4] Lewy, 1904; see also Nageotte, 1922.

of the more rapid multiplication of the cells subjected to the stress, these experiments have completely confirmed the epoch-making essay of W. Roux (1881), by whom the principles of functional differentiation were first clearly stated. From these and other lines of evidence, it appears highly probable that the size and fibre-direction of all the tendons of the body have no direct hereditary

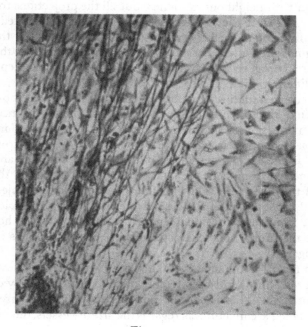

Fig. 211

Portion of a tissue-culture of chick fibroblasts exposed to regional tension (by cultivation as a film in a quadrangular frame). In the region under tension (left) the cells are arranged in fibres parallel to the directions of the tensile force, and are more numerous than in the remainder, where they are scattered and of irregular form. (From Weiss, *Arch. Entwmech.* CXVI, 1929.)

basis, but are determined epigenetically *de novo* in each individual by the stresses and strains to which they are exposed during development. The fact that fibroblasts arrange themselves along lines of mechanical stress, and multiply faster when exposed to tension, automatically accounts for the production of a mechanically adaptive structure.

HEE   28

The fine architecture of bones appears also to be determined in the same way. Here, too, structures which are mechanically adapted in great detail to their functions are not determined hereditarily. On the other hand, the general form of bones is predetermined in great detail by chemo-differentiation. Certain depressions in the surface of avian bones appear to result from mechanical interaction with neighbouring bones, but all the projections from the surface, including the joint-structures, will arise in isolated bones grown in culture media (see p. 225). It may prove that the cartilaginous rudiment is rigidly predetermined, whereas the bony structure, being secondary from the start, is always dependent in its differentiation.

The coarse structure of a bone is, then, a result of chemo-differentiation during the prefunctional period, but function is necessary for the perfection of its finer structure, viz. the orientation of its spicules. Function is also necessary for the normal growth of bones. If one leg of a new-born animal is kept immobile and non-functional, the long bones remain much slenderer than in the used limb of the other side. On the other hand, if a leg is subjected to changed function, as in the case of the hind legs of puppies born without front legs, the hind legs, from the practice of hopping, assume the proportions characteristic of hopping animals such as the kangaroo.[2]

With regard to the blood-system, little is known as to how much of the broad lines of its architecture may be determined by chemo-differentiation. What is certain, however, is that a very great deal of its detailed architecture, as regards the size of vessels, the angles of their branchings, and the courses which they follow, are determined hydrodynamically. The pressure of the blood moulds the vessels in such a way as to offer the least resistance to its flow.[2]

Lastly, instances may be given of functional changes involving cell-form as well as the total size of an organ and the development of its parts. The first case, like so much of the functional differentiation of the blood-vessels, shows the effect of pressure of a contained fluid on the walls of its container. The urinary bladder of a dog of medium size normally evacuates a quarter of a litre of fluid per day. The wall of the bladder is composed of smooth muscle cells and is about half a millimetre thick. By means of a

[1] Fuld, 1901.   [2] Oppel and Roux, 1910.

tube connected with the bladder, large quantities of a neutral fluid can be introduced into it, with the result that its internal pressure

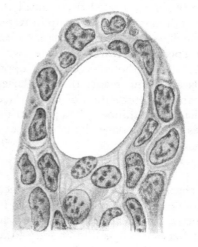
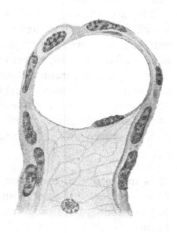

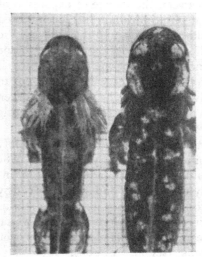

Fig. 212

Functional activity and morpho-genesis in amphibian gills. Below: two salamander larvae; left, reared in conditions of oxygen-deficiency (in water under an atmosphere with 11 per cent. O_2); the gills are long and feathery; right, reared in con-ditions of oxygen-excess (in water under an atmosphere of pure O_2); the gills are short and stumpy. Above: sections of gill-filaments from two similar larvae; right, oxygen-deficiency: epithelium one layer thick, of flattened cells; left, oxygen-excess: epithelium often two layers thick, of rounded cells. (From L. Drastich, *Zeitschr. f. vergl. Physiol.* II, 1925.)

is raised. The quantity of fluid evacuated per day may reach 50 litres under these experimental conditions. As a result of this increased work to which the wall of the bladder has been put, it was found that it had become ten times as thick, that its cells had developed striations very similar to those which characterise heart-muscle, and that the whole bladder pulsated rhythmically.[1] The other case is that of salamander larvae, brought up in water which is deficient in oxygen. Such larvae show much enlarged external gills, while the gills of specimens reared in water with excess of oxygen are extremely small. In the enlarged gills, upon which extra respiratory demands are being made, the capillaries are larger, nearer to the surface, and the epithelium of the surface and the endothelium of the capillaries are thinner, thus permitting of a more rapid diffusion of gases. The converse changes are seen in the reduced gills[2] (fig. 212).

§ 7

It is important to note that no sharp line can be drawn between functional responses of considerable morphogenetic extent, as in the cases just cited, and transitory adjustments of a physiological nature which leave no structural traces, such as a temporary local vaso-dilation. The connexion between the degree of expansion of melanophores and their rate of multiplication has been noted above (p. 426). Further, it should be remembered that one and the same kind of organ can respond by a morphogenetic change to one degree of functional stimulus and not to another. For instance, it appears that only severe demands on muscles will cause them to hypertrophy; movements involving little mechanical strain, even when rapid and prolonged, have no effect—e.g. those of knitting or piano-playing.

It must also be remembered that functional adaptation can only take place within certain limits prescribed by heredity. The thyroid responds very readily to the demands made upon it by increasing or decreasing its supply of hormone and its size. Yet by selection, it has been possible to establish separate genetic strains in pigeons, a high-thyroid strain and a low-thyroid strain, which differ from each other in the size and activity of their thyroids even under identical external conditions[3] (see also p. 409).

Most important of all, it must be borne in mind that functional modification may be very active in one group of animals, and

---

[1] Carey, 1921 B, 1924.   [2] Drastich, 1925.   [3] Riddle, 1929.

negligible or absent in another. For instance, it is impossible for holometabolous insects to produce functional modifications during individual ontogeny in their skeletons. The hard parts of these animals are definitively formed, with all their adaptive details, on emergence from the pupa, and no further growth is possible. The same is true for the development of their muscles and tendons: these must be preformed during the pupa stage so as to permit of perfect function and locomotion of the animal as soon as they are called upon.

There is thus a remarkable contrast between the development of vertebrates and that of higher insects. In the former, prefunctional differentiation lays down a rough sketch of the organism, upon which most of the finer adaptive details are later inserted by means of functional response to the demands made upon the parts. In the latter group, on the other hand, although doubtless some details, such as those of the blood-vessels, may be determined through functional response, the greater part of the structure, including even the finer adaptive details, must be laid down by elaborate chemo-differentiation, unaided by functional response.

There are, of course, other equally fundamental differences in developmental methods between groups. Hormones play a very large part in the later stages of vertebrate morphogenesis; but in insects their rôle appears to be altogether subsidiary. Similarly, the adult form of a vertebrate is determined by changes in proportion of parts which are brought about by differential growth in already functioning organs, and which continue through a large fraction of the life-span; in holometabolous insects, no growth occurs in differentiated parts, and proportions must be definitively fixed during the short pupal period.

The subordination in Ascidians of the period in which the total gradient-field system is the sole form of organisation, as contrasted with its long persistence in Amphibia, is another example, in this case concerning early stages of development, of the differences which may exist between groups as regards their developmental mechanisms.

Chapter XIV

SUMMARY

§ 1

It is now possible to give a brief summary of the chief points which have emerged from our study of development, during which attention was focussed on differentiation and its origin as the central problem.

In the first place, animal development is truly epigenetic, in that it involves a real creation of complex organisation. It is also predetermined, but only in the sense that an egg cannot give rise to an organism of a species different from its parent. The development of each individual is unique. It is the result of the interaction of a specific hereditary constitution with its environment. Alterations in either of these will produce alterations in the end result.

Determination is progressive. In the earliest stages, the egg acquires a unitary organisation of the gradient-field type in which quantitative differentials of one or more kinds extend across the substance of the egg in one or more directions. The constitution of the egg predetermines it to be able to produce a gradient-field of a particular type; however, the localisation of the gradients is not predetermined, but is brought about by agencies external to the egg. The respective rôles of internal predetermination and external epigenetic determination are clearly seen in regard to the bilateral symmetry of the egg. The amphibian egg is predetermined to be able to give rise to a gradient-field system of bilateral type through the establishment of the grey crescent at a particular latitude of one meridian. The particular meridian is not predetermined, but is normally decided by the point of sperm-entry; the precise latitude is determined as a result of the primary axial gradient of the egg, impressed upon it by factors in the ovary. On the other hand, the egg of a radially symmetrical animal like a Hydroid is incapable of developing bilateral symmetry; the predetermined capacity to react to stimuli localised in one meridian is not given in its constitution.

The agencies which determine the position of the various axes involved in the gradient-field system may be of very various nature; they may be factors in the maternal environment (ovarian conditions), biological factors (point of sperm-entry), or external physical factors (as in the determination of the polarity of the egg of *Fucus*). In any case, they are external to the egg. They may also operate at very different times relatively to fertilisation.

A number of chemical processes are set going by fertilisation. These will proceed differently in the quantitatively different environments provided in different parts of the gradient-field system, until qualitative differences are set up. In most cases, these differences are at first not visible, and are presumably of chemical nature; this step in differentiation is therefore spoken of as chemo-differentiation. These chemical differences appear at first to be reversible (e.g. labile determination of the presumptive neural tube region in the Urodele before gastrulation) but after a certain point to become irreversible. From this moment onwards, the organism consists of a mosaic of chemo-differentiated regions, each determined to give rise only to one or a limited number of kinds of structure. These are what we have called partial fields.

§ 2

The attainment of the mosaic stage often takes place under the influence of a dominant region or organiser. This may determine the extent and form of the whole gradient-field within which chemo-differentiation occurs, as in Planarian regeneration, or may interact with a previously established gradient-field orientated in another direction, as in amphibian organiser grafts.

The organiser may exert its effects at a distance, as does the regenerated head on a cut piece of a Planarian, or may supplement such distance effects by more powerful contact effects, as happens when the amphibian organiser comes to underlie a certain portion of the animal hemisphere, and at once determines it irrevocably as a nervous system.

Modifications of the gradients by external agencies will entail alterations in the structures produced. These alterations may consist in changed proportions, or in the total absence of certain regions (temperature-gradient experiments with frogs' eggs, cyclopia in fish, modification of regeneration in Planarians). Here again, there is a predetermined capacity to produce a certain type of structure in certain conditions; but the precise localisation of the structures produced depends upon the form of the gradients in the field-system.

Once the mosaic stage has set in, further differentiation may be brought about by the influence of one point on its neighbours. The classical example of this is the induction of a lens from epidermis by the optic cup.

During the period when the organisation of the developing animal consists of a single field-system, far-reaching regulation is possible; after irreversible chemo-differentiation has occurred, it is not. The precise time at which irreversible chemo-differentiation sets in varies markedly in different groups. In Amphibia it occurs during gastrulation; in Ascidians at fertilisation.

After the establishment of a mosaic of partial fields, it does not follow that all the cells of any given partial field necessarily give rise to the organ characteristic of the field. Thus, more cells are capable of giving rise to the amphibian fore-limb than do in fact give rise to it in normal development. Further, the boundaries of the partial fields overlap: a given group of cells in the limb-rudiment of the chick may contribute to the formation of either a thigh or a shank, according as to whether it is allowed to remain attached to or is isolated from one partial field or the other. Gradients may exist in such fields: the capacity of cells within the fore-limb field to give rise to a limb decreases with their distance from a subcentral portion of the field: the same is true for many other organ-fields.

§ 3

Up to a certain time, regulation is still possible within each of the partial fields; but as development proceeds, each of these becomes split up into progressively smaller fields, each with its own determined fate: for instance, the fields for leg, shank, and foot, within the originally single hind-limb field.

Each area in the mosaic passes from the state of invisible *chemo-differentiation* by the process of *histo-differentiation* to full visible differentiation, and so reaches the functional stage. After the organism as a whole has reached the functional stage, many new morphogenetic agencies come into play. The organism also, through acquiring the power of regeneration, reacquires much of the regulative capacity which it lost in its passage through the mosaic stage.

The type of organisation characteristic of one stage appears to persist, in whole or in part, throughout subsequent stages. Thus, the main gradient-system of the embryo permeates the partial fields of the limb, neural folds, ear, gills, and heart, and determines their axis; and the growth of the lateral line along a particular level of the flank can best be interpreted in terms of a persistent total gradient-field.

Again, a total field-system certainly exists in adult Planarians and appears to reveal its presence in late stages of other groups through the presence of growth-gradients permeating the whole organism.

The persistence into adult life of the partial field-systems of the mosaic stage is shown by the phenomena of regeneration, by the existence of localised growth-gradients within single areas, and notably by phenomena such as those found in newts, where, for instance, indifferent regeneration-buds produced by an amputated limb will produce legs when grafted into a certain area round the leg, while if grafted near the base of the tail they will produce tails.

§ 4

With this, of course, only a start has been made with the scientific analysis of development. It remains for the future to discover such fundamentals as the physiological basis of the field-systems, and the elaborate physico-chemical processes which must be operative at the time when the quantitative differences of the early gradient-field system are being converted into the qualitative differences of the chemo-differentiated mosaic stage.

It is, however, already a good deal to have arrived at this first outline of development on the biological level. To have established the fact that organisations of quite different type succeed one another during development is important. The recognition of the gradient-field system, with its purely quantitative differentials, as the basis of early organisation, is a great step forward, since it provides an adequate formal explanation of many phenomena of regulation which have been considered by various authors, notably by Driesch, as affording proof of vitalistic theories of development.

Further, the epigenetic analysis of development is pointing the way to a large extension of the field of heredity, in the shape of physiological genetics. It is only through a study of development that it will be possible to understand what the term "genetic characters" really stands for—in other words, what are the basic processes involved in the action of a particular Mendelian gene.

Experimental embryology as a separate branch of science was initiated by Roux; in its next phase, in which Driesch, Boveri, Wilson, Herbst, Morgan, Brachet and Jenkinson are outstanding names, a large body of facts was amassed, and the experimental proof of epigenesis provided; in the third phase, Spemann and Harrison are the outstanding figures within the sub-science, while the theories of Child have not only linked the facts of regeneration with those of embryonic differentiation, but have provided a scientific basis for a field hypothesis for early development, thus filling a large gap in the theoretical aspect of the subject. Meanwhile, experimental embryology has been making fruitful contacts with physiology, notably in the field of hormone action, with genetics, and with growth studies.

The fourth stage is now beginning, in which this framework of general principle will be filled in through intensive research, and the whole science deepened by a search for the physico-chemical bases of the empirical biological principles which have been discovered in its earlier stages.

BIBLIOGRAPHY AND INDEX
OF AUTHORS

NOTE. This bibliography and index of authors has been specially designed to facilitate reference both to the text of this book and to the original works in a library. For this purpose, where two or more works by an author are concerned with the same subject, they are referred to together; and where such works are in the same periodical, their references are placed together. This system involves a certain trifling disturbance of the chronological order in some cases, which should, however, present no inconvenience owing to the facility with which the date-figures in bold type can be picked out.

BIBLIOGRAPHY AND INDEX OF AUTHORS

Author	Reference	Subject	Page where quoted
Abeloos, M.	*Comptes Rendus Soc. de Biol.* XCVIII, **1928**, p. 917	Growth-gradient and heterogony in Planarians	368, 372, 425
Abeloos, M.	*La régénération et les problèmes de la morphogénèse*, Paris, **1932**	General	309
Adams, A. E.	*Journ. Exp. Zool.* XL, **1924**, p. 311; LVIII, **1931**, p. 147	Mouth, dependent differentiation	180
Adelmann, H. B.	*Arch. Entwmech.* CXIII, **1928**, p. 704	Eye-cup and lens-formation	187, 237
Adelmann, H. B.	*Journ. Exp. Zool.* LIV, **1929**, pp. 249, 291; LVII, **1930**, p. 223	Determination of eye in Urodela	243, 244, 245
Adler, L.	*Arch. Entwmech.* XLIII, **1917**, p. 343	Effects of over-ripeness of egg on thyroid development	262
Allen, B. M.	*Endocrinology*, VIII, **1924**, p. 639	Hormones and brain development	396
Allen, E.	*Sex and Internal Secretions*, Baltimore and London, **1932**	Sex, general	254, 255
Allen, E. J.	*Phil. Trans. Roy. Soc. Lond.* B, CCXI, **1921**, p. 131	Regeneration, *Procerastea* (Polychaete)	281
Baker, J. R.	*Journ. Anat.* LX, **1926**, p. 374	Asymmetry of gonads, pigs	362
Balinsky, B. I.	*Arch. Entwmech.* CV, **1925**, p. 718; CVII, **1926**, p. 679; CX, **1927**, p. 71	Limb-formation after graft of auditory vesicle or celloidin pill	177, 231
Balinsky, B. I.	*Arch. Entwmech.* CXXII, **1930**, p. 12	Lens fibres, dependent differentiation	189
Baltzer, F.	*Verh. Schweiz. Naturf. Ges.* CI, **1920**, p. 217	Larval hybrids, Amphibia	405
Baltzer, F.	*Rev. Suisse Zool.* XXXVIII, **1931**, p. 361	*Bonellia*, male-determination	140

Author	Reference	Topic	Pages
Bataillon, E.	*Arch. de Zool. Exp. et Gén.* VI, **1910**, p. 101	Artificial parthenogenesis and dorsal meridian in frog's egg	38
Bautzmann, H.	*Arch. Entwmech.* CVIII, **1926**, p. 283	Organiser, Urodele, extent of area in blastula	135, 145, 152, 159
Bautzmann, H.	*Arch. Entwmech.* CX, **1927**, p. 631	Organiser and ventral half-embryo	89
Bautzmann, H.	*Arch. Entwmech.* CXIV, **1928**, p. 177; CXIX, **1929** A, p. 1; CXXVIII, **1933**, p. 665	Notochord, inductive powers	145, 147, 152, 374
Bautzmann, H.	*Naturwiss.* XVII, **1929** B, p. 818; *Sitzber. Ges. Morph. u. Physiol. München*, XXXIX, **1929** C, p. 1	Explantation experiments, general	139
Bautzmann, H., Holtfreter, J., Spemann, H. and Mangold, O.	*Naturwiss.* XX, **1932**, p. 971	Organiser, narcotised, desiccated, etc.	153
Beckwith, C. J.	*Journ. Exp. Zool.* XLIX, **1927**, p. 217	*Amblystoma*, lens-regeneration	187, 239
Bell, E. T.	*Anat. Anz.* XXXI, **1907**, p. 283	Balancer determination	236
Bellamy, A. W.	*Biol. Bull.* XXXVII, **1919**, p. 312; XLI, **1921**, p. 351; *Amer. Journ. Anat.* XXX, **1922**, p. 473	Origin of axial gradient in frog; cyclopia; differential susceptibility	35, 68, 320, 344, 346, 402
Bellamy, A. W. and Child, C. M.	*Proc. Roy. Soc. Lond.* B, XCVI, **1924**, p. 132	Axial gradients in frog's egg	68, 348
Belogolowy, G.	*Arch. Entwmech.* LXIII, **1918**, p. 556	Frog's egg grafted into embryo	328
Benoit, J.	*C.R. Soc. Biol.* CIV, **1930**, p. 1329	Embryonic castration, birds	263
Berrill, N. J.	*Journ. Exp. Zool.* LVIII, **1931**, p. 495	Regeneration and regulation in *Sabella* (Polychaete)	166, 281, 288, 309
Bertalanffy, L. von	*Kritische Theorie der Formbildung*, Berlin, **1928**	General, theoretical	46, 195, 274
Bertalanffy, L. von and Woodger, J. H.	*Modern Theories of Development*, Oxford, **1933**	General, philosophical	11
Beyer, K. M. and Child, C. M.	*Physiol. Zool.* III, **1930**, p. 342	*Planaria*, polarity of lateral pieces	279

BIBLIOGRAPHY AND INDEX OF AUTHORS (*continued*)

Author	Reference	Subject	Page where quoted
Bijtel, J. H.	*Arch. Entwmech.* cxxv, **1931**, p. 448	Formation of tail in Amphibia	28
Bijtel, J. H. and Woerdeman, M. W.	*Proc. Koning. Akad. Weten. Amsterdam*, xxxi, **1928**, p. 1030	Formation of tail in Amphibia	28
Bischler, V.	*Rev. Suisse Zool.* xxxiii, **1926**, p. 431	Regeneration and fields, Urodela	198, 304, 366
Bodenstein, D.	*Arch. Entwmech.* cxxviii, **1933**, p. 564	Insects, moulting hormones	174
Boerema, I.	*Arch. Entwmech.* cxv, **1929**, p. 601	Neural tube, rolling up	249
Bok, S. T.	*Folia Neurobiol.* ix, **1915**, p. 475	Formation of axons and nerve-fibres	381
Bond, C. J.	*Journ. Genet.* x, **1920**, p. 87; xvi, **1926**, p. 253	Asymmetry, polydactyly, birds	362
Bonnet, C.	*Palingénésie Philosophique*. Geneva, **1769**	Preformation	2, 3
Born, G.	*Arch. Mikr. Anat.* xxiv, **1885**, p. 475	Frog's egg, inverted	36
Boveri, T.	*Verh. Phys. Med. Ges. Würzburg*, xxxiv, **1901**, p. 145	Echinoid larvae, polarity and bilateral symmetry	69, 274
Boveri, T.	*Arch. Entwmech.* xvi, **1903**, p. 340	Larval hybrids, Echinoids	405
Boveri, T.	*Ergebnisse über die Konstitution des Zellkerns*, Jena, **1904**; *Zellenstudien*, vi, **1907**	Qualitative differences of chromosomes	404
Boveri, T.	*Zellenstudien*. v. *Ueber die Abhängigkeit der Kerngrösse und Zellengrösse der Seeigel-Larven von den Chromosomenzahl der Ausgangszellen*, Jena, **1905**	Cytoplasmo-nuclear ratio	133, 408
Boveri, T.	*Festschr. R. Hertwig*, iii, **1910**, p. 131	Stratification of substances, *Ascaris*	274
Boveri, T. and Hogue, M. J.	*Sitzber. Phys. Med. Ges. Würzburg*, **1909**, p. 34	Diminution of chromosomes, *Ascaris*	101, 400

Bovet, D.	*Rev. Suisse Zool.* XXXVII, **1930**, p. 83	Regeneration-fields and nerves, Urodele	231, 363
Boycott, A. E., Diver, C., Garstang, S. L. and Turner, F. M.	*Phil. Trans. Roy. Soc. Lond.* B, CCXIX, **1930**, p. 51	Inheritance of sinistrality in snails	71, 81, 411
Brachet, A.	*Arch. de Biol.* XIX, **1903**, p. 1; XXI, **1905**, p. 103; XXXIII, **1923**, p. 343; *Arch. Entwmech.* XXII, **1906**, p. 325; CXI, **1927**, p. 250	Frog, development after injury to egg or one blastomere	92, 94
Brachet, A.	*Arch. Zool. Exp. Gén.* VI, **1910**, p. 1	Development polyspermic frog's egg	132
Brachet, A.	*Arch. de Biol.* XXVI, **1911**, p. 337	Frog's egg, artificial parthenogenesis, dorsal meridian	38, 94
Brachet, A.	*L'œuf et les facteurs de l'ontogénèse*, Paris, **1917**	General	xi
Brambell, F. W. R.	*The Development of Sex in Vertebrates*, London, **1930**	Sex, general	254
Brandt, W.	*Arch. Mikr. Anat. u. Entwmech.* CIII, **1924**, p. 517	Limb-bud grafts, Urodele	225, 359
Brauer, A.	*Sitzber. Preuss. Akad. Wiss.* **1917**, p. 208	Scorpion, double monsters	329
Braus, H.	*Morph. Jahrb.* XXXV, **1906**, p. 509	Perforation of operculum, Anura	180, 429
Brien, H.	*Ann. Soc. Roy. Zool. Belgique*, LXI, **1930**, p. 19	Polarity in *Clavellina* buds	65
Browne, E. N.	*Journ. Exp. Zool.* VII, **1909**, p. 1	Organiser, *Hydra*	164
Bruns, E.	*Arch. Entwmech.* CXXIII, **1931**, p. 682	Regulation, amphibian blastula	94
Buchanan, J. W.	*Journ. Exp. Zool.* XLV, **1926**, p. 141	Bird embryo, axial gradients	350
Buddenbrock, W. von	*Zeitschr. f. vergl. Physiol.* XIV, **1930**, p. 415	Insects, moulting hormones	174
Burr, H. S.	*Journ. Exp. Zool.* XX, **1916**, p. 27	Nasal sac and nasal capsule	175
Bytinski-Salz, H.	*Arch. Entwmech.* CXVIII, **1929**, p. 121; CXXIII, **1931**, p. 518	Neural plate, Urodele, regional inductive powers	145, 374

BIBLIOGRAPHY AND INDEX OF AUTHORS (continued)

Author	Reference	Subject	Page where quoted
Cajal, R. y	Trab. Lab. Invest. Biol. Univ. Madrid, IV, 1906, p. 219.	Nerve-fibre development	393
Carey, E. J.	Amer. Journ. Anat. XXIX, 1921 A, p. 93	Muscle differentiation	432
Carey, E. J.	Amer. Journ. Anat. XXIX, 1921 B, p. 341; XXXII, 1924, p. 475	Bladder muscles, induction of striations	436
Carpenter, R. L.	Journ. Exp. Zool. LXI, 1932, p. 149; LXIV, 1933, p. 287	Neuron multiplication in dorsal root ganglia	383
Carpenter, R. L. and Carpenter, E. C.	Journ. Exp. Zool. LXIV, 1932, p. 187	Neuron multiplication in dorsal root ganglia	383
Carrel, A.	Science, LXXIII, 1931, p. 297	Tissue culture, metaplasia	210, 213
Carrel, A. and Ebeling, A. H.	Journ. Exp. Med. XLIII, 1926, p. 461	Fibroblasts, metaplasia	213
Castelnuovo, G.	Boll. di Zool. III, 1932, p. 291	Temperature gradients, axolotl	43, 339, 344
Champy, C.	Comptes Rendus Soc. Biol. LXXVI, 1914, p. 31	Connective tissue and epithelium; differentiation	179
Champy, C.	Arch. de Morph. Gén. et Exp. IV, 1922, p. 1	Regional fields and hormone action	428, 430
Cheng, T. H.	Zeitschr. f. Zellforsch. u. mikr. Anat. XVI, 1932, p. 542	Asymmetry of gonads	362
Child, C. M.	Journ. Exp. Zool. XVII, 1914, p. 61; XXXIII, 1921 A, p. 240	Frequency of head-regeneration in Planarians, gradients	302, 309
Child, C. M.	Individuality in Organisms, Chicago, 1915 A; Senescence and Rejuvenescence, Chicago, 1915 B; The Origin and Development of the	General, axial gradients	7, 37, 62, 165, 274, 278, 288, 290, 296, 297, 299, 300, 344, 348

BIBLIOGRAPHY AND INDEX OF AUTHORS (*continued*)

Author	Reference	Subject	Pages where quoted
Conklin, E. G.	*Journ. Exp. Zool.* II, 1905, p. 185; *Arch. Entwmech.* XXI, 1906, p. 727	Ascidian, blastomere potencies	97, 198, 217
Conklin, E. G.	"Cellular differentiation", in Cowdry's *General Cytology*, Chicago, 1924	General, mosaic development, polarity, centrifuge experiments	67, 100, 123, 217
Conklin, E. G.	*Journ. Exp. Zool.* LX, 1931, p. 1	Ascidian, centrifuge experiments	123, 217
Conklin, E. G.	*Journ. Morph.* LIV, 1932, p. 69; *Journ. Exp. Zool.* LXIV, 1933, p. 303	*Amphioxus*, blastomere isolation	63, 79, 100, 119, 123
Copenhaver, W. M.	*Journ. Exp. Zool.* XLIII, 1926, p. 321	Determination of heart	235, 360
Cotronei, G.	*Arch. Ital. di Biol.* LXXI, 1921, p. 1	Cyclopia, Amphibia	346, 348
Crampton, H. E.	*Ann. New York Acad. Sci.* VIII, 1894, p. c	Cleavage in sinistral snails	72
Crampton, H. E.	*Arch. Entwmech.* III, 1896, p. 1	*Ilyanassa*, isolated blastomeres	113
Dalcq, A.	*Les Bases Physiologiques de la Fécondation*, Paris, 1928	Fertilisation, parthenogenesis	xi
Dalcq, A.	*Arch. d'Anat. Micr.* XXVIII, 1932, p. 223	Ascidian, egg-fragments	108, 124, 126
Dalcq, A. and Simon, S.	*Arch. de Biol.* XLII, 1931, p. 107; *Protoplasma*, XIV, 1932, p. 497	Chromosomes and development	404
Danchakoff, V. [Dantschakoff, W.]	*Zeitschr. f. Anat. u. Entwgesch.* LXXIV, 1924, p. 401	Chorio-allantoic grafts	186, 199, 205
Danchakoff, V.	*Arch. Entwmech.* CXXVII, 1932, p. 542; *Zeitschr. Zellf. Mikr. Anat.* XVIII, 1933, p. 56	Gonad-structure, differentiation independent of germ-cells	263, 265

Dean, I. L., Shaw, M. E. and Tazelaar, M. A.	Brit. Journ. Exp. Biol. v, 1928, p. 309	Temperature-gradient, frog's egg	43, 320, 342
de Beer, G. R.	Arch. de Zool. Exp. et Gén. LXI, 1922, p. 47	Sponge collar-cell spheres	250
de Beer, G. R.	Introduction to Experimental Embryology, Oxford, 1926	General, elementary	xi
de Beer, G. R.	Biol. Rev. II, 1927, p. 137	General	274
de Beer, G. R.	Quart. Journ. Micr. Sci. LXXIV, 1930, p. 701	Visceral cartilages, morphology	395
de Beer, G. R. and Huxley, J. S.	Quart. Journ. Micr. Sci. LXVIII, 1924, p. 471	Differential inhibition, Aurelia	425
Delage, Y.	Structure du protoplasma, Paris, 1895	General	6
Detwiler, S. R.	Journ. Exp. Zool. XXV, 1918, p. 499; LII, 1929 A, p. 315; LXIV, 1933 A, p. 405	Determination of limb-bud and girdle, Urodele	199, 223, 231, 299
Detwiler, S. R.	Proc. Nat. Acad. Sci. U.S.A. VI, 1920 A, p. 96; Journ. Exp. Zool. XLV, 1926 A, p. 399; XLVIII, 1927 A, p. 1	Neuron-multiplication in dorsal root ganglia	383, 385
Detwiler, S. R.	Journ. Exp. Zool. XXXI, 1920 B, p. 117; XXXV, 1922, p. 115; Journ. Comp. Neur. XXXVIII, 1925 A, p. 461	Attraction of nerve to displaced limb	390, 391
Detwiler, S. R.	Journ. Exp. Zool. XXXVII, 1923 A, p. 339; XXXVIII, 1923 B, p. 293; XLII, 1925 B, p. 333; LII, 1929 B, p. 351; Quart. Rev. Biol. I, 1926 B, p. 61; Journ. Comp. Neur. XLIII, 1927 B, p. 143; XLV, 1928 A, p. 191; L, 1930 A, p. 521	Neuron-multiplication in spinal cord	375, 376, 387, 392
Detwiler, S. R.	Journ. Comp. Neur. XL, 1926 C, p. 465	Resonance theory of nerve-action	391

BIBLIOGRAPHY AND INDEX OF AUTHORS (*continued*)

Author	Reference	Subject	Page where quoted
Detwiler, S. R.	*Naturwiss.* XV, **1927** C, pp. 873, 895	Attraction of nerve to grafted eye	391
Detwiler, S. R.	*Journ. Exp. Zool.* LI, **1928** B, p. 1	No attraction of nerve to grafted tail	392
Detwiler, S. R.	*Journ. Exp. Zool.* LV, **1930** B, p. 319; LVII, **1930** C, p. 183	Innervation of grafted limbs	391
Detwiler, S. R.	*Journ. Comp. Neur.* LIV, **1932**, p. 173; *Proc. Nat. Acad. Sci. U.S.A.* XIX, **1933** B, p. 22	Spinal ganglia, segmentation	393
Detwiler, S. R. and Carpenter, R. L.	*Journ. Comp. Neur.* XLVII, **1929**, p. 427	Innervation of grafted limbs	391
Detwiler, S. R. and McKennon, G. E.	*Biol. Bull.* LIX, **1930**, p. 353	Innervation of grafted limbs	391
Dragomirow, N.	*Arch. Entwmech.* CXXIII, **1930**, p. 206	Eye-cup and lens fibres	190
Dragomirow, N.	*Arch. Entwmech.* CXXVI, **1932**, p. 636; CXXIX, **1933**, p. 522	Eye-cup, regulation	244
Drastich, L.	*Zeitschr. vergl. Physiol.* II, **1925**, p. 632	Gill capillaries, function	436
Drew, A. H.	*Brit. Journ. Exp. Path.* IV, **1923**, p. 46	Connective tissue and differentiation	179, 209, 211
Drew, G. H.	*Journ. Exp. Zool.* X, **1911**, p. 349	Metaplasia, *Pecten.*	213
Driesch, H.	*Anat. Anz.* VIII, **1893**, p. 348	Equality of nuclear division during cleavage	84
Driesch, H.	*Arch. Entwmech.* III, **1896**, p. 362	Orientation of mesenchyme in pluteus	181
Driesch, H.	*Arch. Entwmech.* IV, **1897**, p. 75	*Myzostoma*, organ-forming substances	119
Driesch, H.	*Arch. Entwmech.* IX, **1899**, p. 103	*Tubularia*, reconstitution	287
Driesch, H.	*Arch. Entwmech.* X, **1900**, p. 361	Isolated blastomeres of Echinoid eggs	7, 98, 131

Driesch, H.	*Philosophie des Organischen*, Leipzig, 1921	General, vitalism and mechanism	215, 325
Duesberg, J.	*L'œuf et ses localisations germinales*, Paris, 1928	General, and Ascidian egg centrifuged	124
Dürken, B.	*Zeitschr. Wiss. Zool.* XCIX, 1912, p. 189	Effect of limb-extirpation on brain	389
Dürken, B.	*Zeitschr. Wiss. Zool.* CXV, 1916, p. 58	Conjunctiva and limb-grafts	178
Dürken, B.	*Biol. Zentralbl.* XLV, 1925, p. 541; *Biol. Gen.* VI, 1930, p. 511	Limb development and nervous system	230
Dürken, B.	*Arch. Entwmech.* CVII, **1926**, p. 727	Interplantation experiments	138
Dürken, B.	*Lehrbuch der Experimentalzoologie* (2nd ed.), Berlin, 1928	General	xi
Dürken, B.	*Grundriss der Entwicklungsmechanik*, Berlin, 1929	General, elementary	xi
Ebeling, A. H. and Fischer, A.	*Journ. Exp. Med.* XXXVI, 1922, p. 285	Fibroblasts and epithelium differentiation	179
Eidmann, H.	*Biol. Zentralbl.* XLII, 1922, p. 97	Effects of over-ripeness on gonad differentiation	262
Ekman, G.	*Morph. Jahrb.* XLVII, 1913, p. 419; *Comm. Biol. Soc. Sci. Fenn.* I (3), 1922	Gills and operculum, determination	233, 360
Ekman, G.	*Finska Vetensk. Soc. Forh.* LXIII, 1921, p. 1; *Comm. Biol. Soc. Sci. Fenn.* I (9), 1924, p. 1; *Arch. Entwmech.* CVI, 1925, p. 320	Heart-determination	76, 234, 235, 249
Ekman, G.	*Comm. Biol. Soc. Sci. Fenn.* I (6), 1923	Nose-determination	180, 237
Erdmann, W.	*Arch. Entwmech.* CXXIV, 1931, p. 666	Explantation experiments	139, 203
Eycleshymer, A. C.	*Anat. Anz.* XLVI, 1914, p. 1	Non-regeneration, decapitated *Necturus* embryo	195

BIBLIOGRAPHY AND INDEX OF AUTHORS (*continued*)

Author	Reference	Subject	Page where quoted
Fankhauser, G.	*Arch. Entwmech.* CXXII, **1930**, p. 676	Time of origin of organiser	145
Fauré-Fremiet, E.	*La cinétique du développement*, Paris, **1925**	General	xi
Fell, H. B.	*Arch. f. Zellforsch.* VII, **1928**, pp. 69, 390	Differentiation *in vitro* of ear, cartilage, and bone	205, 250
Fell, H. B. and Robison, R.	*Biochem. Journ.* XXIII, **1929**, p. 767; XXIV, **1930**, p. 1905	Phosphatase activity *in vitro*	205
Filatow, D.	*Rev. Zool. Russe*, I, **1916**, p. 48	Ear-vesicle and cartilage	175
Filatow, D.	*Arch. Mikr. Anat. u. Entwmech.* CIV, **1924**, p. 50; *Arch. Entwmech.* CV, **1925**, p. 475; CVII, **1926**, p. 575	Eye-cup and lens-formation	185, 187
Filatow, D.	*Arch. Entwmech.* CX, **1927**, p. 1	Ear-vesicle and limb-induction	177, 231
Filatow, D.	*Arch. Entwmech.* CXXII, **1930**, p. 546	Ear-vesicle and cartilage, *Acipenser*	177
Filatow, D.	*Zool. Jahrb. (Abt. allg. Zool. Physiol.)*, LI, **1932**, p. 589	Limb-bud differentiation and size	418
Fischel, A.	*Arch. Entwmech.* VII, **1898**, p. 557	Ctenophore, isolated blastomeres	105
Fischel, A.	*Arch. Entwmech.* XLII, **1917**, p. 1	Eye-cup and conjunctiva	178
Fischer, A.	*Comptes Rendus Soc. de Biol.* XCII, **1925**, p. 109	Tissue-culture, metaplasia	213
Flower, S. S.	*Proc. Zool. Soc. Lond.* **1927**, p. 155	Metamorphosis and behaviour	396
Ford, E. B.	*Journ. Genetics*, XX, **1929**, p. 93	Genes and developmental rates	411
Ford, E. B. & Huxley, J.S.	*Brit. Journ. Exp. Biol.* V, **1927**, p.112	Genes and developmental rates	409
Frederici, E.	*Arch. de Biol.* XXXVI, **1926**, p. 465	Anuran blood-rudiment	198
Frew, J. G. H.	*Brit. Journ. Exp. Biol.* VI, **1928**, p. 1	Insect imaginal discs, explantation	174
Friedmann, H.	*Biol. Bull.* LII, **1927**, p. 197	Asymmetry of gonads, birds	362

Author	Reference	Topic	Pages
Fuld, E.	*Arch. Entwmech.* XI, **1901**, p. 1	Function and leg-differentiation	434
Garman, S.	*Amer. Nat.* XXIX, **1895**, p. 1012; XXX, **1896**, p. 232	Fish, asymmetry of genital organs	70
Geinitz, B.	*Zeitschr. f. Ind. Abst. u. Vererb.* XXXVII, **1925** A, p. 117	Organiser and polarity of host tissues, Urodele	149
Geinitz, B.	*Arch. Entwmech.* CVI, **1925** B, p. 357	Organiser, Anuran and Urodele, xenoplastic induction	142
Gilchrist, F. G.	*Physiol. Zool.* I, **1928**, p. 231; *Quart. Rev. Biol.* IV, **1929**, p. 544	Temperature-gradients, Amphibia	39, 43, 137, 342
Godlewski, E.	*Comptes Rendus Soc. de Biol. Réun. Plén.* 24 avril, **1925**	Cytoplasmo-nuclear ratio and cleavage	133
Goerttler, K.	*Arch. Entwmech.* CVI, **1925**, p. 503; *Zeitschr. f. Anat. u. Entwick.* LXXX, **1926**, p. 283	Processes of gastrulation and neurulation	22, 25, 50, 137
Goerttler, K.	*Arch. Entwmech.* CXII, **1927**, p. 517	Dynamic determination	151, 163, 251
Goerttler, K.	*Verh. Anat. Ges.* XXXVII, **1928**, p. 132	Determination of heart-rudiments and prepotency of left side	77, 204
Goetsch, W.	*Verh. Anat. Ges.* XL, **1931**, p. 128	Organiser, structure	154
Goldschmidt, R.	*Arch. Entwmech.* CXVII, **1929**, p. 311	Regeneration, polarity, cell-migration	280, 300
Goldschmidt, R.	*The Mechanism and Physiology of Sex Determination*, London, **1923**	Sex-determination, general	172, 270
Goldschmidt, R.	*Physiologische Theorie der Vererbung*, Berlin, **1927**	General, physiology of gene-action	46, 172, 195, 206, 409
Goodrich, E. S.	*Living Organisms*, Oxford, **1924**	General, heredity and environment	403
Gräper, L.	*Arch. Entwmech.* CXVI, **1929**, p. 382	Chick blastoderm, movements	163
Gray, J.	*Experimental Cytology*, Cambridge, **1931**	General	xi
Groll, O.	*Arch. Mikr. Anat. u. Entwmech.* c, **1924**, p. 385	Eye-cup and conjunctiva	178
Guareschi, C.	*Boll. Ist. Zool. Univ. Roma*, VI, **1928**	Ear-vesicle and cartilaginous capsule	175

456

BIBLIOGRAPHY AND INDEX OF AUTHORS (*continued*)

Author	Reference	Subject	Page where quoted
Guareschi, C.	*Rend. Accad. Naz. Lincei*, XVI, **1932**, p. 345	Frog, cyclopic	349
Gurwitsch, A.	*Arch. Entwmech.* LI, **1922**, p. 383; CXII, **1927**, p. 433	Field hypothesis	274
Guyénot, E.	*Rev. Suisse Zool.* XXXIV, **1927**, pp. 1, 127	Fields and loss of regeneratory power	272, 366
Guyénot, E.	*Comptes Rendus Soc. de Biol.* XCIX, **1928**, p. 27	Fields stimulated by nerves, *Lacerta*	362
Guyénot, E. and Ponse, K.	*Bull. Biol. France et Belgique*, LXIV, **1930**, p. 251	Fields and regeneration	272, 362
Guyénot, E. and Schotté, O.	*Comptes Rendus Soc. de Biol.* LXXXIX, **1923**, p. 491	Regeneration buds	418
Guyénot, E. and Schotté, O.	*Comptes Rendus Soc. de Biol.* XCIV, **1926**, p. 1050	Fields stimulated by nerves, Amphibia	231
Guyénot, E. and Valette, M.	*Comptes Rendus Soc. de Biol.* XCIII, **1925**, p. 1276	Loss of regeneration power after field-extirpation	366
Haldane, J. S.	*The Sciences and Philosophy*, London, **1929**	General, philosophical	215
Hamburger, V.	*Naturwiss.* XV, **1927**, pp. 657, 677; *Arch. Entwmech.* CXIV, **1928**, p. 272; CXIX, **1929**, p. 47	Limb-development and innervation	230, 391, 430
Hämmerling, J.	*Arch. Entwmech.* CX, **1927**, p. 395	Frog's egg and situs inversus	77
Hammond, J.	*Farmer and Stockbreeder Agric. Gaz.* Dec. 10, **1928**	Proportions and feeding, mammals	425
Hammond, J.	*Zeitschr. Züchtungskunde*, IV, **1929**, p. 543	Growth-gradients, mammals	414

457

Author	Reference	Topic	Pages
Harmer, S. F.	Proc. Linn. Soc. Lond. CXLI, 1930, p.69	Polyzoa, polyembryony	328
Harper, E. H.	Biol. Bull. VI, 1904, p. 173	Regeneration and regulation in *Stylaria*	166, 309
Harrison, R. G.	Arch. Entwmech. VII, 1898, p. 430	Embryonic chimaeras	407
Harrison, R. G.	Arch. Mikr. Anat. LXIII, 1904, p. 35	Development of lateral line	355, 357
Harrison, R. G.	Journ. Exp. Zool. IV, 1907 A, p. 239	Aneurogenic limb-bud grafts	378
Harrison, R. G.	Anat. Rec. I, 1907 B, p. 116; Journ. Exp. Zool. IX, 1910, p. 787	Development of axons *in vitro*	378
Harrison, R. G.	Proc. Nat. Acad. Sci. U.S.A. I, 1915, p. 539	Mosaic stage of development and power of regeneration	58, 198
Harrison, R. G.	Journ. Exp. Zool. XXV, 1918, p. 413	Development and determination of limb and limb-girdle, Urodele	199, 223, 229, 231
Harrison, R. G.	Journ. Exp. Zool. XXXII, 1921 A, p. 1; Arch. Entwmech. CVI, 1925 A, p.469	Axes of limb-buds, Urodele	71, 224, 232, 359, 360
Harrison, R. G.	Biol. Bull. XLI, 1921 B, p. 156	Determination of gills	233, 360
Harrison, R. G.	Proc. Nat. Acad. Sci. U.S.A. X, 1924 A, p. 69	Growth-rates of heteroplastically grafted limbs	225
Harrison, R. G.	Journ. Comp. Neur. XXXVII, 1924 B, p. 123	Neural crest and sheath-cells	394
Harrison, R. G.	Journ. Exp. Zool. XLI, 1925 B, p. 349	Balancer, determination	179, 236
Harrison, R. G.	Arch. Entwmech. CXX, 1929, p. 1	Growth-factors, parts of eyes	424
Harvey, E. B.	Biol. Bull. LXIV, 1933, p. 125	Echinoderm eggs, development of fragments	221, 313
Helff, O. M.	Anat. Rec. XXIX, 1924, p. 102; Journ. Exp. Zool. XLV, 1926, p. 1	Perforation of operculum, Anura	180, 429
Helff, O. M.	Physiol. Zool. I, 1928, p. 463	Tympanic membrane, determination	178
Hellmich, W.	Arch. Entwmech. CXXI, 1930, p. 135	Regeneration and resorption of limbs in tadpoles	299, 305
Herbst, C.	Mitt. Zool. Staz. Neapel, XI, 1895, p. 136	Lithium salts and development of sea-urchin larvae	181, 323, 334

BIBLIOGRAPHY AND INDEX OF AUTHORS *(continued)*

Author	Reference	Subject	Page where quoted
Herbst, C.	*Arch. Entwmech.* v, **1897**, p. 649; IX, **1900**, p. 424	Calcium-free water, and separation of blastomeres	401
Herbst, C.	*Formative Reize in der Tierischen Ontogenese*, Leipzig, **1901**	General: formative and directive stimuli	7, 134
Herbst, C.	*Handwörterbuch den Naturwissen-schaften*, III, **1912**, p. 542	General review	175
Herbst, C.	*Arch. Entwmech.* CII, **1924**, p. 130	Salamander pigment-cells, function and multiplication	426
Herlant, M.	*Arch. de Biol.* XXVI, **1911**, p. 103	Polyspermy, frog's eggs, and dorsal meridian	37, 132
Herlitzka, A.	*Arch. Entwmech.* IV, **1896**, p. 624	Two embryos from one newt's egg	53
Hertwig, G.	*Arch. Entwmech.* CXI, **1927**, p. 292	Grafting of haploid limb-buds	364
Hertwig, O.	*Arch. Mikr. Anat.* XLII, **1893**, p. 662	Equality of nuclear division; injury to one blastomere	43, 84, 92
Hertwig, O.	*Sitzber. K. Preuss. Akad. Wiss.*, Berlin, **1897**, p. 14; *Arch. Mikr. Anat.* LXIII, **1904**, p. 643	Frog's egg, centrifuged	40
Hesse, R.	*Zool. Jahrb., Abt. allg. Zool. Physiol.* XXXVIII, **1921**, p. 243	Heart-weight in birds of arctic and temperate regions	432
Hey, A.	*Arch. Entwmech.* XXXIII, **1911**, p. 117	Newt, duplicitas anterior	156
Hiraiwa, Y. K.	*Journ. Exp. Zool.* XLIX, **1927**, p. 441	Mammalian grafts on chorio-allantois	203, 268
His, W.	*Unsere Körperform und das Physio-logische Problem ihrer Entstehung*, Leipzig, **1874**	General, foundations of experimental embryology	8

BIBLIOGRAPHY AND INDEX OF AUTHORS (*continued*)

Author	Reference	Subject	Page where quoted
Huxley, J. S.	*Quart. Journ. Micr. Sci.* LXV, **1921** B, p. 643	*Perophora*, differential inhibition	425
Huxley, J. S.	*Journ. Hered.* XIII, **1923**, p. 349	Amphibian metamorphosis	427
Huxley, J. S.	*Nature*, CXIII, **1924**, p. 276	General, chemo-differentiation	46, 195
Huxley, J. S.	*Proc. Roy. Soc. Lond.* B, XCVIII, **1925**, p. 113	Metamorphosis, Amphibia	424
Huxley, J. S.	*Pubb. Staz. Zool. Napoli*, VII, **1926**, p. 1	Polarity in *Clavellina* winter-buds	65, 198, 287
Huxley, J. S.	*Arch. Entwmech.* CXII, **1927**, p. 480	Temperature-gradient, frog's egg	39, 43, 339, 344
Huxley, J. S.	*Naturwiss.* XVIII, **1930**, p. 265	Gradient-theory and explanted pieces	317
Huxley, J. S.	*Amer. Nat.* LXV, **1931**, p. 289	Growth-gradients	414
Huxley, J. S.	*Problems of Relative Growth*, London, **1932**	General, growth-gradients	206, 225, 366, 367, 370, 422, 424, 425
Huxley, J. S. and de Beer, G. R.	*Quart. Journ. Micr. Sci.* LXVII, **1923**, p. 473	Differential inhibition, *Obelia*	425
Huxley, J. S. and Murray, P. D. F.	*Anat. Rec.* XXVIII, **1924**, p. 385	Keratinisation in chorio-allantois	209
Huxley, J. S. and Wolsky, A.	*Nature*, CXXIX, **1932**, p. 242	Albino eyes in *Gammarus*	411
Hyman, L. H.	*Journ. Exp. Zool.* XX, **1916**, p. 99	Frequency of head-regeneration in *Lumbriculus*	308
Hyman, L. H.	*Biol. Bull.* XL, **1921**, p. 32	Oligochaetes, double monsters	332
Hyman, L. H.	*Biol. Bull.* LII, **1927**, p. 1	Axial gradients, vertebrate embryos	350
Hyman, L. H. and Bellamy, A. W.	*Biol. Bull.* XLIII, **1922**, p. 313	Gradients and electric currents	379

Ingvar, S.	Proc. Soc. Exp. Biol. and Med. XVII, 1920, p. 198	Electric current and direction of axon-growth in vitro	378
Jackson, C. M.	The effects of inanition and malnutrition upon Growth and Structure, London and Philadelphia, 1925	Proportions and feeding, mammals	425
Jenkinson, J. W.	Arch. Entwmech. XXI, 1906, p. 367	Frog development, effects of chemical substances	183
Jenkinson, J. W.	Biometrika, v, 1907, p. 147; VII, 1909 A, p. 148	Point of sperm-entry and bilateral symmetry	15, 37, 38, 39
Jenkinson, J. W.	Experimental Embryology, Oxford, 1909 B	General	xi, 14, 84
Jenkinson, J. W.	Arch. Entwmech. XXXII, 1911, p. 699	Polarity in egg of Echinoids	60
Jenkinson, J. W.	Quart. Journ. Micr. Sci. LX, 1915, p. 61	Development of frog after centrifuging egg	40, 220
Jenkinson, J. W.	Three Lectures on Experimental Embryology. (Posthumous.) Oxford, 1917	General	xi
Jollos, V. and Peterfi, T.	Biol. Zentralbl. XLIII, 1923, p. 286	Cleavage without nucleus, axolotl	132
Jones, W. N.	Ann. of Bot. XXXIX, 1925, p. 359	Seakale roots, polarity in regeneration	297
Juhn, M., Faulkner, G. H. and Gustavsen, R. G.	Journ. Exp. Zool. LVIII, 1931, p. 69	Growth-gradients and feather-regeneration	370
Kaan, H. W.	Journ. Exp. Zool. XLVI, 1926, p. 13	Development and determination of ear	233, 360
Kappers, C. U. A.	Journ. Comp. Neur. XXVII, 1917, p. 261; Brain, XLIV, 1921, p. 125; The Evolution of the Nervous System, Haarlem, 1930	Neurobiotaxis	379, 389
Kathariner, L.	Arch. Entwmech. XII, 1901, p. 597	Gravity and development of frog	36

BIBLIOGRAPHY AND INDEX OF AUTHORS (*continued*)

Author	Reference	Subject	Page where quoted
King, H. D.	*Journ. Exp. Zool.* XII, **1912**, p. 319	External factors and sex-differentiation, Anura	263
Koether, F.	*Arch. Entwmech.* CX, **1927**, p. 578	Newt, double monsters	158
Kopeć, S.	*Arch. Entwmech.* XXXIII, **1911**, p. 1; *Zool. Anz.* XLIII, **1913**, p. 65	Lepidoptera, wing-rudiments grafted on to opposite sex	208
Kornmüller, A. E.	*Fortschr. Neur. Psych.* V, **1933**, p. 419	Bio-electric phenomena, brain	364
Kozelka, A. W.	*Journ. Exp. Zool.* LXI, **1932**, p. 431; *Proc. Soc. Exp. Biol. and Med.* XXX, **1933**, p. 841	Male type of determination of fowl spur-rudiment	269
Krüger, F.	*Arch. Entwmech.* CXXII, **1930**, p. 1	Absence of induction by grafted lens	145, 392
Kuschakewitsch, S.	*Festschr. f. R. Hertwig*, II, **1910**, p. 61	Effect of over-ripeness of egg on gonad differentiation, Anura	261
Kusche, W.	*Arch. Entwmech.* CXX, **1929**, p. 192	Developmental potencies of pieces of blastulae interplanted	139
Lamborn, W. A.	*Proc. Ent. Soc. Lond.* **1914**, p. lxvii	*Papilio*, wings and wing-cases	206
Lankester, E. R.	*Nature*, LI, **1894**, p. 102	Development, response to environment	7
Lebedinsky, N. G.	*Arch. Mikr. Anat. u. Entwmech.* CII, **1924**, p. 101	Limb-development without innervation	430
Lehmann, F. E.	*Arch. Entwmech.* CVIII, **1926**, p. 243; CXIII, **1928** A, p. 123	Determination of neural tube, Urodele	137, 374
Lehmann, F. E.	*Journ. Exp. Zool.* XLIX, **1927**, p. 93	Dorsal root ganglia and myotomes	393
Lehmann, F. E.	*Verh. deutsch. Zool. Ges.* XXXII, **1928** B, p. 267; *Arch. Entwmech.* CXVII, **1929**, p. 312	Neural folds and epidermis, morph. and hist. differentiation	252
Lehmann, F. E.	*Arch. Entwmech.* CXXV, **1932**, p. 566	Newt organiser-grafts	150

BIBLIOGRAPHY AND INDEX OF AUTHORS (*continued*)

Author	Reference	Subject	Page where quoted
Luther, A.	*Comm. Biol. Soc. Scient. Fennicae*, II, **1925**, p. 1	Ear-vesicle and cartilaginous capsule	175
Lyon, E. P.	*Amer. Journ. Physiol.* xv, **1906**, p. xxi	Development of sea-urchin after centrifuging	218
Maas, O.	*Zeitschr. Wiss. Zool.* LXXXII, **1905**, p. 601	Hydroids, isolated blastomeres	95
MacBride, E. W.	*Quart. Journ. Micr. Sci.* LVII, **1911**, p. 235; *Proc. Roy. Soc. Lond. B*, xc, **1918**, p. 323	Hydrocoel and echinoid rudiment	81, 180
Malaquin, A.	*Comptes Rendus Soc. de Biol.* LXXXII, **1919**, p. 433	Organiser, *Salmacina* and *Filigrana* (Polychaetes)	290
Manchot, E.	*Arch. Entwmech.* CXVI, **1929**, p. 689	Determination of eye	244
Mangold, O.	*Arch. Entwmech.* XLVII, **1921** A, p. 249	One newt from two eggs	90
Mangold, O.	*Arch. Entwmech.* XLVIII, **1921** B, p. 505	Separated blastomeres of newt and situs inversus	76
Mangold, O.	*Arch. Mikr. Anat. u. Entwmech.* C, **1924**, p. 198	Plasticity of tissues at early stages, newt	42, 46, 140
Mangold, O.	*Ergebnisse der Biologie*, III, **1928**, p. 151; v, **1929** A, p. 290; VII, **1931** A, p. 193	General, differentiation of vertebrate nervous system, limbs and eye	224, 243, 244
Mangold, O.	*Arch. Entwmech.* CXVI, **1929** B, p. 586	Determination of neural plate	145, 147
Mangold, O.	*Naturwiss.* XIX, **1931** B, p. 475	Eye-rudiments, fish	201
Mangold, O.	*Naturwiss.* XIX, **1931** C, p. 905	Balancer induction	177, 237
Mangold, O.	*Naturwiss.* XX, **1932**, p. 371.	Homoiogenetic induction, regional	147
Mangold, O. and Seidel, F.	*Arch. Entwmech.* CXI, **1927**, p. 593	One newt from two eggs	90, 374

BIBLIOGRAPHY AND INDEX OF AUTHORS (*continued*)

Author	Reference	Subject	Page where quoted
Morgan, T. H.	*Experimental Embryology*, Columbia Univ. Press, **1927**	General	xi, 216
Morgan, T. H.	*Journ. Exp. Zool.* LXIV, **1933**, p. 433	*Ilyanassa*, polar lobe	122
Morgan, T. H. and Boring, A. M.	*Arch. Entwmech.* XVI, **1903**, p. 680	Frog, cleavage	16
Morgan, T. H. and Lyon, E. P.	*Arch. Entwmech.* XXIV, **1907**, p. 147	Sea-urchin egg centrifuged	83, 218
Morgan, T. H. and Spooner, G. B.	*Arch. Entwmech.* XXVIII, **1909**, p. 104	Polarity of centrifuged sea-urchin larva	66, 218
Morrill, C. V.	*Anat. Rec.* XVI, **1919**, p. 265	Situs inversus, fish	76
Mršić, W.	*Arch. Mikr. Anat. u. Entwmech.* XCVIII, **1923**, p. 129; *Arch. Entwmech.* CXXIII, **1930**, p. 301	Trout, effects of over-ripeness of egg	261
Müller, K. J.	*Zeits. Wiss. Zool.* CXLII, **1932**, p. 425	*Clepsine*, duplicitas and asymmetry	72, 114
Murray, P. D. F.	*Proc. Linn. Soc. New South Wales*, LI, **1926**, p. 187	Regional fields in limb-bud, chick	225
Murray, P. D. F.	*Proc. Roy. Soc. Lond.* B, CXI, **1932**, p. 497	Differentiation of blood, chick	205
Murray, P. D. F. and Huxley, J. S.	*Brit. Journ. Exp. Biol.* III, **1925**, p. 9; *Journ. Anat.* LIX, **1925**, p. 379	Grafts of chick embryo on chorio-allantoic membrane	195, 199, 250
Murray, P. D. F. and Selby, D.	*Journ. Exp. Biol.* VII, **1930**, p. 404	Differentiation of pieces of blastoderm on chorio-allantoic membrane	263, 268
Mutz, E.	*Arch. Entwmech.* CXXI, **1930**, p. 210	Organiser, *Hydra*	164
Nageotte, J.	*L'organisation de la matière dans ses rapports avec la vie*, Paris, **1922**	Strains and tendon-formation, etc.	432

Naville, A.	Rev. Suisse Zool. xxxiv, **1927**, p. 269	Loss of power of regeneration, Anura	366
Needham, J.	Chemical Embryology, Cambridge, **1932**	General	xi, 1
Needham, J.	Biol. Rev. viii, **1933**, p. 180	General, biochemical	327
Newman, H. H.	Journ. Exp. Zool. xvi, **1914**, p. 447	Larval hybrids, fish	405
Newman, H. H.	The Biology of Twins, Chicago, **1917**; The Physiology of Twinning, Chicago, **1923**	General, twinning	329, 330
Nicholas, J. S.	Anat. Rec. xxviii, **1924** A, p. 317	Balancer in axolotl	236
Nicholas, J. S.	Journ. Exp. Zool. xl, **1924** B, p. 113; Anat. Rec. xxix, **1925**, p. 108; xxxii, **1926**, p. 218	Postural regulation of limb, and limb-girdle, Urodele	232
Nicholas, J. S.	Proc. Nat. Acad. Sci. U.S.A. xiii, **1927**, p. 695	Fundulus, determination, early stages	195
Nicholas, J. S.	Journ. Exp. Zool. lv, **1930**, p. 1	Urodele, brain determination	175, 374, 387
Nicholas, J. S.	Proc. Soc. Exp. Biol. and Med. xxviii, **1931**, p. 1018	Neuron multiplication in spinal cord	387
Nicholas, J. S. and Rudnick, D.	Proc. Soc. Exp. Biol. and Med. xxix, **1931**, p. 325	Rat embryos on chick chorio-allantois	203
Noble, G. K. and Bradley, H. T.	Journ. Exp. Zool. lxv, **1933**, p. 1	Regeneration, lizard scales	372
Nussbaum, J. and Oxner, M.	Arch. Entwmech. xxx (1), **1910**, p. 74	Nemertine regeneration and metaplasia	211
Ogawa, C.	Journ. Exp. Zool. xxxiv, **1921**, p. 17	Ear-vesicle, self-righting effect, Anura	208
Olmsted, J. D. M.	Journ. Exp. Zool. xxxi, **1920**, p. 369	Taste-buds and innervation, fish	174, 431
Oppel, A. and Roux, W.	Vortr. u. Aufs. ü. Entwmech. x, **1910**	Functional differentiation of blood-vessels	434
Örström, A.	In the press	Sea-urchin, bilateral symmetry	69
Parker, G. H.	Amer. Nat. lxvi, **1932** A, p. 147	Trophic influence of nerves	174, 431

BIBLIOGRAPHY AND INDEX OF AUTHORS (*continued*)

Author	Reference	Subject	Page where quoted
Parker, G. H.	*Humoral Agents in Nervous Activity*, Cambridge, **1932** B	Theoretical	174, 431
Parker, R. C.	*Science*, LXXVI, **1932** A, p. 219; **1932** B, p. 446; *Journ. Exp. Med.* LVIII, **1932** C, p. 713	Fibroblasts, characteristics, tissue-culture, metaplasia	210, 213
Pasquini, P.	*Boll. Ist. Zool. Univ. Roma*, V, **1927**, p. 1	Eye-cup, regulation, Amphibia	244
Pasquini, P.	*Journ. Exp. Zool.* LXI, **1933**, p. 45	Bull-frog, determination of lens	189
Pasteels, J.	*Arch. de Biol.* XLII, **1931**, p. 389	Inequality of cleavage divisions	108
Pasteels, J.	*Arch. de Biol.* LXIII, **1932**, p. 521	Frog, development after injury to egg	94
Pauli, M. E.	*Zeits. Wiss. Zool.* CXXIX, **1927**, p. 483	Insect, determination, early stages	127, 208
Penners, A.	*Arch. Mikr. Anat. u. Entwmech.* CII, **1924**, p. 51	*Tubifex*, double monsters	113
Penners, A.	*Zeitschr. Wiss. Zool.* CXXVII, **1925**, p. 1	*Tubifex*, development of isolated blastomeres	113
Penners, A. and Schleip, W.	*Zeitschr. Wiss. Zool.* CXXX, **1928**, p. 305; CXXXI, **1928**, p. 3	Inverted frog's egg	95
Pflüger, E.	*Pflüger's Arch.* XXXII, **1883**, p. 1	Inverted frog's egg	36
Pressler, K.	*Arch. Entwmech.* XXXII, **1911**, p. 1	Situs inversus, Amphibia	73
Przibram, H.	*Arch. Entwmech.* XLIII, **1917**, p. 1	Growth-rate in regeneration	367
Przibram, H.	*Arch. Entwmech.* XLV, **1919**, p. 1	Regeneration and growth	419
Przibram, H.	*Arch. Mikr. Anat. u. Entwmech.* CII, **1924**, p. 604	Mirror-imaging in reduplications	224
Przibram, H.	*Arch. Entwmech.* CIV, **1925**, p. 434	Growth-rate and temperature	367
Przibram, H.	*Connecting laws in Animal Morphology*, London, **1931** A	General, growth-rates	71

Przibram, H.	*Sitzber. Akad. Wiss. Wien*, IX (vii), **1931** B	Serial heteromorphosis	360
Rand, H. W., Bovard, J. F. and Minnich, D. E.	*Proc. Nat. Sci. U.S.A.* XII, **1926**, p. 565	Organiser, *Hydra*	164
Ranzi, S.	*Pubb. Staz. Zool. Napoli*, IX, **1928**, p. 81; XI, **1931**, p. 104	Cephalopods, determination	208, 250
Raven, C. P.	*Proc. Kon. Akad. Weten. Amsterdam*, XXXIV, **1931** A, p. 554	Neural crest and balancer-induction	177, 236
Raven, C. P.	*Arch. Entwmech.* CXXV, **1931** B, p. 210; CXXIX, **1933** A, p. 179	Neural crest and cartilage	394, 396
Raven, C. P.	*Proc. Kon. Akad. Weten. Amsterdam*, XXXVI, **1933** B, p. 566	Amphibian organiser, glycogen	154
Reagan, F. P.	*Anat. Rec.* XI, **1916**, p. 489	Embryonic castration, birds	263
Reagan, F. P.	*Journ. Exp. Zool.* XXIII, **1917**, p. 85	Ear-vesicle and cartilage	175
Reith, F.	*Zeitschr. Wiss. Zool.* CXXVI, **1925**, p. 181; CXXXIX, **1931**, p. 728; *Arch. Entwmech.* CXXVII, **1932**, p. 283	Insect, determination	127, 208
Ribbert, H.	*Arch. Entwmech.* I, **1894**, p. 69	Compensatory hypertrophy, kidney	431
Richards, O. W.	*Biol. Rev.* II, **1927**, p. 298	Insect genitalia, asymmetry	70
Riddle, O.	*Amer. Nat.* LXIII, **1929**, p. 385	Genetic strains and hormone activity	436
Rienhoff, W. F.	*Bull. Johns Hopkins Hosp.* XXXIII, **1922**, p. 392	Metanephros *in vitro*, self-differentiation	179, 199, 205
Rotmann, E.	*Arch. Entwmech.* CXXIV, **1931**, p. 747; CXXIX, **1933**, p. 85	Urodeles, heteroplastic grafts	142, 424
Roux, W.	*Der Kampf der Teile*, Leipzig, **1881**	General, functional differentiation	34, 425
Roux, W.	*Breslau Ärzt. Zeitschr.* **1884** (also *Ges. Abh.* II, 19, p. 256)	Gravity and development of frog	36
Roux, W.	*Zeitschr. f. Biol.* XXI, **1885**, p. 411	General dependent and self-differentiation; morphological differentiation	9, 249

BIBLIOGRAPHY AND INDEX OF AUTHORS (*continued*)

Santos, F. V.	Biol. Bull. LVII, 1929, p. 188	Organiser, Planarians	165
Sato, T.	Arch. Entwmech. CXXII, 1930, p. 451; CXXX, 1933, p. 19	Lens regeneration in Urodele	238
Schaxel, J.	Genetica, IV, 1922 A, p. 339	Chimaeras in regeneration-buds, Urodele	405
Schaxel, J.	Arch. Entwmech. L, 1922 B, p. 498	Lack of regulation in Mosaic stage	195
Schleip, W.	Die Determination der Primitiventwicklung, Leipzig, 1929	General	xi, 70
Schmidt, G. A.	Arch. Entwmech. CXXII, 1930, p. 663; CXXIX, 1933, p. 1	Two embryos from one frog's egg	53, 94
Schotté, O.	Comptes Rendus Soc. Phys. Hist. Nat. Genève, XLIII, p. 126	Loss of power of regeneration after extirpation of field	366
Schotté, O.	Rev. Suisse Zool. XXXIII, 1926 B, p. 1	Regeneration and sympathetic	420
Schotté, O.	Arch. Entwmech. CXXII, 1930, p. 663	Organiser, plasticity and determination, Anura	94, 143
Schultze, O.	Arch. Entwmech. I, 1894, p. 269	Frog's egg inverted and double monsters	93
Seidel, F.	Biol. Zentralbl. XLVI, 1926, p. 321; XLVIII, 1928, p. 230; XLIX, 1929, p. 577; Arch. Entwmech. CXIX, 1929, p. 322; Verh. deutsch. Zool. Ges. XXXIV, 1931, p. 193	Insects, determination	127, 171, 208, 254
Seidel, F.	Arch. Entwmech. CXXVI, 1932, p. 213		
Sexton, E. W. and Pantin, C. F. A.	Nature, CXIX, 1927, p. 119	Insects, equality of nuclear division during cleavage	88
Singer, C.	Greek Biology and Greek Medicine, Oxford, 1922	Precursor substances in Gammarus	412
Sinnott, E. W. and Hammond, D.	Amer. Nat. LXIV, 1930, p. 509	General, historical	1
		Genetics of proportional growth in gourds	416

BIBLIOGRAPHY AND INDEX OF AUTHORS (*continued*)

Author	Reference	Subject	Page where quoted
Šivickis, P. B.	*Arch. Zool. Ital.* xvi, **1931** A, p. 430; *Arb. 2. Abt. Ungar. Biol. Forsch. Inst.* iv, **1931** B, p. 1; *Vytanto Didziojo Univ. Kaunas*, vii, **1933**, p. 369	Planarians, head-frequency	296
Smith, P. E.	*Amer. Anat. Mem.* xi, **1920**	Hypophysis and pituitary differentiation; melanophore function	179, 198, 426
Spek, J.	*Arch. Entwmech.* cvii, **1926**, p. 54	Ctenophore, isolated blastomeres	105, 108
Spek, J.	*Protoplasma*, ix, **1930**, p. 370	*Nereis*, cleavage, differentiation	132
Spemann, H.	*Verh. Anat. Ges.* xv, **1901** A, p. 61	Eye-cup and lens-formation, conjunctiva, Amphibia	178, 183
Spemann, H.	*Arch. Entwmech.* xii, **1901** B, p. 61; xv, **1902**, p. 448; xvi, **1903**, p. 551	Development of half-blastulae, two larvae from one egg, Urodele	53, 88, 156, 239
Spemann. H.	*Zool. Jahrb. Suppl.* vii, **1904**, p. 429	Duplicitas anterior and cyclopia	350
Spemann, H.	*Zool. Anz.* xxviii, **1905**, p. 419	Eye-cup and lens-formation	183, 187, 237, 238
Spemann, H.	*Verh. Ges. deutsch. Natur. u. Ärzte*, lxxviii, **1906** A, p. 189	Ear-vesicle and posture	209
Spemann, H.	*Verh. deutsch. Zool. Ges.* xvi, **1906** B, p. 195	Organiser, Urodele	134
Spemann, H.	*Arch. Entwmech.* xxx, **1910**, p. 437	Ear-vesicle, self-righting effect	199, 208
Spemann, H.	*Zool. Jahrb. Suppl.* xv (3), **1912** A, p. 1.	Determination of eye, Amphibia	245, 247
Spemann, H.	*Zool. Jahrb. (Abt. f. allg. Zool. u. Physiol.)* xxxii, **1912** B, p. 1	Eye-cup and lens-formation	186, 187

Author	Reference	Subject	Pages
Spemann, H.	*Verh. deutsch. Zool. Ges.* xxiv, **1914**, p. 216; *Zeitschr. Wiss. Zool.* CXXXII, **1928**, p. 105	Equality of nuclear division during cleavage	43, 86, 88
Spemann, H.	*Sitzber. Ges. Nat. Freunde Berlin*, IX, **1916**, p. 306	Organiser, Urodele	134, 158
Spemann, H.	*Arch. Entwmech.* XLIII, **1918**, p. 448; *Naturwiss.* VII, **1919**, p. 581	Plasticity and determination in early stages, Urodele	46, 73, 134, 158, 244
Spemann, H.	*Arch. Entwmech.* XLVIII, **1921**, p. 533	Heteroplastic grafts, Urodele	142,143,274,405,406,407
Spemann, H.	*Brit. Journ. Exp. Biol.* II, **1925**, p. 493; *Proc. Roy. Soc. Lond. B,* CII, **1927**, p. 177	Organiser, Urodele	147
Spemann, H.	*Naturwiss.* XVII, **1929**, p. 287	Organiser, Urodele, desiccated	153
Spemann, H.	*Arch. Entwmech.* CXXIII, **1931**, p. 389	Organiser, regional potencies	42, 43, 147, 149, 151
Spemann, H.	*Rev. Suisse Zool.* XXXIX, **1932**, p. 307; *Naturwiss.* XXI, **1933**, p. 115	Anuran grafts in Urodeles	142, 143, 321
Spemann, H. and Bautzmann, E.	*Arch. Entwmech.* CX, **1927**, p. 557	Regulation in newt blastulae	242
Spemann, H. and Falkenberg, H.	*Arch. Entwmech.* XLV, **1919**, p. 371	Duplicitas anterior and situs inversus, Urodele	75
Spemann, H. and Geinitz, B.	*Arch. Entwmech.* CIX, **1927**, p. 129	Organiser, Urodele, infection of other tissues	152
Spemann, H. and Mangold, H.	*Arch. Mikr. Anat. u. Entwmech.* C, **1924**, p. 599	Organiser, Urodele	12, 50, 135
Spemann, H. and Schotté, O.	*Naturwiss.* XX, **1932**, p. 463	Anuran grafts in Urodeles	142, 143, 321
Spirito, A.	*Boll. Ist. Zool. Univ. Roma,* VI, **1928**	Amphibian eye-cup, self-differentiation	244
Spurling, R. G.	*Anat. Rec.* XXVI, **1923**, p. 41	Mosaic stage and regeneration, chick	58, 198
Stauffacher, H.	*Jenaisch. Zeitschr. Nat.* XXVIII, **1894**, p. 196	*Cyclas*, egg-polarity	62

BIBLIOGRAPHY AND INDEX OF AUTHORS (*continued*)

Author	Reference	Subject	Page where quoted
Steinitz, E.	*Arch. Entwmech.* XX, 1906, p. 537	Eye-cup and cranium, Amphibia	175
Stella, E.	*Arch. Zool. Ital.* XVIII, 1932, p. 133	Eye-determination, Urodele	244
Sternberg, H.	*Arch. Mikr. Anat. u. Entwmech.* CIII, 1924, p. 259	Ear-vesicle, differentiation, Amphibia	199
Stevens, N. M.	*Arch. Entwmech.* XXVII, 1909, p. 622	Nematode, blastomere potencies	101
Stockard, C. R.	*Amer. Journ. Anat.* X, 1910, p. 369; XXVIII, 1921, p. 115	Duplicitas anterior and situs inversus, fish	76, 330, 344, 402
Stöhr, P.	*Arch. Mikr. Anat. u. Entwmech.* CII, 1924, p. 426; *Arch. Entwmech.* CVI, 1925, p. 409	Development of heart, Amphibia	204, 235, 249, 360
Stone, L. S.	*Journ. Exp. Zool.* XLIV, 1926, p. 95; *Arch. Entwmech.* CXVIII, 1929, p. 40; *Anat. Rec.* LI, 1932, p. 267	Neural crest and cartilage, Amphibia	394, 395
Strangeways, T. S. P.	*Tissue Culture*, Cambridge, 1924	Growth and differentiation	209
Strangeways, T. S. P. and Fell, H. B.	*Proc. Roy. Soc. Lond.* B, XCIX, 1926, p. 340; C, 1926, p. 273	Differentiation *in vitro* of eye and limb-bud, chick	205, 250
Streeter, G. L.	*Journ. Exp. Zool.* III, 1906, p. 543; IV, 1907, p. 431; XVI, 1914, p. 149	Ear-vesicle and self-righting effect, Amphibia	199, 208
Stultz, W. A.	*Proc. Soc. Exp. Biol. and Med.* XXIX, 1931, p. 178	Hind-limb field, Urodele	224, 229
Sumner, F. B. and Wells, N. A.	*Journ. Exp. Zool.* LXIV, 1933, p. 377	Fish melanophores, physiological state and rate of multiplication	426
Suzuki, S.	*Arch. Entwmech.* CXIV, 1928, p. 371	Amphibia, injuries to blastopore rim	21

Swett, F. A.	*Anat. Rec.* xxii, **1921**, p. 183	Duplicitas anterior and situs inversus, fish	76
Swett, F. A.	*Journ. Exp. Zool.* xxxvii, **1923**, p. 207	Limb-discs, regional potencies, Urodele	223
Swett, F. A.	*Anat. Rec.* xxvii, **1924**, p. 273	Regeneration of abnormal limbs	197
Tanaka, Y.	*Genetics*, ix, **1924**, p. 479	Maternal inheritance, silkworm	412
Taylor, C. V. and Tennent, D. H.	*Yearbook, Carnegie Inst. Wash.* xxiii, **1924**, p. 201	Induced polarity and cleavage-planes, Echinoderm eggs	83, 101
Taylor, C. V., Tennent, D. H. and Whitaker, D. M.	*Yearbook, Carnegie Inst. Wash.* xxv, **1925**, p. 249	Induced polarity and gastrulation, Echinoderm eggs	83, 313
Taylor, C. V. and Whitaker, D. M.	*Science*, lxv, **1926**, p. 308	Induced polarity and cleavage-planes, Echinoderm eggs	83, 84, 101
Tazelaar, M. A.	*Quart. Journ. Micr. Sci.* lxxii, **1928**, p. 419	Temperature-gradients, bird	344
Tazelaar, M. A., Huxley, J. S. and de Beer, G. R.	*Anat. Rec.* xlvii, **1930**, p. 1	Temperature-gradients, frog's egg	43
Tello, F.	*Vort. u. Aufs. ü. Entwmech.* xxxiii, **1923**	Development of nerves	393
Tennent, D. H.	*Papers Tortugas Lab. Carnegie Inst. Wash.* v, **1914**, p. 127; *Carnegie Inst. Wash. Publ.* No. 312, **1922**, p. 42	Time of appearance of paternal characters in larval hybrids	408
Thompson, D'Arcy W.	*Growth and Form*, Cambridge, **1917**	General	368
Titlebaum, A.	*Proc. Nat. Acad. Sci. U.S.A.* xiv, **1928**, p. 245	*Chaetopterus*, double monsters	114
Tokura, R.	*Folia Anat. Japon.* iii, **1925**, p. 173	Ear-determination, Anura	233, 360
Tornier, G.	*Arch. Entwmech.* xx, **1906**, p. 76	Limbs, reduplication, Amphibia	327
Tung, T.-C.	*Arch. de Biol.* xliv, **1933**, p. 1	Point of sperm-entry and symmetry	39
Twitty, V. C.	*Journ. Exp. Zool.* L, **1928**, p. 319	Epidermis, polarity of cilia beat	236
Twitty, V. C.	*Journ. Exp. Zool.* lv, **1930**, p. 43	Growth-rate of grafted eyes, Urodele	424

BIBLIOGRAPHY AND INDEX OF AUTHORS (*continued*)

Author	Reference	Subject	Page where quoted
Twitty, V. C. and Schwind, J. L.	*Journ. Exp. Zool.* LIX, **1931**, p. 61	Growth-rates of grafted eyes and lenses, Urodele	424
Ubisch, L. von	*Zeitschr. Wiss. Zool.* CVI, **1913**, p.409	Hydrocoel and Echinoid rudiment	180
Ubisch, L. von	*Zeitschr. Wiss. Zool.* CXXIII, **1924**, p. 37; CXXIX, **1927**, p. 213	Eye-cup and lens-formation	186
Ubisch, L. von	*Zeitschr.Wiss.Zool.* CXXIV, **1925**, p.457	Sea-urchin eggs, fused	103
Ubisch, L. von	*Arch. Entwmech.* CXVII, **1929**, p. 80	Lithium and development of sea-urchin half-embryos	323
Ubisch, L. von	*Arch. Entwmech.* CXXIV, **1931**, p. 181	Sea-urchin larvae, heteroplastic grafts	181
Uhlenhuth, E.	*Biol. Bull.* XLII, **1922**, p. 143	Thyroid in Urodele metamorphosis	427
Umanski, E.	*Zool. Anz.* XCVI, **1932** A, p. 299	Organiser, bird	161
Umanski, E.	*Zool. Anz.* XCVII, **1932** B, p. 286	Regeneration-bud, organiser-action	153
Versluys, J.	*Biol. Gen.* III, **1927**, p. 385; IV, **1928**, p. 617	Resonance theory of nerve-action	391
Vogt, W.	*Verh. Anat. Ges.* XXXI, **1922**, p. 53; XXXV, **1926** A, p. 62	Process of gastrulation and neurulation	22, 42
Vogt, W.	*Sitzber. Ges. Morph. u. Phys. München*, XXXV, **1923**, p. 22	Dynamic determination	149
Vogt, W.	*Sitzber. Ges. Morph. u. Phys. München*, XXXVII, **1926** B	Grey crescent in Urodele	15
Vogt, W.	*Verh. deutsch. Zool. Ges.* XXXII, **1928** A, p. 26	Labile determination, Urodele	136, 137
Vogt, W.	*Verh.Anat.Ges.*XXXVII,**1928**B,p.139; *Rev.Suisse Zool.*XXXIX, **1932**,p.309	Lateral temperature-gradients	39, 43, 342
Vogt, W.	*Arch. Entwmech.* CXX, **1929**, p. 384	Maps of presumptive regions	18, 21

Vogt, W.	*Verh. Anat. Ges.* XL, **1931**, p. 141	Lack of regulation in mosaic phase	195
Wachs, H.	*Arch. Entwmech.* XXXIX, **1914**, p. 384; XLVI, **1920**, p. 328	Lens-regeneration, Urodeles	238
Waddington, C. H.	*Nature,* CXXV, **1930**, p. 924; CXXXI, **1933** A, p. 275; *Phil. Trans. Roy. Soc. Lond.* B, CCXXI, **1932**, p. 179; *Journ. Exp. Biol.* X, **1933** B, p. 38; *Arch. Entwmech.* CXXVIII, **1933** C, p. 502	Organiser, bird	100, 136, 159, 161, 163
Waddington, C. H., Needham, J. and Needham, D.	*Nature,* CXXXII, **1933**, p. 239	Organiser, liquid extract, Urodele	12, 154
Waddington, C. H. and Schmidt, G. A.	*Arch. Entwmech.* CXXVIII, **1933**, p. 522	Organiser, bird, heteroplastic grafts	160, 161, 310
Waddington, C. H. and Waterman, A. J.	*Journ. Anat.* LXVII, **1933**, p. 355	Mammalian embryos, *in vitro*	206
Warburg, O.	*Über Stoffwechsel der Tumoren,* Berlin, **1926**	Respiration of normal and cancerous cells	214
Warynsky, S. and Fol, H.	*Rec. Zool. Suisse,* I, **1884**, p. 1	Situs inversus, chick	78
Waterman, A. J.	*Anat. Rec.* LII, **1932**, p. 31	Mammalian embryos, grafts on omentum	201
Weigmann, R.	*Zool. Anz.* LXIX, **1926**, p. 1; *Zeitschr. Wiss. Zool.* CXXIX, **1927**, p. 48	Position of blastopore dorsal lip in inverted eggs, frog	38, 152
Weismann, A.	*Kontinuität des Keimplasmas,* **1885**	Germ-plasm theory	4
Weiss, P.	*Arch. Mikr. Anat. u. Entwmech.* CII, **1924** A, p. 635	Innervation of grafted limbs	391
Weiss, P.	*Arch. Mikr. Anat. u. Entwmech.* CIV, **1924** B, p. 673	Determination of grafted limbs	364
Weiss, P.	*Arch. Mikr. Anat. u. Entwmech.* CIV, **1925**, p. 359	Limb-regeneration without skeleton	304

BIBLIOGRAPHY AND INDEX OF AUTHORS (*continued*)

Author	Reference	Subject	Page where quoted
Weiss, P.	*Arch. Entwmech.* CIX, **1927** A, p. 584	Lung grafted on to limb-stump	304
Weiss, P.	*Arch. Entwmech.* CXI, **1927** B, p. 317	Limb from tail regeneration-bud	272
Weiss, P.	*Morphodynamik. Abh. Theor. Biol.* XXIII, **1927** C; *Entwicklungsphysiologie der Tiere*, Berlin, **1930**	General, fields	xi, 274
Weiss, P.	*Biol. Gen.* IV, **1928**, p. 605	Resonance theory of nerve-action	391
Weiss, P.	*Arch. Entwmech.* CXVI, **1929**, p. 438; *Amer. Nat.* LXVII, **1933**, p. 322	Functional differentiation; tension, etc.	432
Wessel, E.	*Arch. Entwmech.* CVII, **1926**, p. 481	Duplicitas cruciata, Urodele	159
Wetzel, G.	*Arch. Mikr. Anat.* XLVI, **1895**, p. 654	Duplicitas, inverted frog's egg	93
Wetzel, R.	*Arch. Entwmech.* CVI, **1925**, p. 463; CXIX, **1929**, p. 188	Chick, blastoderm, early stages	159, 163
Whitaker, D. M.	*Biol. Bull.* LXI, **1931**, p. 61	*Fucus*, determination of polarity	60
Wilhelmi, H.	*Arch. Entwmech.* XLVI, **1920**, p. 210; XLVIII, **1921**, p. 517	Situs inversus, Urodele	78
Willier, B. H.	*Journ. Exp. Zool.* XLVI, **1927**, p. 409	Gonad determination, bird	206
Willier, B. H.	*Physiol. Zool.* III, **1930**, p. 201	Adrenal differentiation, chick	199
Willier, B. H.	Chap. IV in *Sex and Internal Secretions*, ed. E. Allen, Baltimore and London, **1932**	Gonad, differentiation	255, 261, 263
Willier, B. H. and Rawles, M. E.	*Anat. Rec.* XLVIII, **1931** A, p. 277	Heart and liver differentiation, chick	179
Willier, B. H. and Rawles, M. E.	*Journ. Exp. Zool.* LIX, **1931** B, p. 429	Developmental potencies, chick blastoderm	268
Wilson, C. B.	*Quart. Journ. Micr. Sci.* XLIII, **1900**, p. 97	Nemertines, egg-polarity	62

Author	Reference	Topic	Page
Wilson, E. B.	*Arch. Entwmech.* XVI, **1903**, p. 411	Nemertines, isolated blastomeres	98
Wilson, E. B.	*Journ. Exp. Zool.* I, **1904** A, p. 1	*Dentalium*, determination	110, 120
Wilson, E. B.	*Journ. Exp. Zool.* I, **1904** B, p. 197	*Patella*, isolated blastomeres	114
Wilson, E. B.	*The Cell*, New York (3rd Ed.), **1925**	General	xi
Wilson, E. B.	*Arch. Entwmech.* CXVII, **1929**, p. 180	*Chaetopterus*, egg-fragments	120, 172
Wilson, H. V.	*Journ. Exp. Zool.* V, **1907**, p. 245; XI, **1911**, p. 280	Restitution of sponge and hydroid fragments	65, 281
Witschi, E.	*Verh. Naturforsch. Ges. Basel*, XXXIV, **1922**, p. 33; *Proc. Soc. Exp. Biol. and Med.* XXVII, **1930**, p. 475	Frogs, development of double monsters from over-ripe eggs; cancer	96
Witschi, E.	*Verh. Naturforsch. Ges. Basel*, XXXVIII, **1927**, p. 367	Two frogs' eggs within one membrane	93
Witschi, E.	Chap. v in *Sex and Internal Secretions*, ed. E. Allen, Baltimore and London, **1932**	Gonad, differentiation, general	255, 256, 261
Witschi, E.	*Journ. Exp. Zool.* LXV, **1933** A, p. 213	Sex-differentiation, Urodeles	259
Witschi, E.	*Amer. Journ. Anat.* LII, **1933** B, p. 461	Gonad-differentiation, toads	258, 259, 260
Woerdeman, M. W.	*Arch. Entwmech.* CXVI, **1929**, p. 220	Amphibian eye-field	244
Woerdeman, M. W.	*Ann. Soc. Roy. Sci. Méd. et Nat. Bruxelles*, **1932**, p. 17	Eye and lens-fibre determination	190, 243
Woerdeman, M. W.	*Proc. Kon. Akad. Wetensch. Amsterdam*, XXXVI, **1933** A, p. 189; **1933** B, p. 423	Amphibian organiser and eye, and glycogen activities	154, 190
Woerdeman, M. W.	*Proc. Kon. Akad. Wetensch. Amsterdam*, XXXVI, **1933** C	Tumour-tissue and induction	153
Wolff, C. F.	*Theoria Generationis*, Halle, **1759**	Epigenesis	6, 8
Wolff, G.	*Arch. Entwmech.* I, **1895**, p. 380; XII, **1901**, p. 307; *Festschr. R. Hertwig*, III, **1910**, p. 67	Lens-regeneration, newt	187, 237, 238
Woodger, J. H.	*Biological Principles*, London, **1928**	General, philosophical	10

BIBLIOGRAPHY AND INDEX OF AUTHORS (*continued*)

Author	Reference	Subject	Page where quoted
Yamane, J.	*Arch. Entwmech.* CXXI, **1930**, p. 598	Spinal cord and nerve-development	378, 381, 393
Yatsu, N.	*Journ. Coll. Sci. Tokyo*, XXVII (17), **1910**	Nemertines, egg-fragments	120
Yatsu, N.	*Annotat. Zool. Japon.* VIII, **1912** A; *Journ. Coll. Sci. Tokyo*, XXXII (3), **1912** B	Ctenophores, isolated blastomeres and egg-fragments	108
Zeleny, C.	*Journ. Exp. Zool.* I, **1904**, p. 293	Nemertines, isolated blastomeres	98
Zoja, R.	*Arch. Entwmech.* I, **1895**, p. 578; II, **1896**, p. 1	Hydroids, isolated blastomeres	97
Zur Strassen, O.	*Arch. Entwmech.* VII, **1898**, p. 642	Nematodes, fusion of eggs	101

APPENDIX

1. Exogastrulation in Amphibia.

The work of Holtfreter (*Arch. Entwmech.* CXXIX, 1933, p. 669; *Biol. Zentralbl.* LIII, 1933, p. 404) on this subject only appeared after this book was in page proof. It is, however, so important that we have decided to summarize it in an appendix; and have taken the opportunity of adding some other points that had been overlooked.

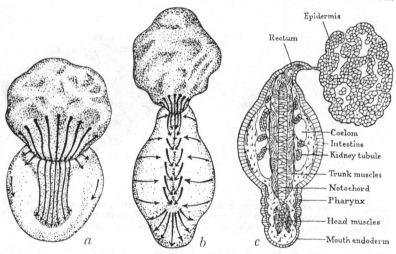

Fig. 213

Exogastrulation in Amphibia (axolotl). Diagrams showing (*a, b*) the mass-movements of the organiser-region and endoderm; *c*, the structure of an exo-embryo. Note wrinkled amorphous epidermis, exogastrulated endo-mesodermal portion inside-out. (From Holtfreter, *Biol. Zentralbl.* LIII, 1933.)

Holtfreter discovered that by the simple procedure of removing the early blastulae of axolotls from their membranes and placing them in Ringer solution of about 0·35 per cent. strength they could be made to exogastrulate—i.e. the presumptive endoderm and mesoderm is evaginated instead of being invaginated, leaving the presumptive ectoderm as a hollow sac. Stages in the process are shown in figs. 213, 214. It is of interest to note that the tendency to constriction in the marginal zone (Ch. III, p. 42) still manifests itself, leading to a waist between the ectoderm and the endo-mesoderm from the earliest stages of gastrulation. Later the waist becomes still further narrowed to a stalk, which can be easily severed, and may break of its own accord.

All the mass-movements of the different regions involved in normal gastrulation (p. 43) are still operative in the exogastrulae, though their mutual interactions in the altered circumstances are a little different, as indicated by the arrows in figs. 213, 215. For

Fig. 214

Exogastrulated axolotl embryo, 8 days old. On the right, the epidermis; on the left, the exogastrulated endo-mesoderm: note pharynx with inverted gill-pouches (on left). (From Holtfreter, *Biol. Zentralbl.* LIII, 1933.)

instance, the organiser-region, forming the dorsal side of the marginal zone (p. 41), stretches out as a tongue on the dorsal side of the evaginated mass, and subsequently becomes sunk in a groove and finally overgrown by the endoderm. This confirms the view that the dynamics of gastrulation are predetermined in the various local regions of the germ.

The final result of exogastrulation is what we may call an exo-embryo (Holtfreter's *Exokeim*). This consists of two very distinct parts. The ectoderm has flattened down to an irregular wrinkled mass with the blastocoel largely obliterated. It shows no medullary differentiations, notably no trace of neural tube or even of local thickening to form neural plate tissue.

The endo-mesoderm on the other hand bears a considerable resemblance to an embryo, showing well-marked regions—a head and gill-region with gill-clefts, a trunk-region, and a tail-region. However, it is entirely abnormal in its detailed structure. It is morphologically inside-out; its outer layer is endodermal, and this contains a more or less solid mass of notochord, somites, mesenchyme and cartilage (figs. 213, 216).

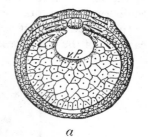

The endodermal epithelium is, as in the normal embryo, polarised: but its outer surface corresponds with that which bounds the gut-lumen in normal ontogeny. This may be compared with the fact noted on p. 250, that spheres composed of gastral layer only (collar-cells), arising in sponge dissociation experiments, have the collars directed outwards, whereas in normal animals they face the gastral cavity. In both sponges and amphibia, one surface of the epithelium orients itself towards the most favourable environment, whether this be an internal lumen or the external medium.

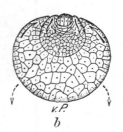

a

b

Fig. 215

Diagrams of transverse sections through Urodele embryos, showing the structure and directions of movement of parts in *a*, normal embryo, *b*, exo-embryo. *v.P.* site of original vegetative pole. (From Holtfreter, *Arch. Entwmech.* cxxix, 1933.)

Exogastrulation gives us a method by which ectoderm can be totally separated from endo-mesoderm from the first onset of the gastrulation-process. In addition, it provides pseudo-embryos, containing all the derivatives of the endo-mesoderm, in which we can be certain that no nervous tissue is present; and further, in the inversion of the endodermal and mesodermal layers, it provides a natural experiment in abnormal spatial relations which it would be impossible to duplicate artificially. (See also p. 252 for a comparable case in insects.)

As might be expected, conclusions of considerable importance have been arrived at by analysis of the results. In the first place, we have the complete failure of the nervous system to differentiate in the ectodermal portion. The only difference observable between presumptive epidermis and presumptive neural plate is that the latter shows an autonomous tendency to elongation in the direction of the egg's major axis. We shall later return to the absence of ectodermal differentiations.

In marked contrast with this incapacity of the ectoderm is the capacity of the endo-mesoderm for self-differentiation. We have in

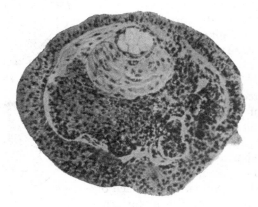

Fig. 216

Transverse section through trunk-region of exogastrulated endo-mesoderm, axolotl. Note superficial endoderm surrounding notochord, myotomes, connective tissue, and (below on the right) heart. (From Holtfreter, *Arch. Entwmech.* CXXIX, 1933.)

the first place typical notochord. Then the mesoderm produces the following derivatives: in the anterior region, head-musculature, mesenchyme and cartilage; in the trunk-region, somitic mesoderm, pro- and meso-nephric tubules with coelomic funnels and associated with gonads, smooth gut-musculature and (empty) hearts capable of rhythmical contraction, empty endothelial sacs, masses of blood-cells, coelomic spaces, and connective tissue. The endoderm shows equal powers of self-differentiation, and gives rise to buccal cavity, pharynx with endodermal portion of the gill-clefts (the visible apertures on the surface of the exo-embryo of course corresponding to the normal internal apertures leading out of the pharynx), thyroid (probably), oesophagus, stomach, lungs, liver, pancreas, intestine and rectum. The various sections of the gut are characterised by the same histological peculiarities as in the normal animal, e.g. ciliation of the oesophagus and typical glands in the

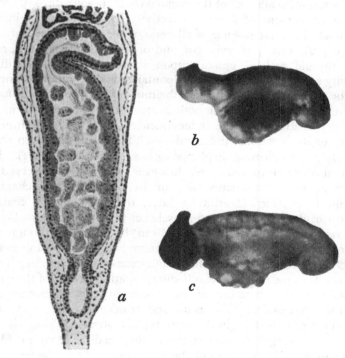

Fig. 217

Self-differentiation in exo-embryos, axolotl. Extrusion of yolk-cells, *a*, in normal embryo, into lumen of gut; *b* and *c*, in exo-embryos, from the surface. (From Holtfreter, *Arch. Entwmech.* cxxix, 1933.)

stomach. Perhaps the most remarkable self-differentiation is that of the small intestine. In the normal axolotl larva of a certain definite age, a number of yolk-rich cells belonging to this region partially degenerate and become detached into the lumen and are subsequently digested by the remainder of the epithelium (fig. 217). In the exo-embryos, this same process of degeneration and detach-

ment occurs at a corresponding stage, though of course the cells
here are detached into the surrounding medium (fig. 217 *b* and *c*).
The process occurs only in the central region of the gut corresponding to the small intestine. This determination of a timed
degeneration recalls that of the isolated chick mesonephros (p. 205).

The attainment of functional activity by many tissues in the
demonstrable total absence of all nervous tissue is of great interest
(cf. p. 430). The epidermis, gut, and pronephros tissue, and probably thyroid vesicles, embark upon active secretion, and ciliary
activity sets in where expected. Spontaneous rhythmic movements
of the outward-facing gut-endothelium occur regularly, brought
about by the underlying smooth musculature, thus providing the
first demonstration of the independence of this tissue of innervation for its functional differentiation. The heart may also reach
this stage, confirming explantation experiments (p. 203). The
striated skeletal musculature, however, was never observed to
contract, either spontaneously, or in reaction to mechanical,
chemical or electrical stimuli: later, the degenerative changes
typical of denervated striped muscle set in. Thus the histological
differentiation of skeletal muscle can be reached (though not
maintained) in the total absence of nervous connexions, but not its
functional activity. This confirms and extends other work (p. 431).

In spite of the remarkable self-differentiating powers of the endo-
mesoderm, the structure of the exo-embryo is abnormal in a
number of respects. The head and trunk-musculature, though
differentiating histologically into typical striated fibres, is not
arranged in a regular metameric plan, and the direction of the fibres
is irregular. The cartilages of the head (no cartilage appears to be
formed in the trunk-region) are massed together in a single irregular
lump. The total absence of the cartilages arising from the neural
crest (p. 394) is doubtless largely responsible for the lack of regular
arrangement of the head-musculature, as well as for its small size.
The gonad appears not as a long ridge, but as a series of small cell-
masses in close connexion with the pronephric tubules. The liver
tissue is always very small in amount, and late in appearance; no
gall-bladder has been noted.

The tail-region is of particular interest. A caudal zone of activity
is present in notochord and trunk-musculature, and a conical tail-

bud arises in a more or less typical way; but it never becomes large, its growth is soon arrested, and it is finally resorbed, in spite of the absence of any degenerative signs in its tissues. Holtfreter ascribes this (and also the absence of regular segmental arrangement of the trunk-muscles) to the absence of the neural tube and especially to the absence of the mesenchyme derived from the neural crest, which is known (see pp. 193, 396) to have the tendency for directive outgrowth. As the neural tube and crest are first induced, by the chorda-mesoderm, and then supply material necessary for tail-mesoderm differentiation, we have here an interesting case of mutual induction on the part of an organiser and of that which it organises. The same interaction is apparent in the head-region, where neural crest material, originally induced by the prechordal portion of the organiser, appears to be necessary for the proper anatomical differentiation of the tissues (muscle and cartilage) derived from this region. (See also p. 181 for a comparable case of mutual dependence in sea-urchins.)

Among other special points may be mentioned the fact that teeth, even partial or rudimentary, are never found in exo-embryos, showing that the presence of ectoderm is necessary for their initiation. Taste-buds, however, do differentiate in the pharynx, thus demonstrating that the view sometimes maintained of their derivation from immigrant ectoderm is incorrect (see p. 498). The presence of gill-clefts shows that their initial determination proceeds from the pharyngeal ectoderm and is quite independent of the presence of ectoderm. The fact that they later disappear, however, suggests that contact with ectoderm is needed for their maintenance.

Blood-tissue is rarely found in exo-embryos, apparently because its primary site of origin lies far back in the ventro-caudal region, and from here it often tends to become included within the ectodermic vesicle.

The ciliary beat on the surface of the ectoderm is also of interest. In the normal embryo this is directed in an orderly way, in a predominantly antero-posterior direction (see p. 236). In wholly isolated ectodermic vesicles, however, it is completely irregular, indicating that a polarity or polarized gradient-field is normally imposed upon the epidermis from the underlying endo-mesodermal tissues. This is beautifully demonstrated by cases in which exogastrulation is not complete, but a portion of the ectoderm has been underlain by endo-mesoderm and has been organised. In such portions, the direction of ciliary beat is regular and normal, while remaining irregular over the rest of the epidermis (fig. 218).

This brings us to a more general consideration of partial exogastrulation. Total exogastrulation is a comparatively rare occur-

Fig. 218

Induction of polarity in epidermis by underlying organiser. The direction of cilia-beat (indicated by arrows) of non-underlain epidermis is irregular and chaotic; that of epidermis underlain by organiser-tissue is regular and polarised. (From Holtfreter, *Biol. Zentralbl.* LIII, 1933.)

rence: in the majority of cases, exogastrulation only proceeds to a certain point, and then the remainder of the endo-mesoderm is invaginated under the ectoderm. All gradations are to be found from a minimal invagination to a normal embryo. The first step is the presence of some blood- and yolk-cells in the ectodermic vesicle. When they are present, they induce a smooth two-layered epithelium, with normal tempo of differentiation, in place of the irregularly folded and wrinkled epidermis derived from wholly isolated ectoderm (see figs. 13 and 219). This appears to be due largely to the formation of mesenchymatous vesicles containing fluid, which produce normal tension in the ectodermic vesicle.

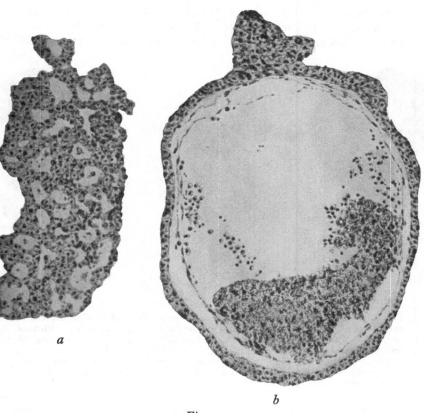

a

b

Fig. 219

Organisation of epidermis. *a*, wrinkled, irregular epidermis of axolotl exoembryo not underlain by organiser; *b*, two-layered epithelium induced by presence of underlying connective tissue and blood-cells. (From Holtfreter, *Arch. Entwmech.* CXXIX, 1933.)

Neural formations are never induced in such conditions, but are always formed if any of the ectoderm comes to be underlain by chorda-mesoderm. If only a narrow portion of chorda-mesoderm is invaginated, the actual neural plate may be much narrower than the presumptive neural region (see p. 155). Further, it does not differentiate into a complete but undersized nervous system, but into tail neural tube only: the organiser, in other words, has regional properties. If only a slight degree of invagination occurs, the invaginated material is presumptive caudal tissue, and the induced structure is a tail (fig. 220 *a*). This then grows out as quite a normal

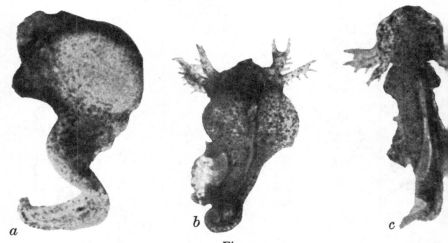

a   *b*   *c*

Fig. 220

Partial exogastrulation and progressive organising effects of varying degrees of organiser-invagination. *a*, slight invagination (of tail-organiser only) producing only a tail; *b*, medium invagination (of tail- and trunk-organiser) producing tail and trunk ending anteriorly with gills; *c*, nearly complete invagination (tail-, trunk-, and head-organiser) producing an embryo which is complete except for the diminutive length of the head and size of the eyes. (From Holtfreter, *Biol. Zentralbl.* LIII, 1933.)

tail, confirming the conclusions reached above (p. 193) as to the co-operation of mesoderm and neural crest mesenchyme in normal tail-elongation. With progressive increase in the amount of tissue invaginated, there is a progressive increase in the amount of organisation of the ectoderm, first trunk-structures appearing, then gills, and finally head-structures (fig. 220 *b* and *c*). The direction of ciliary beat is normal on the organised portions of the ectoderm of such partial embryos (e.g. with tail only), irregular on the unorganised portions. When the embryos are almost but not quite complete they are cyclopic (see pp. 245, 350).

There is one further interesting point to be mentioned. When only somitic mesoderm is invaginated, the resultant neural tube is very small and is abnormal in cross-section; only in the presence of notochord material will a full-sized and normally constructed neural tube be induced (see p. 374).

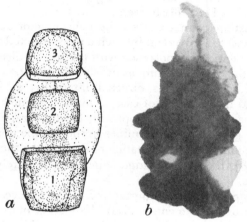

Fig. 221

Regional inductive powers of exogastrulated organiser. *a*, diagram showing experiment of placing pieces of undetermined epidermis on the anterior (1), middle (2), and posterior (3) regions of the organiser (chorda-mesoderm) of exo-embryo of axolotl. *b*, tail induced from epidermis in position 3. (From Holtfreter, *Biol. Zentralbl.* LIII, 1933.)

A further set of important experiments was made by taking undetermined ectoderm from normal early gastrulae and laying it on different regions of the chorda-mesoderm of the exogastrulated endo-mesoderm in the exogastrulation experiments. This must be done in early stages, before the presumptive chorda-mesoderm has disappeared under the surface of the endoderm (see fig. 221).

The results are striking. From whatever presumptive region the pieces of ectoderm were taken, they differentiate in accordance with the regional properties of the organiser on which they are lying. If placed on the anterior (head-organiser) part of the chorda-mesoderm, the ectoderm is induced to form brain with eyes, nasal pits, and ear-vesicles. If placed on the trunk chorda-mesoderm, it produces a normal spinal cord, which becomes displaced below the surface. And if the piece of ectoderm is placed on the protruded tail-region, the chorda-mesoderm grows into the ectoderm, induces a neural tube, and then the two tissues in co-operation grow out as a typical tail. The regional differentiation of the chorda-mesoderm into head-organiser, trunk-organiser, and tail-organiser, is here clearly seen (see p. 147).

An interesting point concerns the behaviour of any endoderm which happens to be overlain by such a piece of ectoderm. Instead of the epithelium being polarised with the free or distal ends of its cells facing the outer medium, as over the rest of the surface of the exogastrula, its polarity is directed internally, as in a normal embryo; and in a number of cases a miniature gut-lumen is produced. Thus, in normal development, the topographical arrangement of the germ-layers determines in the gut the normal polarity of its epithelium, and the formation of its lumen, although as noted above (p. 483) the determination appears to be purely mechanical in its nature.

But perhaps the most important of the facts revealed by these exogastrulation experiments concern the absence of neural differentiation in the ectoderm. This is all the more striking in view of the indications which previous work has given (see pp. 50, 136, 203) of the existence of a labile determination of neural folds, as evidenced by experiments of removal or inactivation of the organiser-region, and of explantation and interplantation of portions of blastulae (see figs. 18, 62, 63).

It might be held, and is held by Holtfreter (*loc. cit.*), that these new results show that an invaginated organiser is indispensable for the determination of neural folds; that these results dispose of the hypothesis of a labile determination (and therefore of a "double assurance") of the neural folds; and that the conclusions drawn from previous experiments are erroneous. While realising the strength of this argument, it is as well to consider the possibilities that the non-appearance of neural differentiations in the exogastrulation experiments may be due to other causes. In this connexion, four points may be called to mind:

i. So far as it goes, there *is* in the exogastrulation experiments a determination for the presumptive neural fold region to stretch more than the neighbouring presumptive epidermis (p. 484), and this determination must be independent of an invaginated organiser.

ii. On comparing the experiments in which the organiser-region is removed or inactivated with the exogastrulation experiments, it will be noticed that there is a difference in the distance between the presumptive neural fold region and the organiser, and in the time during which the latter could act on the former at any given distance. While previous experiments have led to the view that there exists a gradient-field under the influence of the organiser before invagination, in which the presumptive neural fold region undergoes labile determination, the conditions of the exogastrulation experiment are such that it might be argued that the gradient-field is deformed, or even not formed.

iii. The labile determination in question would be part of the general effect of an uninvaginated organiser working from a distance as the centre of an individuation-field (see p. 310). It is held by Holtfreter that neural fold formation can result only by contact with an underlying organiser ("evocation", Waddington and Needham). But it is clear from the experiments on newts (see p. 149), in which trunk-organiser is grafted at head-level and induces the formation of head-structures, that the tissues there are under the influence of an action exerted from a distance by the host-organiser acting as the centre of an individuation-field. The same conclusions emerge from experiments on birds (see p. 162). There is therefore evidence for the existence of individuation-fields, in both Amphibia and birds, which is not disproved by the exogastrulation experiment.

iv. The variations between the results of explantation of portions of blastulae in inorganic media, and those of interplantation of similar portions into living embryos (see pp. 139, 317), show that the reactivity of the tissues (i.e. their differentiation into epidermis, neural tube or notochord) is markedly affected by environmental changes. It is possible that the lack of differentiation of neural structures in the exogastrulation experiments is to be explained on such lines as these.

2. Lateral temperature-gradients and gradient-fields in Amphibia.

Gilchrist (*Journ. Exp. Zool.* LXVI, 1933, p. 15) has recently re-attacked the question (raised on p. 342) as to whether the gradient-system in the amphibian embryo can be directly altered and deformed by means of lateral temperature-gradients so as to give rise to asymmetrically developed neural folds, independently of the effects of alteration of growth of the invaginated organiser. A new method introduced consists in applying a lateral temperature-gradient for a certain (not too great) length of time, and then to reverse the sign of the gradient for an equal length of time. Eggs of the Urodele *Triturus* thus treated, and heated first on the right during the late blastula and then on the left during the early gastrula stages, show abnormally large neural folds on the right side.

Gilchrist draws the conclusion that the processes of neural plate determination (in what we should call the primary gradient-system) take place during the late blastula, for which reason the temperature-gradient to which they are then exposed is able to bring about a larger development on the heated side. This view receives support from the results of other experiments in which the temperature-gradient is applied earlier: heated on the right in the early blastula, and on the left in the late blastula. In such cases the neural folds are larger on the left.

We are, however, not informed as to whether the invaginated gut-roof is symmetrical or not, and it would still be possible to hold that the late blastula stage is the critical time for neural plate determination, not because of any effect on the gradient-system, but because the processes of invagination of the organiser (gut-roof) are also susceptible to modification by temperature at this period. In this case, the effects of the temperature-gradient on the neural folds would be indirect, and exerted via the organising action of the gut-roof. Against this, however, it must be mentioned that Gilchrist presents evidence of the early blastula stage as being that at which the embryo is most susceptible to temperature-gradients for the production of abnormalities in the subsequent processes of gastrulation.

3. Gradients and prelocalisation in Polychaetes (*Nereis*).

The work of Spek (*Protoplasma*, IX, 1930, p. 370; and Parts IV and V of Gellhorn's *Lehrbuch der allgemeinen Physiologie*, 1931) on this subject presents such a number of features of interest, that it deserves special mention here. To the technique of *intra vitam* staining of the early stages of the developing egg of *Nereis*, he has added the use of indicators enabling him to detect changes of pH in different parts of the embryo. The unfertilised egg of *Nereis* contains a large germinal vesicle (nucleus) surrounded by a number (about 30) of drops of fat arranged around it in the equatorial plane; in addition, there is, scattered through the egg, a large number of small albuminous droplets containing a lemon-yellow pigment. This pigment is found to turn violet in acid media and therefore acts as a natural indicator.

Fertilisation results in the formation of the polar bodies at the animal pole, as well as in a number of changes in the cortical regions of the cytoplasm which do not directly concern us here. What is of great interest, however, is the fact that soon after the polar bodies are extruded, the fat drops and albumen droplets undergo a re-arrangement, as a result of which they leave a clear zone or "pole-plasm" (see p. 113) at the animal pole and become concentrated in the vegetative hemisphere of the egg (see p. 119 for a comparable case of re-arrangement and prelocalisation of egg-contents in Ascidians). The albumen droplets in the equatorial zone take on a peculiar colour when stained *intra vitam*, and the whole egg shows a clear stratification along the main axis.

The egg is in this condition when the first two (meridional) cleavage-divisions take place, leaving four blastomeres, each possessing the characteristic stratification. There next occurs a remarkable change in the pH of different regions along the main axis, as shown by the natural and by the experimentally added indicators. The region of the animal pole shows a shift to the alkaline, while the region of the vegetative pole shows a shift to the acid side of the scale.

The whole phenomenon, as Spek says, presents the appearance of a natural experiment of cataphoresis. The changes in pH at the two ends of the main axis of the egg reflect a gradient in electrical potential, due to the accumulation of ions of opposite sign at the two poles, and resulting in the segregation of the egg-contents and in their stratified re-arrangement.

The next (third, equatorial) cleavage separates the animal cytoplasm (mostly clear, "alkaline") from the vegetative cytoplasm (granular, "acid"): the former goes to the formation of the first quartet of micromeres ($1 a$ to $1 d$), while the latter forms the macromeres ($1 A$ to $1 D$). The droplets which become included in the micromeres are those which lay in the equatorial zone of the egg, and, as we shall see, are destined to become included in the primary trochoblasts (see also p. 132).

The forces which were at work prior to the equatorial cleavage in segregating the egg-contents, continue to function after that cleavage-division in both micromeres and macromeres. The result is that in the micromeres, the albumen droplets which have become included in these cells are concentrated at their most vegetative end; while in the macromeres, what little clear cytoplasm there is, is situated at their most animal end, the fat and albumen drops being still further concentrated vegetatively.

At the next (fourth) cleavage the droplets in the micromeres find themselves in the primary trochoblasts ($1 a^2$ to $1 d^2$), while the clear cytoplasm becomes included in the apical cells ($1 a^1$ to $1 d^1$). At the same time, the clear cytoplasm in the macromeres becomes incorporated in the micromeres of the second quartet ($2 a$ to $2 d$), while the fat drops, etc. remain in the macromeres ($2 A$ to $2 D$).

In general, therefore, it appears that those regions of the cytoplasm which are characterised by an alkaline reaction give rise to the ectoderm, while those regions with an acid reaction become endoderm. Facts such as these throw an interesting light on the inhibition of ectoderm in lithium-induced exogastrulae of Echinoids (see p. 336).

The free-swimming trochophore larva provides evidence of a ventro-dorsal gradient, for the ventrally-situated first (ectodermal) and second (mesodermal) somatoblasts are particularly alkaline, while, in the endoderm, a similar gradient from ventral to dorsal side may be observed.

We do not yet know whether any of the various egg-contents which we have seen are distributed to various blastomeres during the cleavage of *Nereis* may be regarded as organ-forming substances rather than raw materials (see p. 217): further experiments, involving isolation of blastomeres and centrifugalisation, combined with study of *p*H indicators, will have to decide on this point. As an example, however, of the stratification of substances (perhaps, of potencies, see p. 102) resulting from the action of an electrical gradient, these observations and experiments are of great value.

4. Organiser-properties in living and dead tissues.

In amplification of the statement on p. 153 that certain tissues which possess no capacity to act as organisers when alive may show this capacity when they are killed, we may refer to further recent experiments by Holtfreter (*Naturwissenschaften*, XXI, 1933, p. 766). He has found that the property to induce the formation of a secondary embryo or parts of it in an amphibian gastrula are possessed by the following: *all* parts of uncleaved amphibian eggs that have been boiled to a state of hardness; *all* parts of an amphibian gastrula that have been preserved for six months in 70 per cent. alcohol, treated with xylol, embedded in paraffin and brought back to water; boiled pieces of muscle of the Annelid *Enchytraea* and of the molluscs *Planorbis* and *Limnea*; heat-coagulated cell-free extracts of the crustacean *Daphnia* and of the pupæ of moths; pieces of all organs so far tested of the stickleback, fresh or boiled; living pieces of larval amphibian liver, brain and retina, and of adult liver, ovary, and heart; living pieces of liver, kidney, testis and other organs of lizards, birds, and mice; coagulated bird embryo-extract; extract of killed calf's liver, and boiled pieces of several mammalian organs; pieces of liver, brain, kidney, thyroid and tongue of a fresh human corpse.

No vegetable material was found to possess organising properties, but in the animal kingdom the chemical substance which forms the basis of the organising action is clearly widespread and may probably be regarded as universal. The fact that it is absent (in the living state) from all regions of the vertebrate embryo except the organiser, while it is present in a variety of organs in the adult, is noteworthy and is probably to be regarded as an adaptation to a specialised mode of development in which organiser action by contact (p. 310) is employed.

5. Development of the amphibian mouth (see also p. 179).

The recent work of Ströer (*Arch. Entwmech.* cxxx, 1933, p. 131), on mouth-development in *Amblystoma*, may be referred to as an excellent example of detailed experimental analysis. He finds, by means of grafting experiments, that the mouth-region is composed of both ectodermal and endodermal portions, of which the ectodermal alone is capable of forming teeth. The presumptive mouth-ectoderm is dependent on underlying endoderm for the realisation of its potencies: without this, it develops into epidermis. Ectoderm from the ventral region of the abdomen grafted in place of the presumptive mouth-ectoderm does not react with the underlying endoderm to produce a mouth (contradicting Adams; see p. 179). On the other hand, pieces of the mouth-inducing region of the endoderm (anterior wall of fore-gut) introduced into the blastula may produce a mouth in interaction with the ectoderm of the heart region. Thus we do not know the limits, either in space or time, of the ectoderm-field capable of reacting to form mouth-epithelium. When presumptive mouth-endoderm grafted into the blastula does not succeed in inducing the ectodermic portions of a mouth, it develops into endodermic portions only (portions of buccal cavity, pharynx, oesophagus, with taste-buds, but without teeth).

An interesting point is that if only a small piece of ventral ectoderm is grafted into the presumptive mouth-region, it is caught up in the invagination process carried out by the remaining presumptive mouth-ectoderm and apparently "infected" with its qualities, for it then differentiates into true mouth-epithelium.

Another point is that the determination of the presumptive mouth-ectoderm to produce teeth takes place rather earlier than its determination to become differentiated mouth-epithelium.

The only region capable of inducing mouth-formation is the anterior wall of the fore-gut. This was proved by implanting pieces of various regions of the developing egg into blastulae. Pieces of anterior neural plate or neural fold have no mouth-inducing capacity. Neither neural tube nor neural crest material is necessary for tooth-formation, so that the mesodermal portion of the teeth must be derived not from mesectoderm but mesendoderm. Taste-buds are produced by pure endodermal grafts (see p. 487).

INDEX

Bold type denotes pages on which figures will be found.